ntry University

hone 024 7688 7555

Micro-organisms and Earth systems – advances in geomicrobiology

There is growing awareness that important environmental transformations are catalysed, mediated and influenced by micro-organisms, and such knowledge is having an increasing influence on disciplines other than microbiology, such as geology and mineralogy. Geomicrobiology can be defined as the study of the role that microbes have played and are playing in processes of fundamental importance to geology. As such, it is a truly interdisciplinary subject area, necessitating input from physical, chemical and biological sciences. The book focuses on some important microbial functions in aquatic and terrestrial environments and their influence on 'global' processes and includes state-of-the-art approaches to visualization, culture and identification, community interactions and gene transfer, and diversity studies in relation to key processes. Microbial involvement in key global biogeochemical cycles is exemplified by aquatic and terrestrial examples. All major groups of geochemically active microbes are represented, including cyanobacteria, bacteria, archaea, microalgae and fungi, in a wide range of habitats, reflecting the wealth of diversity in both the natural and the microbial world. This book represents environmental microbiology in its broadest sense and will help to promote exciting collaborations between microbiologists and those in complementary physical and chemical disciplines.

Geoffrey Michael Gadd is Professor of Microbiology and Head of the Division of Environmental and Applied Biology in the School of Life Sciences at the University of Dundee, UK.

Kirk T. Semple is a Reader in the Department of Environmental Science at Lancaster University, UK.

Hilary M. Lappin-Scott is Professor of Environmental Microbiology in the School of Biosciences, University of Exeter, UK.

Symposia of the Society for General Microbiology

Managing Editor: Dr Melanie Scourfield, SGM, Reading, UK
Volumes currently available:

43   Transposition
45   Control of virus diseases
47   Prokaryotic structure and function – a new perspective
51   Viruses and cancer
52   Population genetics of bacteria
53   Fifty years of antimicrobials: past perspectives and future trends
54   Evolution of microbial life
55   Molecular aspects of host–pathogen interactions
56   Microbial responses to light and time
57   Microbial signalling and communication
58   Transport of molecules across microbial membranes
59   Community structure and co-operation in biofilms
60   New challenges to health: the threat of virus infection
61   Signals, switches, regulons and cascades: control of bacterial gene expression
62   Microbial subversion of host cells
63   Microbe–vector interactions in vector-borne diseases
64   Molecular pathogenesis of virus infections

SIXTY-FIFTH SYMPOSIUM OF THE
SOCIETY FOR GENERAL MICROBIOLOGY
HELD AT KEELE UNIVERSITY SEPTEMBER 2005

Edited by
G. M. Gadd, K. T. Semple & H. M. Lappin-Scott

micro-organisms and earth systems – advances in geomicrobiology

Published for the Society for General Microbiology

CAMBRIDGE UNIVERSITY PRESS
Cambridge, New York, Melbourne, Madrid, Cape Town,
Singapore, São Paulo

Cambridge University Press
The Edinburgh Building, Cambridge CB2 2RU, UK

Published in the United States of America by
Cambridge University Press, New York

www.cambridge.org
Information on this title: www.cambridge.org/9780521862221

First published 2005

Printed in the United Kingdom at the University Press, Cambridge

A catalogue record for this publication is available from the British Library

ISBN 13   978 0 521 86222 6 hardback
ISBN 10   0 521 86222 1 hardback

Typeface   Sabon (Adobe) 10·5/13·5 pt   System   QuarkXPress™   [SGM]

Front cover illustration: Microbial colonization of red sandstone: extensive colonization of the outer weathered layers is evident by varied microbial communities, including cyanobacteria, microalgae, bacteria and fungi. The variety of pigmentation reflects this as well as accompanying microbe-mediated mineral transformations such as secondary mineral formation. Dissolution reactions can enhance physical weathering – places where weakened outer layers have fallen off are clearly evident. Characterization of microbial communities, their activities and roles is a significant challenge for geomicrobiologists. Image recorded in Hawkstone Park, Shropshire, UK, by G. M. Gadd (© 2004).

CONTENTS

Contributors   vii

Editors' Preface   xi

M. Wagner and M. W. Taylor
Isotopic-labelling methods for deciphering the function of uncultured
micro-organisms   1

L. A. Warren
Biofilms and metal geochemistry: the relevance of micro-organism-induced
geochemical transformations   11

N. D. Gray and I. M. Head
Minerals, mats, pearls and veils: themes and variations in giant sulfur bacteria   35

**D. W. Hopkins, B. Elberling, L. G. Greenfield, E. G. Gregorich, P. Novis,
A. G. O'Donnell and A. D. Sparrow**
Soil micro-organisms in Antarctic dry valleys: resource supply and utilization   71

V. R. Phoenix, A. A. Korenevsky, V. R. F. Matias and T. J. Beveridge
New insights into bacterial cell-wall structure and physico-chemistry: implications
for interactions with metal ions and minerals   85

J. Coombs and T. Barkay
Horizontal gene transfer of metal homeostasis genes and its role in microbial
communities of the deep terrestrial subsurface   109

L. G. Benning, V. R. Phoenix and B. W. Mountain
Biosilicification: the role of cyanobacteria in silica sinter deposition   131

K. H. Nealson and R. Popa
Metabolic diversity in the microbial world: relevance to exobiology   151

D. B. Nedwell
Biogeochemical cycling in polar, temperate and tropical coastal zones:
similarities and differences   173

G. M. Gadd, M. Fomina and E. P. Burford
Fungal roles and function in rock, mineral and soil transformations   201

K. Pedersen
The deep intraterrestrial biosphere   233

J. A. Raven, K. Brown, M. Mackay, J. Beardall, M. Giordano, E. Granum, R. C. Leegood, K. Kilminster and D. I. Walker
Iron, nitrogen, phosphorus and zinc cycling and consequences for primary productivity in the oceans   247

J. R. Lloyd
Mechanisms and environmental impact of microbial metal reduction   273

M. Krüger and T. Treude
New insights into the physiology and regulation of the anaerobic oxidation of methane   303

N. Clipson, E. Landy and M. Otte
Biogeochemical roles of fungi in marine and estuarine habitats   321

P. C. Bennett and A. S. Engel
Role of micro-organisms in karstification   345

Index   365

CONTRIBUTORS

Barkay, T.
Department of Biochemistry and Microbiology, Cook College, Rutgers University,
76 Lipman Dr., New Brunswick, NJ 08901, USA

Beardall, J.
School of Biological Sciences, Monash University, Clayton, VIC 3800, Australia

Bennett, P. C.
Department of Geological Sciences, The University of Texas at Austin, Austin, TX 78712,
USA

Benning, L. G.
Earth and Biosphere Institute, School of Earth and Environment, University of Leeds, UK

Beveridge, T. J.
Department of Molecular and Cellular Biology, College of Biological Science,
University of Guelph, Guelph, Ontario, Canada N1G 2W1

Brown, K.
Plant Research Unit, Division of Environmental and Applied Biology, School of Life
Sciences, University of Dundee at SCRI, Scottish Crop Research Institute, Invergowrie,
Dundee DD2 5DA, Scotland, UK

Burford, E. P.
Division of Environmental and Applied Biology, Biological Sciences Institute, School of Life
Sciences, University of Dundee, Dundee DD1 4HN, Scotland, UK

Clipson, N.
Department of Industrial Microbiology, University College Dublin, Belfield, Dublin 4, Ireland

Coombs, J.
Department of Biochemistry and Microbiology, Cook College, Rutgers University,
76 Lipman Dr., New Brunswick, NJ 08901, USA

Elberling, B.
Institute of Geography, University of Copenhagen, Øster Voldgade 10, DK-1350,
Copenhagen K., Denmark

Engel, A. S.
Department of Geology and Geophysics, Louisiana State University, Baton Rouge,
LA 70803, USA

Fomina, M.
Division of Environmental and Applied Biology, Biological Sciences Institute, School of Life
Sciences, University of Dundee, Dundee DD1 4HN, Scotland, UK

Gadd, G. M.
Division of Environmental and Applied Biology, Biological Sciences Institute, School of Life
Sciences, University of Dundee, Dundee DD1 4HN, Scotland, UK

Giordano, M.
Department of Marine Science, Università Politecnica delle Marche, 60131 Ancona,
Italy

Granum, E.
Department of Animal and Plant Sciences, University of Sheffield, Sheffield S10 2TN, UK

Gray, N. D.
School of Civil Engineering and Geosciences, Institute for Research on the Environment
and Sustainability and Centre for Molecular Ecology, University of Newcastle, Newcastle
upon Tyne NE1 7RU, UK

Greenfield, L. G.
School of Biological Sciences, University of Canterbury, Private Bag 4800, Christchurch,
New Zealand

Gregorich, E. G.
Agriculture Canada, Central Experimental Farm, Ottawa, Canada K1A 0C6

Head, I. M.
School of Civil Engineering and Geosciences, Institute for Research on the Environment
and Sustainability and Centre for Molecular Ecology, University of Newcastle, Newcastle
upon Tyne NE1 7RU, UK

Hopkins, D. W.
School of Biological and Environmental Sciences, University of Stirling, Stirling FK9 4LA,
Scotland, UK

Kilminster, K.
School of Plant Biology, University of Western Australia, M090 35 Stirling Highway, Crawley,
WA 6009, Australia

Korenevsky, A. A.
Department of Molecular and Cellular Biology, College of Biological Science,
University of Guelph, Guelph, Ontario, Canada N1G 2W1

Krüger, M.
Federal Institute for Geosciences and Resources (BGR), Stilleweg 2, D-30655 Hannover,
Germany, and Max-Planck-Institute for Marine Microbiology, Celsiusstrasse 1, D-28359
Bremen, Germany

Landy, E.
School of Biomedical and Molecular Sciences, University of Surrey, Guildford GU2 7XH,
UK

Leegood, R. C.
Department of Animal and Plant Sciences, University of Sheffield, Sheffield S10 2TN, UK

Lloyd, J. R.
The Williamson Research Centre for Molecular Environmental Studies and the School
of Earth, Atmospheric and Environmental Sciences, University of Manchester, Manchester
M13 9PL, UK

Mackay, M.
Plant Research Unit, Division of Environmental and Applied Biology, School of Life Sciences, University of Dundee at SCRI, Scottish Crop Research Institute, Invergowrie, Dundee DD2 5DA, Scotland, UK

Matias, V. R. F.
Department of Molecular and Cellular Biology, College of Biological Science, University of Guelph, Guelph, Ontario, Canada N1G 2W1

Mountain, B. W.
Institute of Geological and Nuclear Sciences, Wairakei Research Centre, Taupo, New Zealand

Nealson, K. H.
Department of Earth Sciences, University of Southern California, Los Angeles, CA 90089-0740, USA

Nedwell, D. B.
Department of Biological Sciences, University of Essex, Colchester CO4 3SQ, UK

Novis, P.
Manaaki Whenua – Landcare Research, PO Box 69, Lincoln 8152, New Zealand

O'Donnell, A. G.
Institute for Research on Environment and Sustainability, University of Newcastle upon Tyne, Newcastle upon Tyne NE1 7RU, UK

Otte, M.
Department of Botany, University College Dublin, Belfield, Dublin 4, Ireland

Pedersen, K.
Deep Biosphere Laboratory, Department of Cell & Molecular Biology, Göteborg University, Box 462, SE-405 30 Goteborg, Sweden

Phoenix, V. R.
Department of Molecular and Cellular Biology, College of Biological Science, University of Guelph, Guelph, Ontario, Canada N1G 2W1

Popa, R.
Department of Earth Sciences, University of Southern California, Los Angeles, CA 90089-0740, USA

Raven, J. A.
Plant Research Unit, Division of Environmental and Applied Biology, School of Life Sciences, University of Dundee at SCRI, Scottish Crop Research Institute, Invergowrie, Dundee DD2 5DA, Scotland, UK

Sparrow, A. D.
School of Biological Sciences, University of Canterbury, Private Bag 4800, Christchurch, New Zealand, and Department of Natural Resources and Environmental Sciences, University of Nevada, 1000 Valley Rd, Reno, NV 89512, USA

Taylor, M. W.
Department of Microbial Ecology, University of Vienna, Althanstr. 14, A-1090 Vienna, Austria

Treude, T.
Max-Planck-Institute for Marine Microbiology, Celsiusstrasse 1, D-28359 Bremen, Germany

Wagner, M.
Department of Microbial Ecology, University of Vienna, Althanstr. 14, A-1090 Vienna, Austria

Walker, D. I.
School of Plant Biology, University of Western Australia, M090 35 Stirling Highway, Crawley, WA 6009, Australia

Warren, L. A.
School of Geography and Earth Sciences, McMaster University, 1280 Main St. West, Hamilton, Ontario, Canada L8S 4K1

EDITORS' PREFACE

The science of the environment encompasses a huge number of biological, chemical and physical disciplines. For several years, scientists have been interested in large-scale environmental processes/phenomena, such as soil formation, global warming and global elemental cycling. Until recently, the role and impact of micro-organisms on these 'global' environmental processes has been largely ignored or, at best, underestimated. However, there is growing awareness that important environmental transformations are catalysed, mediated and influenced by micro-organisms, and such knowledge is having an increasing influence on disciplines other than microbiology, such as geology and mineralogy. Geomicrobiology can be defined as the study of the role that microbes have played and are playing in processes of fundamental importance to geology. As such, it is a truly interdisciplinary subject area, necessitating input from physical, chemical and biological sciences, in particular combining the fields of environmental and molecular microbiology together with significant areas of mineralogy, geochemistry and hydrology. As a result, geomicrobiology is probably the most rapidly growing area of microbiology at present. It is timely that this topic should be the subject of a Plenary Symposium volume of the Society for General Microbiology (SGM) to emphasize and define this important area of microbiological interest, and help to promote exciting collaborations between microbiologists and other environmental and Earth scientists.

This Symposium arose from the Environmental Microbiology Group of the SGM and presents a snapshot of some key areas of geomicrobiology written by experts in their respective fields. The book focuses on some important microbial functions in aquatic and terrestrial environments and their influence on 'global' processes and includes state-of-the-art approaches to visualization, culture and identification, community interactions and gene transfer, as well as diversity studies in relation to key pathways. Novel approaches for the study of diversity and function of microbial communities are highlighted and applied to key environmental problems, such as community interactions in biofilms, and the microbiology of surface, sub-surface and extreme environments. Microbial involvement in key global biogeochemical cycles is exemplified by aquatic and terrestrial examples, and includes metal and mineral transformations and development, element cycling in marine and estuarine systems, primary production, and anaerobic methane oxidation. All major groups of geochemically active microbes are represented, including cyanobacteria, bacteria, archaea, microalgae and fungi, in a wide range of habitats both aquatic and terrestrial, aerobic and anaerobic, and benign and extreme, reflecting the wealth of diversity in both the natural and the microbial world. It should also be appreciated that many of the natural processes discussed also have application in applied contexts such as agriculture and plant productivity, environmental exploitation and resource utilization and microbial treatment of pollution. This book will truly represent environmental microbiology in its broadest sense and we hope that it will have broad appeal, not only to environmental microbiologists, but also to environmental scientists, geologists, geochemists, Earth scientists, ecologists and environmental biotechnologists. A beneficial outcome may be the promotion of exciting collaborations between microbiologists and those in complementary physical and chemical disciplines. Only through interdisciplinary scientific approaches will the roles of micro-organisms in Earth systems be better clarified, appreciated and understood.

We would like to thank all the authors for enthusiastically supporting this project despite their obvious heavy work commitments. We also wish to thank Diane Purves (University of Dundee) for expert assistance with manuscripts and author liaison, and all at the SGM office, particularly Melanie Scourfield, Josiane Dunn and Janet Hurst, for their efficient help and support with the Symposium organization and production of this volume.

Geoffrey M. Gadd
Kirk T. Semple
Hilary M. Lappin-Scott

Isotopic-labelling methods for deciphering the function of uncultured micro-organisms

Michael Wagner and Michael W. Taylor

Department of Microbial Ecology, University of Vienna, Althanstr. 14, A-1090 Vienna, Austria

INTRODUCTION

With the benefit of hindsight, the last 20 years in microbial ecology will probably be referred to as the census period that dramatically changed our perception of bio-diversity within the three domains of life. Bacteria and archaea are no longer viewed as groups of peculiar and morphologically simple organisms that show relatively little diversification despite their long evolutionary history, but have now been recognized to harbour a perplexing number of novel phylogenetic lineages (Rappé & Giovannoni, 2003). Current estimates assume that the number of prokaryotic species ranges in the millions and thus vastly exceeds the fewer than 10 000 described prokaryotic species that have been isolated to date in pure culture (Curtis et al., 2002). This dramatic para-digm shift was only made possible by the development of cultivation-independent molecular approaches for surveying microbial diversity in nature. Whilst it is now evident that most prokaryotes cannot be cultured easily, due to their living in complex communities and their intimate metabolic links with both their abiotic and biotic environments, the powerful arsenal of techniques at our disposal enables us to see beyond the 'cultured few' and gain valuable insights into the realm of uncultured micro-organisms (Wagner, 2004). It is now relatively straightforward to determine the species richness of natural microbial communities by comparative sequence analysis of environmentally retrieved 16S rRNA gene sequences (Olsen et al., 1986; Schloss & Handelsman, 2004). Furthermore, even the abundance of bacteria in environmental samples can be determined easily by using quantitative PCR (Skovhus et al., 2004) or hybridization formats such as fluorescence in situ hybridization (FISH) (Wagner et al., 2003). Due to these technological advances, many novel and often numerically

SGM symposium 65: Micro-organisms and Earth systems – advances in geomicrobiology.
Editors G. M. Gadd, K. T. Semple & H. M. Lappin-Scott. Cambridge University Press. ISBN 0 521 86222 1 ©SGM 2005

important bacterial and archaeal species have been identified in various ecosystems. However, for most of these players, their ecophysiology and contribution to the functioning of the ecosystem remain hidden.

Given the above, one of the biggest challenges in contemporary microbial ecology is to develop strategies that enable us (i) to directly investigate metabolic properties of defined but uncultured micro-organisms, and (ii) to identify those uncultured organisms that are responsible for defined processes within their natural environment. These two criteria highlight important differences in approaches that are currently taken in the study of microbial community function. In a microbial world dominated by uncultured representatives, those investigators who approach matters from a diversity standpoint typically target organisms of interest (e.g. those that are highly abundant in a particular ecosystem) and endeavour to find out what metabolic traits these species possess. An alternative, equally valid approach is to recognize a process occurring in the environment (e.g. sulfate reduction or denitrification) and try to establish which organisms are responsible. Both approaches are served equally well by the methods outlined in this chapter. Specifically, we provide an overview of a group of recently developed methods that exploit the addition of isotope-labelled substrates to complex microbial communities in order to bridge the gap between microbial community structure and function.

IDENTIFICATION OF PLAYERS THAT INCORPORATE LABELLED SUBSTRATES

One of the most elegant ways to better understand the ecophysiology of uncultured bacteria is to expose them to isotope-labelled substrates and, subsequently, to link their identification with a measurement of substrate incorporation into their biomass. This concept was first realized by extracting phospholipid fatty acids from complex microbial communities after incubation with ^{13}C-labelled substrates, and identifying active groups of micro-organisms by detection of labelled signature compounds using isotope-ratio mass spectrometry (Boschker *et al.*, 1998). However, this approach is not suited for the identification of uncultured bacteria, because their phospholipid fatty-acid composition is unknown. Consequently, several methods, which are presented in more detail below, were developed to exploit other, more informative biomarkers or even to allow simultaneous organism identification and detection of substrate incorporation at the single-cell level. A common feature of all these methods is that it is possible to fine-tune the experimental setup such that very specific questions regarding the ecophysiology of selected micro-organisms can be answered. This is achieved by performing parallel experiments using modified incubation conditions (pH, temperature, presence/absence of different electron acceptors, presence/absence of specific inhibitors for selected microbial groups etc.) under which the microbial biomass encounters the labelled substrate. Alternatively, this modus operandi enables microbial ecologists to hunt for novel micro-organisms responsible for defined functions in the environment. However, with the exception of the so-called $HetCO_2$-microautoradiography (MAR) approach (Hesselsoe *et al.*, 2005), all of these methods are dependent on the availability of commercial isotope labelling for the compounds of interest. Furthermore, all mentioned techniques fail to detect micro-organisms that convert labelled compounds without using them for the synthesis of biomass or storage compounds.

DNA stable-isotope probing (DNA-SIP) is based on the incorporation of stable isotope-labelled substrates into the DNA of substrate-consuming micro-organisms (reviewed recently by Dumont & Murrell, 2005). To date, DNA-SIP has only been performed with ^{13}C-labelled substrates. In DNA-SIP, the heavy DNA of the substrate consumers is separated from the light DNA of the other community members by buoyant density-gradient centrifugation with added ethidium bromide. The identity of the active bacteria is revealed subsequently by amplification, cloning and comparative sequence analysis of 16S rRNA genes or selected functional genes from the heavy DNA. Probing the light- and heavy-DNA fractions with specific PCR primers can be applied to show rapidly whether a given bacterial player has incorporated the offered substrate. One of the major advantages of DNA-SIP is that large genomic-DNA fragments of the active community fraction can be isolated and subsequently analysed by the environmental-genomics approach (DeLong, 2002, 2005). On the other hand, the DNA of active community members will only be labelled sufficiently for DNA-SIP if the added compounds are highly enriched in ^{13}C, and if relatively high concentrations of labelled substrate are offered over extended time periods (up to several weeks) so that slow-growing micro-organisms are also able to replicate their DNA and to divide. It is essential to be aware that such incubation conditions might induce major shifts in the composition of the resident microbial community, perhaps comparable to the biases reported previously for enrichment cultures (Wagner *et al.*, 1993). Compared to the rapid-growing r-strategists, K-strategists that are adapted to low substrate concentrations might thus be heavily underrepresented or even missing (if the applied substrate concentration is inhibitory to them) in the heavy-DNA fraction. Furthermore, due to the long incubation times required for DNA-SIP, primary consumers will transfer label to other community members by cross-feeding. Therefore, not all genomes found in the heavy DNA are from bacteria capable of using the labelled compound (although, on a more positive note, this effect could also be exploited to obtain indications of metabolic interactions among different community members).

Several of the limitations of DNA-SIP can be overcome by RNA-SIP (Manefield *et al.*, 2002, 2005). In microbial cells, synthesis rates of RNA (including 16S rRNA) are significantly higher than rates of DNA replication; consequently, if DNA is replaced by

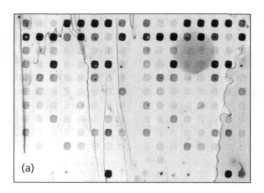

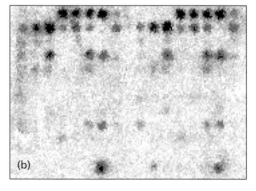

Fig. 1. Isotope-array experiment to identify butyrate-oxidizing bacteria under aerobic conditions in activated sludge from a nutrient-removal wastewater-treatment plant. (a) Fluorescence scan of the array; (b) radioactivity scan of the same array. Pictures kindly provided by A. Loy, M. Schloter and M. Hesselsoe.

RNA as a biomarker, much shorter incubation times are required to obtain sufficient amounts of labelled material. ^{13}C-labelled RNA can be separated from light RNA via caesium trifluoroacetate (CsTFA) density-gradient centrifugation, although it has been observed that labelled RNA with a specific buoyant density is found over a number of fractions in CsTFA gradients. To solve this problem, 16S rRNA from each density fraction must be amplified separately by RT-PCR and subsequently analysed by a fingerprinting method such as denaturing-gradient gel electrophoresis (DGGE). Increasing band intensities in the fingerprint pattern of certain 16S rDNA fragments from high-density fractions throughout the duration of the labelled-substrate pulse are then used to demonstrate that the 16S rRNA of a particular bacterium has become increasingly labelled during the duration of the experiment. Subsequently, the respective DGGE band is cut from the gel, reamplified by PCR and sequenced in order to identify the substrate-consuming micro-organisms. The disadvantage of this procedure is that it is very time-consuming and dependent on the resolving power of DGGE, which is insufficient in highly complex microbial communities. In future studies, the identification of bacteria with a specific function could be accelerated and improved in resolution by hybridizing heavy DNA or RNA obtained by SIP to diagnostic rRNA-targeted microarrays (Guschin *et al.*, 1997; Loy *et al.*, 2002, 2005; Wilson *et al.*, 2002).

RNA as a biomarker is also exploited by the so-called isotope-array approach (Adamczyk *et al.*, 2003) but, in contrast to SIP, the microbial communities are incubated with radioactively (and not ^{13}C)-labelled compounds prior to nucleic-acid extraction. To date, exclusively ^{14}C-labelled compounds have been used for isotope-array analysis. Again, as RNA is targeted, only relatively short incubation times are required. After community RNA extraction, the RNA is labelled covalently with a fluorescent dye and hybridized to a suitable rRNA-targeted microarray by using standard procedures. Visualization of the microarray by using a fluorescence scanner reveals the community structure, whilst recording the radioactivity of each microarray spot by use of a beta-imager shows which community member has incorporated radioactive label into its rRNA and thus consumed the offered substrate (Fig. 1). Due to limited resolution of commercially available beta-imagers, larger probe-spot sizes are required for optimal results compared with standard microarrays. An advantage of the isotope array compared with RNA-SIP is that it is PCR-independent and thus offers the potential for quantitative analysis. The ratio of radioactive signal to fluorescence signal, which can be determined easily for each hybridized probe spot, reflects the incorporation rate of the labelled substrate into the RNA of the detected bacterial population(s) and can thus be used to compare their metabolic activities within a complex microbial community. On the other hand, and in contrast to SIP, the nature of this format makes it impossible to identify novel substrate-consuming micro-organisms whose 16S rRNA sequences are not yet known (i.e. you can only find what you are looking for). In the future, the isotopc array should be adapted to the use of stable isotopes so as to avoid radioactive lab work and also render it adaptable to the analysis of microbial communities colonizing humans. For this purpose, beta-imaging must be replaced by detection methods such as time-of-flight secondary-ion mass spectrometry.

In contrast to the aforementioned methods, structure and function of microbial communities can be analysed by a combination of FISH and MAR at the single-cell level (Lee *et al.*, 1999; Ouverney & Fuhrman, 1999; Daims *et al.*, 2001; Nielsen *et al.*, 2002, 2003b) (Fig. 2), as long as the environmental sample is suited for FISH analysis. After incubation of the microbial communities with radioactively labelled substrate, community members are identified by FISH with rRNA-targeted oligonucleotide probes. In parallel, the incorporation of radioactive substrate into the biomass of bacterial cells is visualized with a radiation-sensitive silver halide emulsion, which is placed on top of the radiolabelled organisms and subsequently processed by standard photographic procedures. Excited silver ions will precipitate as metallic silver and

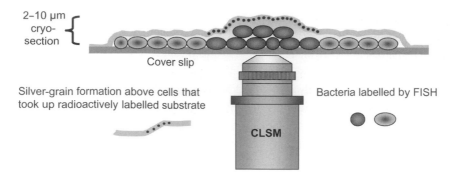

Fig. 2. Schematic representation of the FISH-MAR approach. CLSM, Confocal laser-scanning microscope.

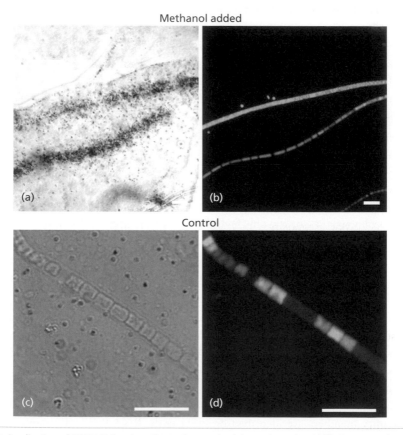

Fig. 3. Application of FISH-MAR to investigate the ecophysiology of uncultured filamentous micro-organisms. The filaments showed uptake of methanol in MAR experiments (a) and were identified simultaneously by FISH as members of the '*Gammaproteobacteria*' (b). Control experiments with pasteurized biomass showed no unspecific uptake or binding of labelled methanol (c, d). Pictures kindly provided by K. Stoecker and H. Daims.

appear as black grains after development of the film (Fig. 3). As FISH requires perme-abilization of the bacterial cell envelope, radioactive substrates that are taken up by a bacterial cell, but not incorporated into macromolecules, will not result in silver-grain formation. It should, however, be noted that MAR cannot differentiate between con-version of labelled organic substrates into intracellular storage products and active substrate metabolism. The recently developed HetCO$_2$-MAR approach (Hesselsoe *et al.*, 2005) allows this limitation to be overcome by utilizing the tendency of meta-bolically active heterotrophic bacteria to assimilate significant amounts of CO$_2$. Thus, ^{14}CO$_2$ is added as an activity marker to complex microbial communities and, after suitable incubation periods, active autotrophic and heterotrophic community members can be identified by FISH-MAR. HetCO$_2$-MAR also avoids difficulties associated with obtaining many isotope-labelled substrates, as this approach requires that only the CO$_2$ is labelled and other added substrates are not.

Major advantages of FISH-MAR are that incubation times are comparatively short, because every labelled macromolecule of the monitored bacterial cells contributes to the detection of substrate incorporation. Furthermore, the amount of substrate incor-poration into specific uncultured bacterial cells can be quantified by counting the number of silver grains on top of the cells if bacteria with known specific radioactivity are used as an internal standard. This technique can be applied to measure *in situ* substrate affinities (K_s) of uncultured bacteria and to study physiological differences within naturally occurring single populations of bacteria (Nielsen *et al.*, 2003a). In future studies, it should be possible to modify the FISH-MAR method such that radio-actively labelled substrates are replaced by stable isotope-labelled compounds and the stable-isotope composition of FISH-identified microbial cells is measured by secondary-ion mass spectrometry (Orphan *et al.*, 2001, 2002).

The last few years have seen the development of an impressive battery of tools that measure function of uncultured micro-organisms by the addition of labelled substrates. After initial, proof-of-principle type studies, several of these tools are now applied (almost) routinely in microbial ecology laboratories and they will be used increasingly in time-course experiments to uncover metabolic networks in multi-component microbial communities (Lueders *et al.*, 2004). Together with quantitative FISH-based analysis of co-occurring microbial populations (Daims *et al.*, 2005), these approaches will shed new light on the complex network of interactions existing in microbial communities.

OUTLOOK

We have provided here a brief overview of cutting-edge methods for the analysis of microbial-community function. As a field that necessarily emphasizes the importance of technological advances, microbial-community ecology should continue to benefit significantly from the development and application of isotopic-labelling methods. We predict that this area should prove a productive meeting ground for process- and diversity-focused microbial ecologists.

The approaches outlined above should prove especially fruitful when used in combination with another emerging subdiscipline of microbial ecology, that of environmental genomics. Much has been made recently of the power of environmental genomics, exemplified by the massive sequencing efforts undertaken in the Sargasso Sea (Venter *et al.*, 2004). However, such studies tell us more about potential, rather than actual, community function because, for many of the retrieved genes, annotation provides no indication of a possible function of the encoded protein. Even for those genes where annotation is meaningful, no information on actual expression in the environment is apparent. Environmental proteomics offer a partial solution to this quandary (Wilmes & Bond, 2004), but this field is in its infancy and widespread application is surely several years away. The real beauty of environmental genomics, then (at least from a community-function perspective), may be its role in the formulation of hypotheses (Wagner, 2005). Strong hints about organismal metabolism can be obtained via meta-genome sequencing, hints that can be evaluated by using the isotopic-labelling techniques described in this chapter. The synergy derived from mixing these approaches could be profound: imagine a situation where one discovers a 'new' organism of presumed importance in an environment (e.g. any of a dozen or more candidate phyla with currently no cultured representatives), but nothing is known of its metabolism. FISH-MAR, SIP and isotope arrays all offer potential avenues for investigation, yet where would one start? Possible carbon sources alone number in the hundreds and it is simply not feasible to test all of these with isotopic methods. However, if one has genomic information for such organisms (as obtained by bacterial artificial chromosome or shotgun sequencing from a mixed sample), then valuable clues as to suitable target substrates should be forthcoming. The need for environmental genomics is arguably reduced when one approaches community function from the other side, i.e. starting with a process and identifying the key players. Here, the critical step is in defining incubation conditions where only the organisms capable of the function(s) of interest are active.

Since its first coupling with molecular-phylogenetic techniques in the late 1990s, isotopic labelling has provided important insights into the function of specific, un-cultured micro-organisms. Continuing technological advances and further integration with environmental-genomics and -proteomics approaches can only enhance the value of this area of microbial ecology.

REFERENCES

Adamczyk, J., Hesselsoe, M., Iversen, N., Horn, M., Lehner, A., Nielsen, P. H., Schloter, M., Roslev, P. & Wagner, M. (2003). The isotope array, a new tool that employs substrate-mediated labeling of rRNA for determination of microbial community structure and function. *Appl Environ Microbiol* **69**, 6875–6887.

Boschker, H. T. S., Nold, S. C., Wellsbury, P., Bos, D., de Graaf, W., Pel, R., Parkes, R. J. & Cappenberg, T. E. (1998). Direct linking of microbial populations to specific biogeochemical processes by ^{13}C-labelling of biomarkers. *Nature* **392**, 801–805.

Curtis, T. P., Sloan, W. T. & Scannell, J. W. (2002). Estimating prokaryotic diversity and its limits. *Proc Natl Acad Sci U S A* **99**, 10494–10499.

Daims, H., Nielsen, J. L., Nielsen, P. H., Schleifer, K.-H. & Wagner, M. (2001). In situ characterization of *Nitrospira*-like nitrite-oxidizing bacteria active in wastewater treatment plants. *Appl Environ Microbiol* **67**, 5273–5284.

Daims, H., Lücker, S. & Wagner, M. (2005). daime, a novel image analysis program for microbial ecology and biofilm research. *Environ Microbiol* (in press).

DeLong, E. F. (2002). Microbial population genomics and ecology. *Curr Opin Microbiol* **5**, 520–524.

DeLong, E. F. (2005). Microbial community genomics in the ocean. *Nat Rev Microbiol* **3**, 459–469.

Dumont, M. G. & Murrell, J. C. (2005). Stable isotope probing – linking microbial identity to function. *Nat Rev Microbiol* **3**, 499–504.

Guschin, D. Y., Mobarry, B. K., Proudnikov, D., Stahl, D. A., Rittmann, B. E. & Mirza-bekov, A. D. (1997). Oligonucleotide microchips as genosensors for determinative and environmental studies in microbiology. *Appl Environ Microbiol* **63**, 2397–2402.

Hesselsoe, M., Nielsen, J. L., Roslev, P. & Nielsen, P. H. (2005). Isotope labeling and microautoradiography of active heterotrophic bacteria on the basis of assimilation of ^{14}CO$_2$. *Appl Environ Microbiol* **71**, 646–655.

Lee, N., Nielsen, P. H., Andreasen, K. H., Juretschko, S., Nielsen, J. L., Schleifer, K.-H. & Wagner, M. (1999). Combination of fluorescent in situ hybridization and micro-autoradiography – a new tool for structure-function analyses in microbial ecology. *Appl Environ Microbiol* **65**, 1289–1297.

Loy, A., Lehner, A., Lee, N., Adamczyk, J., Meier, H., Ernst, J., Schleifer, K.-H. & Wagner, M. (2002). Oligonucleotide microarray for 16S rRNA gene-based detection of all recognized lineages of sulfate-reducing prokaryotes in the environment. *Appl Environ Microbiol* **68**, 5064–5081.

Loy, A., Schulz, C., Lücker, S., Schöpfer-Wendels, A., Stoecker, K., Baranyi, C., Lehner, A. & Wagner, M. (2005). 16S rRNA gene-based oligonucleotide microarray for environmental monitoring of the betaproteobacterial order "*Rhodocyclales*". *Appl Environ Microbiol* **71**, 1373–1386.

Lueders, T., Wagner, B., Claus, P. & Friedrich, M. W. (2004). Stable isotope probing of rRNA and DNA reveals a dynamic methylotroph community and trophic interactions with fungi and protozoa in oxic rice field soil. *Environ Microbiol* **6**, 60–72.

Manefield, M., Whiteley, A. S., Griffiths, R. I. & Bailey, M. J. (2002). RNA stable isotope probing, a novel means of linking microbial community function to phylogeny. *Appl Environ Microbiol* **68**, 5367–5373.

Manefield, M., Griffiths, R. I., Leigh, M. B., Fisher, R. & Whiteley, A. S. (2005). Functional and compositional comparison of two activated sludge communities remediating coking effluent. *Environ Microbiol* **7**, 715–722.

Nielsen, J. L., Juretschko, S., Wagner, M. & Nielsen, P. H. (2002). Abundance and phylogenetic affiliation of iron reducers in activated sludge as assessed by fluorescence in situ hybridization and microautoradiography. *Appl Environ Microbiol* **68**, 4629–4636.

Nielsen, J. L., Wagner, M. & Nielsen, P. H. (2003a). Use of microautoradiography to study *in situ* physiology of bacteria in biofilms. *Rev Environ Sci Biotechnol* **2**, 261–268.

Nielsen, J. L., Christensen, D., Kloppenborg, M. & Nielsen, P. H. (2003b). Quantification of cell-specific substrate uptake by probe-defined bacteria under *in situ* conditions by microautoradiography and fluorescence *in situ* hybridization. *Environ Microbiol* **5**, 202–211.

Olsen, G. J., Lane, D. J., Giovannoni, S. J., Pace, N. R. & Stahl, D. A. (1986). Microbial ecology and evolution: a ribosomal RNA approach. *Annu Rev Microbiol* **40**, 337–365.

Orphan, V. J., House, C. H., Hinrichs, K.-U., McKeegan, K. D. & DeLong, E. F. (2001). Methane-consuming archaea revealed by directly coupled isotopic and phylogenetic analysis. *Science* **293**, 484–487.

Orphan, V. J., House, C. H., Hinrichs, K. U., McKeegan, K. D. & DeLong, E. F. (2002). Multiple archaeal groups mediate methane oxidation in anoxic cold seep sediments. *Proc Natl Acad Sci U S A* **99**, 7663–7668.

Ouverney, C. C. & Fuhrman, J. A. (1999). Combined microautoradiography-16S rRNA probe technique for the determination of radioisotope uptake by specific microbial cell types in situ. *Appl Environ Microbiol* **65**, 1746–1752.

Rappé, M. S. & Giovannoni, S. J. (2003). The uncultured microbial majority. *Annu Rev Microbiol* **57**, 369–394.

Schloss, P. D. & Handelsman, J. (2004). Status of the microbial census. *Microbiol Mol Biol Rev* **68**, 686–691.

Skovhus, T. L., Ramsing, N. B., Holmström, C., Kjelleberg, S. & Dahllöf, I. (2004). Real-time quantitative PCR for assessment of abundance of *Pseudoalteromonas* species in marine samples. *Appl Environ Microbiol* **70**, 2373–2382.

Venter, J. C., Remington, K., Heidelberg, J. F. & 20 other authors (2004). Environmental genome shotgun sequencing of the Sargasso Sea. *Science* **304**, 66–74.

Wagner, M. (2004). Deciphering the function of uncultured microorganisms. *ASM News* **70**, 63–70.

Wagner, M. (2005). The community level: physiology and interactions of prokaryotes in the wilderness. *Environ Microbiol* **7**, 483–485.

Wagner, M., Amann, R., Lemmer, H. & Schleifer, K.-H. (1993). Probing activated sludge with oligonucleotides specific for proteobacteria: inadequacy of culture-dependent methods for describing microbial community structure. *Appl Environ Microbiol* **59**, 1520–1525.

Wagner, M., Horn, M. & Daims, H. (2003). Fluorescence *in situ* hybridisation for the identification and characterisation of prokaryotes. *Curr Opin Microbiol* **6**, 302–309.

Wilmes, P. & Bond, P. L. (2004). The application of two-dimensional polyacrylamide gel electrophoresis and downstream analyses to a mixed community of prokaryotic microorganisms. *Environ Microbiol* **6**, 911–920.

Wilson, K. H., Wilson, W. J., Radosevich, J. L., DeSantis, T. Z., Viswanathan, V. S., Kuczmarski, T. A. & Andersen, G. L. (2002). High-density microarray of small-subunit ribosomal DNA probes. *Appl Environ Microbiol* **68**, 2535–2541.

Biofilms and metal geochemistry: the relevance of micro-organism-induced geochemical transformations

Lesley A. Warren

School of Geography and Earth Sciences, McMaster University, 1280 Main St. West, Hamilton, Ontario, Canada L8S 4K1

INTRODUCTION

This chapter is intended to provide a brief overview of the key concepts underlying the emerging area of environmental microbial metal geochemistry, rather than an exhaustive synthesis. The reader is referred to the following more comprehensive reviews on the biogeochemistry of metals (Warren & Haack, 2001), metal–mineral reactions (Brown & Parks, 2001), emerging molecular-level geochemical techniques (O'Day, 1999; Brown & Sturchio, 2002) and a recent synthesis of how genetic expression in the environment can underpin geochemical reactions (Croal et al., 2004). The relevance of micro-organisms to metal behaviour arises from the overlap of the biosphere with the geosphere and the transformations that occur because of their interactions. Micro-organisms have evolved in intimate association with the rocks, soils and waters (i.e. geosphere) in which they find themselves. In order to grow and survive, they have adapted to these environments and use the inorganic components to drive their metabolic machinery; the myriad functional pathways by which they do so ensure that they influence a number of key elemental cycles in the process. As a consequence, many important geochemical processes are ultimately shaped by life, rather than strict geochemical equilibria, a fact that is increasingly recognized as strict geochemical principles fail to constrain observed environmental behaviour.

Trace-metal behaviour in the environment is of increasing global concern as water and soil contamination with these toxic substances continues and the detrimental effects on ecosystems and human health emerge. While investigation into metal behaviour has spanned many disciplines reflecting discipline-specific foci and approaches to the topic

SGM symposium 65: Micro-organisms and Earth systems – advances in geomicrobiology.
Editors G. M. Gadd, K. T. Semple & H. M. Lappin-Scott. Cambridge University Press. ISBN 0 521 86222 1 ©SGM 2005

(e.g. biology, chemistry, geochemistry, toxicity, physics), what coalesces from this broad, substantial literature is that the controls on aqueous metal behaviour supersede individual discipline boundaries; rather, it is often dynamic, complex, biological (principally micro-organism driven) and geochemical linkages that drive geochemistry (Reysenbach & Shock, 2002; Newman & Banfield, 2002; Warren & Kauffman, 2003; Warren, 2004).

Probing microbial–mineral–metal interactions has provided substantive evidence of microbial shaping of metal fate (e.g. Ehrlich, 2002; Holden & Adams, 2003; Islam *et al.*, 2004), and has clearly alluded to the need to understand how microbial activity and geochemistry interact to ultimately determine metal impact. It is increasingly evident that microbial activity can substantially impact aqueous geochemical behaviour in ways not predicted by classic geochemical models, a serious issue for contaminated environments where bacteria flourish such as acid mine drainage (AMD) or contaminated subsurface environments where migration of contaminants to groundwater supplies represents a serious health hazard.

AQUEOUS METAL GEOCHEMISTRY: FUNDAMENTALS

At the most fundamental level, the single most important predictor of metal behaviour, i.e. mobility within the geosphere, bioavailability, bioaccumulation and toxicity, is the solution or dissolved (operationally defined typically as less than 0·45 μm or, more rigorously, <0·1 μm) concentration of the element (e.g. Martell *et al.*, 1988; Campbell & Tessier, 1989; Hare, 1992; Campbell, 1995; Unz & Shuttleworth, 1996; Warren *et al.*, 1998; Hassler *et al.*, 2004). Typically, the greater the solution concentration of a metal, the more likely are the negative impacts associated with that contaminant, reflecting greater mobility, bioavailability and toxicity. Thus, trace-element geochemistry has focused on understanding the processes by which metals partition between the solution and solid compartments within environmental systems and the controls on the processes involved (Warren & Haack, 2001; and references therein). While precipitation of metals within metal-bearing minerals such as sulfides, carbonates or hydroxides can play a role, principally in sedimentary systems where concentrations of metals can reach saturation with respect to a given mineral phase, it is now well established that metal solid–solution partitioning is often dominantly controlled by interfacial reactions occurring between charged functional groups at solid surfaces and solution metal species, collectively referred to as sorption reactions (Jenne, 1968; Dzombak & Morel, 1990; Stumm & Morgan, 1996; Brown & Parks, 2001; Brown & Sturchio, 2002). In the majority of cases, most metal–solid interactions can be considered dynamically reversible, such that changes in certain geochemical conditions can lead to release of metals back into solution. A further, defining caveat to the framework of metal–solid interactions is that neither metal(s) nor solid phases can be considered collectively, as the behaviour observed will be predicated upon which specific metal(s) and solid phase(s) are involved, reflecting relative affinities, differing mechanisms of sequestration and controls on reactivity.

Aqueous geochemistry has served as the foundation discipline for much of the environmental metal literature spanning freshwater and marine metal behaviour, subsurface metal migration and the bioavailability and toxicity of metals to biota (e.g. Lee *et al.*, 2004; Martino *et al.*, 2004; Basta *et al.*, 2005). The defining controls of pH, redox status and ionic strength (Allard *et al.*, 1987; Johnson, 1990; Brown & Parks, 2001; Hassler *et al.*, 2004) on metal partitioning reflect their influence on: (i) formation/dissolution of key solid phases for sequestering trace metals (e.g. carbonates, sulfides and oxyhydroxide minerals, organic matter), as well as metal speciation and associated behaviour; and (ii) sorption reactions.

The nature and extent of a mineral's importance for metal sequestration in any given system will reflect its respective relative abundance and spatial distribution in the geosphere. The heterogeneous distribution of key mineral phases reflects the differing conditions and controls on their formation and dissolution and also provides some delineation of which minerals are likely to occur and play a role in metal behaviour across differing geochemical environments. Minerals can sequester metals through sorption at their surfaces as well as through direct precipitation or solid–solution formation within a mineral (Stumm & Morgan, 1996; Brown & Parks, 2001). The mechanism by which a metal becomes associated with a solid will determine its relative sensitivity to remobilization and the conditions under which release may occur. Sorption reactions can be considered reversible: metal uptake to solids is not a permanent phenomenon, as small fluctuations in either pH or solution metal(s) concentration can often lead to release of metals from a solid surface (Warren & Haack, 2001). Of key importance to note is that both the relative affinities of different solid phases for different metal ions, as well as the controls on their reactivity, differ. Thus, predicting metal solid–solution behaviour requires information on the types and abundances of solid phases present, the specific metal(s) and respective concentrations involved, as well as system geochemical conditions.

Important solid phases

In general, there are four major solid phases that can be considered most relevant for metal sequestration in aqueous systems: (i) carbonates, (ii) oxyhydroxide minerals (Fe and Mn); (iii) organic matter (live and dead) and (iv) sulfidic minerals (Tessier *et al.*, 1979). Many studies have shown that these typically comprise the majority of solid phases involved in metal uptake (Tessier & Campbell, 1988; Warren & Zimmerman, 1994; Brown *et al.*, 1999; Filgueiras *et al.*, 2002). Another final sedimentary pool is often referred to as the refractory or mineralized component (Tessier *et al.*, 1979), reflecting metals held within crystalline mineral lattices. However, metals associated with this refractory pool are not likely to be released under normal geochemical fluctuations and thus its utility is usually for determining the absolute total mass of metal associated with a given sediment. Changes in pH, redox or ionic strength can affect metal uptake to the previous four fractions, and thus much research has focused on understanding metal associations with these specific sedimentary pools, which commonly occur as heterogeneous mixtures in environmental sediments and soils.

Carbonate minerals [e.g. limestone, $CaCO_3(s)$, or dolostone, $Ca,MgCO_3(s)$] are typically found in circumneutral to alkaline pH environments, where their concentrations can effect significant metal uptake, especially for certain elements which show high affinity for carbonates, such as Cd and U. Carbonate minerals are susceptible to acid dissolution, and thus decreasing pH can lead to dissolution and subsequent release of carbonate-associated metals back into solution.

In contrast, sulfidic minerals such as pyrite [$FeS(s)$] tend to be associated with anoxic sedimentary systems where the necessary reducing conditions are present for their formation and/or preservation. Sulfidic minerals are commonly formed as a consequence of sulfate reduction in anoxic sediments, which leads to a build-up of reduced S in the sediment pore waters, which, when sufficient concentrations build up to saturate with respect to various metal sulfide mineral phases, can then precipitate as a metal sulfide [e.g. $CdS(s)$, $CuS(s)$, $FeS(s)$; Warren *et al.*, 1998]. Once formed, sulfide minerals tend to be fairly robust, even in the presence of oxygen, to oxidative dissolution, except where direct microbial catalysis is involved. Sulfidic minerals are also a key component in mine-associated waste or tailings rock, driving the production of AMD.

The Fe and Mn oxyhydroxides [e.g. Fe(III) oxides, $FeOOH(s)$ and Mn(III,IV) oxides, $MnOOH(s)$] are highly metal-reactive and are typically widespread throughout most environmental systems, making them a dominant solid phase controlling metal behaviour in the environment. Oxyhydroxides of Fe and Mn are redox-sensitive, with the more oxidized form of both elements, Fe(III) and Mn(III,IV), forming a solid phase, while the reduced forms, Fe(II) and Mn(II), are soluble. Thus changes in redox conditions will profoundly affect the concentrations of these solids and associated metal behaviour. Although it should be noted that reductive dissolution of both oxyhydroxides proceeds much more quickly with microbial catalysis than without and that, while abiotic oxidation of Fe(II) to Fe(III) [and associated hydrolysis and formation of $FeOOH(s)$ solid particles] occurs spontaneously above pH 3 in the presence of oxygen, abiotic Mn(II) oxidation is extremely slow below pH 9 (e.g. months to years; Morel & Hering, 1993; Stumm & Morgan, 1996).

Natural organic matter (NOM), both living (e.g. micro-organisms) and dead (both labile and refractory, e.g. humic and fulvic acids), is an efficient, but often dynamically reversible, metal sequestration pool (Gustafsson *et al.*, 2003; Smiejan *et al.*, 2003; Jacob & Otte, 2004; Boullemant *et al.*, 2004). Sequestration by NOM can be especially important in systems of high NOM concentration such as wetlands and eutrophic lakes, typically through sorption reactions (discussed subsequently) associated with their highly electrically charged surfaces. Decomposition of organic matter often leads to release of associated metals into solution and thus the ability of the NOM pool in any given system to hold onto metals will reflect the relative magnitude of decompositional processes.

Sorption reactions

The profound importance of interfacial reactions in controlling the ultimate partitioning of metals between the solid and solution phases is well recognized (Brown & Parks, 2001; Warren & Haack, 2001; and references therein). These interfacial or sorption reactions involve attraction or bonding between a charged solution metal ion and a charged functional group on the surface of the particle (Fig. 1a) and are viewed to proceed analogously to complexation reactions involving charged dissolved species in solution (Buffle, 1988). Mineral surface reactivity is described by physical characteristics such as size (surface area to volume), crystallinity and nature and density of reactive binding sites (Brown *et al.*, 1999; Martínez & McBride, 1998). Organic matter reactivity reflects both these same geochemical particle characteristics and, for live cells, metabolic pathway and level of activity, which can shape external and internal microgeochemical environments, influence mineral formation and dissolution and alter surface charge characteristics and thus potential reactivity (e.g. Ferris *et al.*, 1989, 1999; Nelson *et al.*, 1995; Parmar *et al.*, 2000; Philip *et al.*, 2000; Templeton *et al.*, 2003a, b; Brown *et al.*, 2004).

This particle surface charge is imparted by surface acid functional groups that variously deprotonate or protonate depending on their pK_a values. Commonly, mineral surface functional groups are hydroxyl groups, which often display a spectrum of pK_a values reflecting the specific hydroxyl group co-ordinational environment on the particle (e.g. edge or defect sites within the crystal structure). Micro-organisms also carry a surface charge associated with their cell wall, exopolysaccharide (EPS) or sheath, arising from constituent functional groups (e.g. carboxylic, phosphoryl, amine as well as hydroxyl; Cox *et al.*, 1999). Given the lower pK_a values particularly associated with carboxylic groups ($pK_a \sim 4$), micro-organisms almost always carry a net negative surface charge in most aqueous solutions. Microbial ability to sorb cationic metal ions makes them a highly relevant sorbent, especially in low-pH environments where important mineral

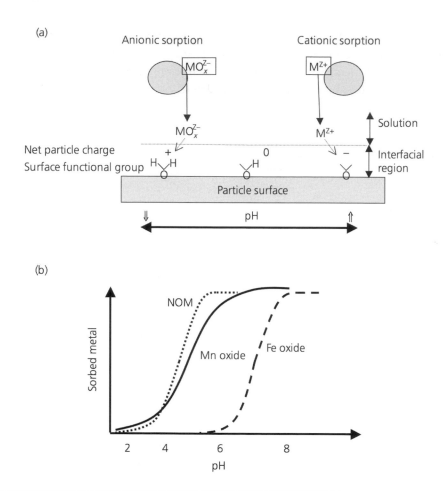

Fig. 1. Interfacial reactions involving particle-surface functional groups and solution ions play a defining role in controlling metal partitioning between solid and solution phases, ultimately influencing resulting metal behaviour. (a) Particle surfaces in aqueous systems carry a net overall charge reflecting the acid–base characteristics of their functional surface groups, dependent on system pH conditions. Generally, as pH decreases, greater anionic sorption occurs as particle surface charge becomes net positive, reflecting greater uptake of protons. In contrast, as pH increases, a greater net negative charge exists on particle surfaces so that cationic solution species such as divalent metal ions (e.g. Cd^{2+}) sorb more effectively to the negatively charged particle. The pH at which individual particles (e.g. different minerals, organic matter, dead or alive) become negatively or positively charged is dependent on the structural composition of the functional groups involved. Typically, most mineral functional groups are hydroxyl groups (as shown), although organic matter can have several types of functional groups with widely differing pK_a values (e.g. carboxylic versus amine groups; Cox et al., 1999). (b) Generalized sorption edges shown for three important metal sorbents in natural systems. Most cationic elements show a characteristic exponential increase in sorption over a relatively small pH range for a given solid, reflecting pH–pK_a relationships for that solid. Note that the pH at which the 'edge' effect occurs for a given solid will vary depending on the metal involved. Mn oxyhydroxides and organic matter are typically negatively charged at much lower pH values than Fe oxides, reflecting the differences in the acid–base characteristics of their respective surface functional groups, and thus can sorb cationic species at much lower pH values than can Fe oxyhydroxides, which, in contrast, are often effective anionic sorbents at low pH.

sorbents (e.g. Fe oxyhydroxides) are less effective. Further, cells need be neither viable nor intact to act as metal sequesters (Ferris *et al.*, 1989), as the functional groups associated with their cell-wall structures can still sorb metals (often with higher sorptive capacities) when the cells die or lyse (Urrutia, 1997).

Most particles in solution carry a charge and, as that surface becomes more net negatively or positively charged (many mineral surfaces are amphoteric), their relative abilities to sorb cationic or anionic species, respectively, increases. Thus, the overwhelming importance of pH in regulating trace-metal sorption across systems has been widely accepted (Stumm, 1992; Stumm & Morgan, 1996). As pH increases, there is a greater net deprotonation of particle surfaces and associated sorption of cationic metal species, partitioning them to the solid, and less mobile, bioavailable and toxic, phase. In contrast, as pH decreases, there is greater sorption potential for anionic species reflecting the greater positive charge associated with particle surfaces (the exact pH at which this occurs is solid dependent). Further, adding to the complexity in dealing with heterogeneous solid and metal mixtures as is observed in the environment, 'edge' effects are observed for metal sorption to a given solid (Fig. 1b), whereby sorption increases substantially over a relatively narrow pH range as the pH of the system moves above the pK_a for dissociation of the specific solid's surface acid–base sites (Dzombak & Morel, 1990). However, these pH edges are both solid dependent, reflecting the different pH values at which a surface becomes net negatively or positively charged (Fig. 1b), and metal dependent, reflecting differential affinities of specific elements (often correlated to a metal's ability to hydrolyse). As is evident from Fig. 1, organic matter, e.g. microorganisms, and Mn oxyhydroxides are able to sorb cationic species at much lower pH values than Fe oxyhydroxides (reflecting their different pH_{zpc} values, or pH at which their surfaces shift between net negative and net positive overall charge; Fig. 1). This makes them of key importance in low-pH systems, whilst Fe oxyhydroxides would be more important for anionic species under lower pH conditions. Typically, these sorption edges are also element specific for a given solid phase, indicating that selective affinity occurs. This negates the possibility of a generalized model for metal sorption independent of either the metal or solid involved.

In brief, the role of ionic strength as a control in metal solid–solution partitioning derives from its influence on the potential electrostatic attractions that occur in solution, as well as on the nature of the interfacial region of the solid surface (Puls *et al.*, 1991; Lores & Pennock, 1998). The most labile fraction of solid-associated metals is often referred to as 'exchangeable', referring to those metals that are electrostatically attracted to particle surfaces, but not necessarily bonded to a specific site on the surface or to a specific sedimentary component within a mixed sediment pool. This fraction is the most easily released back into solution with any changes in solution conditions. In freshwater systems, increasing ionic strength increases both solution interactions, thereby decreasing element activity and compressing the electrical double layer surrounding the particle, and the associated sorptive uptake of metals (Drever, 1997).

The role of redox status

Changes in redox status can affect both solids and metals that are redox sensitive, influence partitioning between the solid and solution pools and thus change metal behaviour. Oxyhydroxide minerals are particularly sensitive to changes in redox status, such that reducing conditions can lead to their reductive dissolution and the associated release of any sorbed metals into solution. Oxidizing conditions would favour (especially for Fe) solid formation and associated sequestration of metals. Reductive dissolution of Fe and Mn oxyhydroxides is commonly driven by micro-organism Fe and Mn reduction coupled to organic matter degradation (see equation below; note that 'CH$_2$O' represents organic matter), where reduction of the oxidized Fe and Mn in the oxyhydroxide mineral leads to dissolution of the solid and release of any associated metals from the oxyhydroxide into solution:

$$4Fe(III)OOH(s) + CH_2O + 8H^+ \rightarrow 4Fe(II) + CO_2 + 7H_2O$$

Typically, reductive dissolution of both oxyhydroxides, as well as oxidative dissolution of sulfidic minerals or organic matter, effectively proceeds only with microbial catalysis: abiogenic reaction rates are extremely slow (e.g. Morel & Hering, 1993; Stumm & Morgan, 1996; O'Day et al., 2004). Further, some redox-active (and often therefore bioactive) metals such as U, Cr and As show profoundly different behaviour depending on their redox status [e.g. U(VI) or U(IV), Cr(VI) or Cr(III), As(V) or As(III)]. For instance, Cr speciation is controlled by a shifting array of processes that include redox transformations, precipitation/dissolution and sorption/desorption reactions, all of which are intimately tied to both oxidation state and the geochemical status of a system (Bartlett, 1991; Richard & Bourg, 1991; Baruthio, 1992). In its VI oxidation state, Cr commonly occurs as the ligand CrO_4^{2-}, which is highly mobile, toxic and soluble (Felter & Dourson, 1997) and is generally controlled by sorption–desorption reactions at lower concentrations. Cr(VI), as chromate, is sorbed to minerals under lower pH conditions, when surface sites are positively charged. For example, Zachara et al. (1989) have shown that the degree of chromate sorption is greater in acidic soils and in subsurface media containing iron oxyhydroxides and clays. The reduced form of Cr, Cr(III), is insoluble in aqueous environments. Precipitation of $Cr(OH)_3$ or (Cr,Fe)-$(OH)_3$ compounds commonly controls its behaviour.

THE OVERLAP BETWEEN MICROBIOLOGY AND METAL GEOCHEMISTRY

The role of emerging technologies

It is only a small step across discipline divides to see the very real and often significant overlap between microbial activity in the environment and metal geochemistry. In large part, the recognition that microbial activity can affect metal geochemistry has grown with our ability to examine the linkages at the appropriate scale, i.e. the micron level. Our ability to effectively probe and 'image' at the appropriate scale of resolution, using such techniques as X-ray absorption spectroscopy (XAS; Brown & Sturchio, 2002), atomic force microscopy (AFM) and scanning transmission X-ray microspectroscopy (STXM; O'Day, 1999), has increased our understanding of the mechanisms involved in solid–metal interactions. In addition, perhaps the most powerful tool which revolutionizes our ability to examine micro-organisms in an environmental context has been the development of culture-independent molecular-biological tools for phylogenetic characterization (e.g. Pace, 1997; Kaeberlein *et al.*, 2002), which have permitted evaluation of community genetic diversity.

Relevance of microbial functional metabolism

Micro-organisms can influence metal behaviour through a number of processes, reflecting the direct link between types and rates of metabolic activity and geochemical conditions, processes and reaction rates (Fig. 2). The occurrence of micro-organisms in almost every environment investigated (e.g. Bennett *et al.*, 2000; Bond *et al.*, 2000; Chapelle *et al.*, 2002; Takai *et al.*, 2004) hints at their widespread influence on geochemical processes. Unlike higher eukaryotic organisms, micro-organisms are not restricted by geographical barriers (Finlay, 2002). Micro-organisms seek energy and carbon to survive and grow. In the environment, they select for geochemical conditions which are favourable and commonly catalyse geochemical processes for their growth (Nealson & Stahl, 1997; Nealson, 2003). While thermodynamics sets energetic constraints on microbial activity, i.e. organisms must live within the realm of reactions that are thermodynamically favourable, it is increasingly clear that the extent of this domain of possible reactions in the geosphere is not yet completely described (e.g. Spear *et al.*, 2005).

Beyond the simple aerobic versus anaerobic demarcation in metabolism, there are myriad potential metabolic pathways (i.e. electron donors and acceptors) with relevance for metal behaviour through which micro-organisms can selectively shape geochemical processes dependent on the particular redox couples that they catalyse. Such processes as nitrification and denitrification, methanogenesis, methanotrophy, sulfur/iron/manganese oxidation and reduction (Holt & Leadbetter, 1992; Nealson &

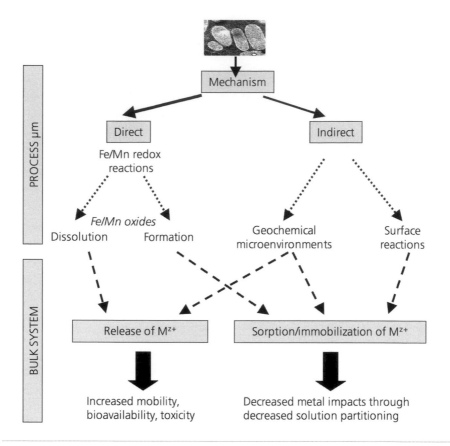

Fig. 2. Microbial influence on metal geochemistry can occur through both indirect and direct mechanisms. In both instances, microscale influence by microbial metabolic activity can scale to bulk-system impacts for metal behaviour, depending on the level of metabolic activity. Indirect mechanisms refer to general changes in the local geochemical environment associated with metabolic activity that then directionally affect metal solid–solution partitioning. Direct mechanisms refer to those specifically catalysed by microbes, e.g. Fe oxidation will lead to the formation of Fe oxyhydroxides and will likely lead to metal sequestration, while Fe reduction will dissolve Fe oxyhydroxides, liberating any associated metals into solution, increasing their mobility, bioavailability and likely impact.

Stahl, 1997) and hydrogen-based metabolism (Chapelle *et al.*, 2002; Spear *et al.*, 2005) can all substantively affect metal speciation and thus behaviour (Fig. 3).

In addition to the ecological factors of sufficient nutrients, water and protection from predation, probably one of the most important parameters defining metabolic habitats is oxygen concentration and/or flux. This is also a defining geochemical parameter controlling the oxidative state of reactive geochemical components and thus segregating differing geochemical processes within the environment. Thus, it is clear that aerobic and anaerobic organisms will select for differing environments on the basis of

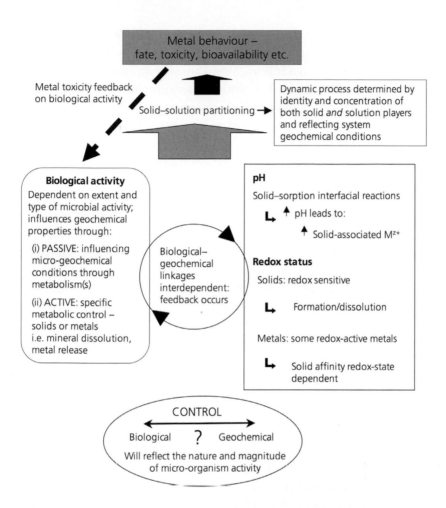

Fig. 3. Understanding the controls of metal geochemistry is increasingly interpreted as a dynamic array reflecting the interactions between microbial activity and the important controls on metal solid–solution partitioning. Feedback mechanisms of local geochemical conditions on existence, activity and growth of micro-organisms also occur. The relative dominance of microbial or classic geochemical controls on metal behaviour will reflect the extent of microbial activity in a given system or at a given time.

oxygen concentration, which similarly stratifies geochemically important solids and/or forms of elements important in metal behaviour. Thus, metal impacts will be selectively dependent on both the types of metabolism and levels of activity occurring in any micro-environment. Regardless of the specific metabolic guild involved, the common and widespread occurrence of microbial control on important geochemical processes is increasingly demonstrated (e.g. biogenic Fe oxide formation: Chafetz *et al.*, 1998; Newman & Banfield, 2002; Haack & Warren, 2003).

As field investigations (e.g. Smiejan *et al.*, 2003; Brown *et al.*, 2004; Labrenz & Banfield, 2004; Meylan *et al.*, 2004) and associated laboratory experimentation to probe mechanisms and effects (e.g. Warren & Ferris, 1998; Parmar *et al.*, 2000; Moreau *et al.*, 2004; Tani *et al.*, 2004a, b; Templeton *et al.*, 2003a, b) demonstrate, micro-organisms influence trace-metal behaviour through both passive and active mechanisms (Fig. 2). Passive or indirect mechanisms are viewed as those associated with general metabolic activity, i.e. where the particular metabolic pathways and levels of activity associated with a given microbial community induce changes in geochemical conditions within the immediate micro-geochemical environment. These can then influence metal behaviour, e.g. increasing pH would generally favour sorption of more metals to the solid compartment. Active or direct mechanisms are defined as those that reflect direct catalysis by micro-organisms of geochemical processes that control metal behaviour, e.g. reduction of Fe resulting in the dissolution of Fe oxide minerals and the associated release of metals into solution. Increasing evidence indicates the ability of many micro-organisms to directly metabolize many bioactive (redox-active) elements such as U, Cr and As (e.g. Lores & Pennock, 1998; Lovley & Anderson, 2000; Liu *et al.*, 2002), which alters the behaviour of these metals in the process.

Micro-organisms can also profoundly influence metal behaviour through associated impacts on mineral formation and dissolution (Konhauser *et al.*, 1993; Bennett *et al.*, 2000; Haack & Warren, 2003; Templeton *et al.*, 2003a; Moreau *et al.*, 2004). However, it is also clear that the geochemical conditions in which micro-organisms find themselves can influence their viability, activity and growth and, commonly, interactions between microbial behaviour and geochemistry exert feedback on each other (Warren & Haack, 2001; Warren & Kauffman, 2003). Thus, there is the potential for geochemical feedback on biological processes that will in turn influence the metal behaviour observed, which clearly indicates that the linkages between microbial activity and geochemical behaviour are complex and dynamic (Fig. 3).

Biofilms: existence, structure and function

It is increasingly clear that many micro-organisms self-organize into diverse biofilm communities (Davies *et al.*, 1998; Branda *et al.*, 2005), which can produce complex and dynamic internal geochemical conditions based on the particular array of metabolic processes occurring within the biofilm (Little *et al.*, 1997). Biofilms form at interfaces between solid and solution phases, overlapping with the same dynamic reactive zone that controls metal partitioning. Biofilms can also form on particles suspended in solution (Leppard *et al.*, 2003, 2004; Roberts *et al.*, 2004) or at the bed-sediment surface (Fig. 4a; Vigneault *et al.*, 2001; Haack & Warren, 2003). In either scenario, these biologically controlled layers can selectively control metal–solid interactions depending on the particular geochemical conditions created by the biofilm's specific metabolic array (Fig. 4b). Biofilms must provide a selective advantage for the organisms that form them. Typically, biofilms are thought to provide structural protection, through the associated EPS matrix, from predation and/or impacts of potentially toxic contaminants (Lawrence *et al.*, 1995, 1998; Neu & Lawrence, 1997; Davies *et al.*, 1998).

Mixed-community biofilms exhibit stratified metabolic functions reflecting the energetic array of reactants and products which link these microbial communities together. Further, minerals are commonly associated with biofilms (e.g. Douglas & Beveridge, 1998; Ferris *et al.*, 1999; Haack & Warren, 2003; Templeton *et al.*, 2003a; Moreau *et al.*, 2004), either through passive biomineralization, where the EPS and/or cells of the biofilm act as nucleation templates for minerals to form, or actively, where specific minerals are catalysed through specific functions of the biofilm. In either scenario, minerals are commonly associated with biofilms as either bioreactants and/or bioproducts of microbial metabolism.

Biofilm metabolic links to metals

Since biofilms represent a concentrated cell mass with significant metabolic activity, their occurrence in the environment can lead to a significant biological overlay on geochemically reactive processes relevant to metal behaviour that will be governed by the ecological controls on biofilm formation and function. While it is generally held that most systems are at or near equilibrium with respect to acid–base reactions (e.g. pH), it is very clear that, without microbial catalysis of important redox geochemical reactions, many would not occur over relevant timescales due to the extremely slow kinetics of many of these reactions in the geosphere. Another clearly important factor to consider is that geochemical processes that proceed abiogenically (geochemically controlled) often do so under different environmental conditions and controls and with potentially different geochemical outcomes from biogenically micro-organism-catalysed processes (e.g. Warren & Ferris, 1998; Haack & Warren, 2003; Tani *et al.*, 2004a, b).

Environmental biofilm metal geochemistry

Biofilms are common in metal-contaminated and/or low-pH systems (Southam & Beveridge, 1992; Ledin & Pedersen, 1996; Nordstrom & Southam, 1997; Edwards *et al.*, 1999, 2000; Hunt *et al.*, 2001; Leveille *et al.*, 2001; Vigneault *et al.*, 2001; Baker *et al.*, 2004; Labrenz & Banfield, 2004), indicating that these extreme conditions provide favourable conditions for adapted microbial communities to flourish. They also indicate that there is significant potential for biofilms in these environments to impact metal geochemistry. As yet, there are not many studies that specifically evaluate the linkage between metal behaviour and biofilm metabolic functional activity. However, acid mine drainage (AMD) has long been recognized as one of the earliest environ-

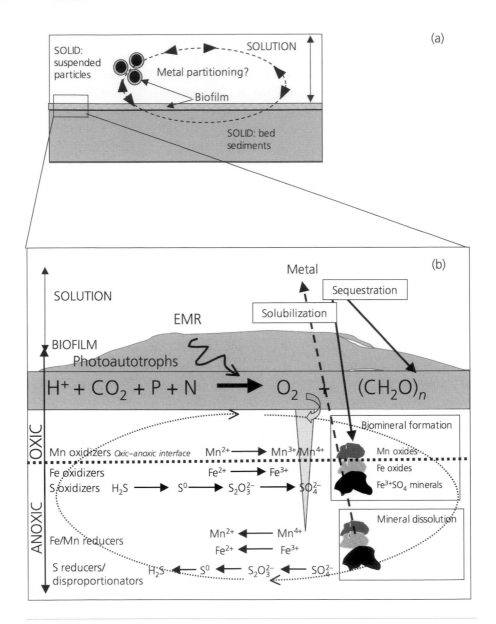

Fig. 4. Biofilms form at solid–solution interfaces, either at bed-sediment solution interfaces or at suspended particle surfaces (a), thus overlapping with the reactive interface for metal solid–solution partitioning. These biologically controlled interfacial structures can profoundly influence the nature, extent and types of geochemical processes that occur, particularly with respect to those that influence metal partitioning in a manner dependent on the particular array of metabolic functions expressed within a given biofilm (b). As shown in (b), in this case a biofilm with phototrophs occurring at the surface driving O_2 fluctuations within the surface biofilm, a generalized microbial biofilm is typically a stratified community that will show depth-dependent metabolic pathways typically as a function of oxygen status, which will also strongly influence the types of geochemical processes and the nature of both solid and solution elements that can occur at any given depth within the biofilm

ments for evidence of microbially controlled geochemical processes (e.g. Ledin & Pedersen, 1996; Nordstrom & Southam, 1997), providing some evidence of the shift in geochemical behaviour with micro-organism activity. AMD is strongly driven by microbially induced weathering and oxidation of sulfidic (e.g. pyrite, pyrrhotite) minerals exposed during the mining of base metals and coal (Fig. 5). The oxidation of sulfide-containing waste rock is significantly catalysed by iron- and sulfur-oxidizing bacteria that are attached or in close proximity to the sulfide minerals. This oxidation produces large amounts of acid. Recent investigation of AMD-associated biofilm metal dynamics (Haack & Warren, 2003) indicates that these biological structures are efficient metal sequesterers (Fig. 4b), through bacterially catalysed biomineralization reactions that would not be predicted to occur abiogenically.

While studies are emerging that provide direct evidence of both biomineralization by biofilms (e.g. Haack & Warren, 2003; Templeton *et al.*, 2003b; Labrenz & Banfield, 2004; Moreau *et al.*, 2004) and metal uptake by biofilms (e.g. Templeton *et al.*, 2001, 2003a; Vigneault *et al.*, 2001; Haack & Warren, 2003), as yet relatively few have integrated metal geochemistry dynamics with microbial community dynamics. However, those studies that have been done demonstrate that the following two major processes influence biofilm metal uptake. Firstly, the process of biofilm-associated mineral formation, whether passive or active, often plays a key role in metal sequestration (e.g. Nelson *et al.*, 1995, 1999; Ferris *et al.*, 1999). However, subsequent dissolution of biominerals, which tend to be small and amorphous, can lead to dynamic temporal fluctuations in biofilm metal content as mineral formation and dissolution are cyclically catalysed (Haack & Warren, 2003). Secondly, the organic fraction of the biofilm can also play a substantive role in metal sequestration through sorption to EPS as well as to cell walls (live and/or dead cells) and intracellular metal uptake, accumulation and storage (Beveridge, 1989; Schorer & Eisele, 1997; Lünsdorf *et al.*, 1997; Vigneault *et al.*, 2001; Haack & Warren, 2003; Brown *et al.*, 2004; Meylan *et al.*, 2004).

('CH_2O' represents organic carbon molecules; EMR, electromagnetic radiation). In an interlinked fashion, the micro-organisms dependent on the geochemical components for their growth and survival and the geochemical components influenced by the microbial catalysis of reactions driving their consumption or production, a biofilm is a geochemical reactor that is strongly driven by biology and overlaps directly with the reactive metal zone between solids and solution. The nature of the metabolic processes that occur within each functional metabolic stratum can profoundly influence key geochemical processes controlling metal behaviour, particularly those reactions involving the formation and dissolution of metal reactive minerals such as Fe and Mn oxyhydroxides (as shown in b). Thus, biofilm metal behaviour is likely to vary across systems, reflecting the specific interplay of geochemistry and microbiology, as well as for a given biofilm structure in both a depth-dependent and a temporal manner. If the biofilm has a phototrophic component (as exemplified in b), this will show shifts in respiration dominance (and therefore O_2 saturation) over diel timescales.

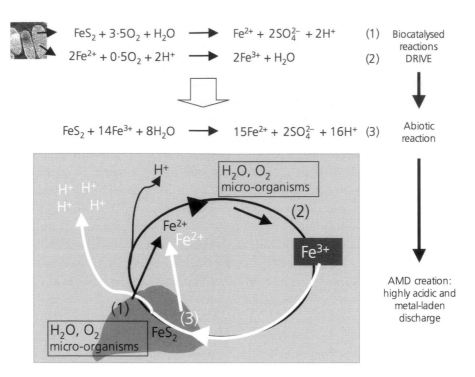

Fig. 5. The weathering of tailings, metal-sulfide waste rock material, is catalysed through bacterial action (the diagram indicates that reactions 1 and 2 are biogenic and drive reaction 3, an abiogenic process that liberates significant acid). Abiogenic oxidation/weathering of these reduced sulfidic minerals, even when exposed to O_2, are exceedingly slow. The bacterium *Thiobacillus ferrooxidans* catalyses oxidation of $FeS_{(x)}$ minerals through direct oxidation of $FeS_{(x)}$ (pyrite, FeS_2, used here) and minerals (equation 1) and through the oxidation of Fe^{2+} to Fe^{3+} (equation 2, increasing rates of Fe^{2+} oxidation up to five orders of magnitude; Singer & Stumm, 1970). This biogenically produced Fe^{3+} then abiogenically catalyses further (more rapid) abiogenic weathering of $FeS_{(x)}$ (equation 3). It is reaction 3 that drives significant acid production associated with AMD. In the process, acid is liberated, metals are typically solubilized and resulting AMD is both acidic and metal laden. The diagram indicates that the microbe-driven processes of (1) and (2) (top left half of diagram) drive abiogenic Fe^{3+}-subsequent weathering of the $FeS_{(x)}$ minerals.

CONCLUSIONS

Trace-metal behaviour is controlled by a complex and shifting array of processes that include redox transformations, precipitation/dissolution and sorption/desorption reactions, all of which are intimately tied to the geochemical status of a system and can be selectively influenced by microbial activity in a community-dependent manner, reflecting the particular functional array involved. Our understanding of how biology and geochemistry integrate in such a dynamic array of potential interactions is, at best, rudimentary. However, it is clear that, in many environmental instances, an overlap occurs between microbial activity and metal behaviour, reflecting the physical over-lap of microbial biofilm habitat and reactive metal zones in the environment. Micro-organisms influence their immediate micro-geochemical environment as well as directly catalyse geochemical transformations required for their survival and growth. In both instances they can profoundly influence the metal behaviour observed. However, micro-organisms are also intimately affected by their environment and thus feedback is exerted in both directions as dynamic and fluctuating geochemical conditions influence metabolic pathways, rates and times of expression. Thus, it is evident that under-standing and predicting geochemical behaviour requires an understanding of the often complex and dynamic linkages between ecological controls on where micro-organisms exist, how they make their living and their level of metabolic activity, and the geo-chemical processes they use to sustain themselves. Further, microbial plasticity ensures that organisms not only exploit available redox couples, but they also respond to microscopic variations in geochemical conditions, whilst simultaneously creating heterogeneity from the output of their own metabolic activity (Finlay, 2002). Thus, system-dependent and site-dependent microscale spatial and temporal heterogeneity needs to be investigated in order to constrain at the geosphere level the complex interactions of micro-organisms and geochemistry that result in the ultimate behaviour observed.

The challenge for this emerging field is to define the relevant controls in an integrated ecological and geochemical space and to develop sensitive techniques for their detection and accurate quantification. Emerging techniques are providing new insight into 'which organisms are present' in different environments, a fundamental question that was, until recently, difficult to answer. The extension of this question to 'what are they doing' is the current focus and the future of this area, as we seek to address mechanistically how microbial metabolism or activity links directly to geochemical reactions and shapes outcomes of relevant processes (Croal et al., 2004; Warren, 2004). As recog-nition of the need to examine the types and levels of metabolic activity occurring in the environment (e.g. proteomics, metallomics) has grown in appreciation of the challenge of quantifying microbial links to geochemical reactions, increasing methodological development has occurred. It is still a challenge to characterize metabolic pathways with microbial strains only identified on the basis of 16S rRNA oligonucleotide sequences, as so few environmental micro-organisms are currently in culture. This means that, while an exponentially growing database of environmental strains is genetically identified, the vast majority of these novel organisms remain undescribed in terms of their metabolic pathways. Because of this hurdle, other techniques are focused on evaluation of gene expression in the environment (dependent on their character-ization) or on using isotopes to track community diversity, structure and function. The field of molecular ecology is providing exciting new studies that track both molecular diversity and function, with apparently promising selectivity, through both labelling and probing of environmental processes using stable isotopes (e.g. Engel et al., 2004;

Griffiths *et al.*, 2004; Lueders *et al.*, 2004; Manefield *et al.*, 2004; Pearson *et al.*, 2004; Andersen *et al.*, 2005; Lu *et al.*, 2005). Until such time as gene expression (i.e. functional activity) for any specific organism in the environment is easily identifiable and quantifiable, the use of isotopes as an additional tool to probe actively occurring geochemical transformations linked to microbial activity appears highly promising.

ACKNOWLEDGEMENTS

This chapter presents a synthesis of ideas that have grown, coalesced and crystallized predominantly from the many discussions and interactions I have had with my students over the last 5 years. I would like especially to thank Dr Elizabeth Haack, Luc Bernier, Derek Amores, Corrie Kennedy, Tara Nelson and Lisa Melymuk. I would also like to acknowledge substantial funding support from NSERC, CFI, OIT, McMaster University and Noranda/Falconbridge Ltd.

REFERENCES

Allard, B., Hakansson, K. & Karlsson, S. (1987). The importance of sorption phenomena in relation to trace element speciation and mobility. In *Speciation of Metals in Water, Sediment and Soil Systems*, Lecture Notes in Earth Sciences, no. 11, pp. 99–112. Edited by L. Landner. Berlin: Springer.

Andersen, J. S., Lam, Y. W., Leung, A. K. L., Ong, S. E., Lyon, C.-E., Lamond, A. I. & Mann, M. (2005). Nucleolar proteome dynamics. *Nature* **433**, 77–83.

Baker, B. J., Lutz, M. A., Dawson, S. C., Bond, P. L. & Banfield, J. K. (2004). Metabolically active eukaryotic communities in extremely acidic mine drainage. *Appl Environ Microbiol* **70**, 6264–6271.

Bartlett, R. J. (1991). Chromium cycling in soils and water: links, gaps, and methods. *Environ Health Perspect* **92**, 17–24.

Baruthio, F. (1992). Toxic effects of chromium and its compounds. *Biol Trace Elem Res* **32**, 145–153.

Basta, N. T., Ryan, J. A. & Chaney, R. L. (2005). Trace element chemistry in residual-treated soil: key concepts and metal bioavailability. *J Environ Qual* **34**, 49–63.

Bennett, P. C., Hiebert, F. K. & Roberts Rogers, J. (2000). Microbial control of mineral-groundwater equilibria: macroscale to microscale. *Hydrogeol J* **8**, 47–62.

Beveridge, T. J. (1989). Role of cellular design in bacterial metal accumulation and mineralization. *Annu Rev Microbiol* **43**, 147–171.

Bond, P. L., Smriga, S. P. & Banfield, J. F. (2000). Phylogeny of microorganisms populating a thick, subaerial, predominantly lithotrophic biofilm at an extreme acid mine drainage site. *Appl Environ Microbiol* **66**, 3842–3849.

Boullemant, A., Vigneault, B., Fortin, C. & Campbell, P. G. C. (2004). Uptake of neutral metal complexes by a green alga: influence of pH and humic substances. *Aust J Chem* **57**, 931–936.

Branda, S. S., Vik, A., Friedman, L. & Kolter, R. (2005). Biofilms: the matrix revisited. *Trends Microbiol* **13**, 20–26.

Brown, G. E., Jr & Parks, G. A. (2001). Sorption of trace elements on mineral surfaces: modern perspectives from spectroscopic studies, and comments on sorption in the marine environment. *Int Geol Rev* **43**, 963–1073.

Brown, G. E., Jr & Sturchio, N. C. (2002). An overview of synchrotron radiation applications to low temperature geochemistry and environmental science. *Rev Miner Geochem* **49**, 1–115.

Brown, G. E., Jr, Henrich, V. E., Casey, W. H. & 11 other authors (1999). Metal oxide surfaces and their interactions with aqueous solutions and microbial organisms. *Chem Rev* **99**, 77–174.

Brown, G. E., Jr, Yoon, T. H., Johnson, S. B., Templeton, A. S., Trainor, T. P., Bostick, B. C., Kendelewicz, T., Doyle, C. S. & Spormann, A. M. (2004). The role of organic molecules and microbial organisms in metal ion sorption processes. *Geochim Cosmochim Acta* **68**, A160.

Buffle, J. (1988). *Complexation Reactions in Aquatic Systems: an Analytical Approach.* Chichester: Ellis Horwood.

Campbell, P. G. C. (1995). Interactions between trace metals and aquatic organisms: a critique of the free-ion activity model. In *Metal Speciation and Bioavailability in Aquatic Systems*, IUPAC Series on Analytical and Physical Chemistry of Environmental Systems, pp. 45–102. Edited by A. Tessier & D. Turner. Chichester: Wiley.

Campbell, P. G. C. & Tessier, A. (1989). Geochemistry and bioavailability of trace metals in sediments. In *Aquatic Ecotoxicology*, pp. 124–148. Edited by A. Boudou & F. Ribeyre. Boca Raton, FL: CRC Press.

Chafetz, H. S., Akdim, B., Julia, R. & Reid, A. (1998). Mn- and Fe-rich black travertine shrubs; bacterially (and nanobacterially) induced precipitates. *J Sediment Res* **68**, 404–412.

Chapelle, F. H., O'Neill, K., Bradley, P. M., Methe, B. A., Ciufo, S. A., Knobel, L. L. & Lovley, D. R. (2002). A hydrogen-based subsurface microbial community dominated by methanogens. *Nature* **415**, 312–315.

Cox, J. S., Smith, D. S., Warren, L. A. & Ferris, F. G. (1999). Characterizing heterogeneous bacterial surface functional groups using discrete affinity spectra for proton binding. *Environ Sci Technol* **33**, 4514–4521.

Croal, L. R., Gralnick, J. A., Malasarn, D. & Newman, D. K. (2004). The genetics of geochemistry. *Annu Rev Genet* **38**, 175–202.

Davies, D. G., Parsek, M. R., Pearson, J. P., Iglewski, B. H., Costerton, J. W. & Greenberg, E. P. (1998). The involvement of cell-to-cell signals in the development of a bacterial biofilm. *Science* **280**, 295–298.

Douglas, S. & Beveridge, T. J. (1998). Mineral formation by bacteria in natural microbial communities. *FEMS Microbiol Ecol* **26**, 79–88.

Drever, J. I. (1997). *The Geochemistry of Natural Waters: Surface and Groundwater Environments*, 3rd edn. London: Prentice-Hall.

Dzombak, D. A. & Morel, F. M. M. (1990). *Surface Complexation Modeling: Hydrous Ferric Oxide.* New York: Wiley.

Edwards, K. J., Gihring, T. M. & Banfield, J. F. (1999). Seasonal variations in microbial populations and environmental conditions in an extreme acid mine drainage environment. *Appl Environ Microbiol* **65**, 3627–3632.

Edwards, K. J., Bond, P. L., Gihring, T. M. & Banfield, J. F. (2000). An archaeal iron-oxidizing extreme acidophile important in acid mine drainage. *Science* **287**, 1796–1799.

Ehrlich, H. L. (2002). How microbes mobilize metals in ores: a review of current understandings and proposals for further research. *Miner Metallurg Process* **19**, 220–224.

Engel, A. S., Porter, M. L., Stern, L. A., Quinlan, S. & Bennett, P. C. (2004). Bacterial diversity and ecosystem function of filamentous microbial mats from aphotic (cave) sulfidic springs dominated by chemolithoautotrophic "*Epsilonproteobacteria*". *FEMS Microbiol Ecol* **51**, 31–53.

Felter, S. P. & Dourson, M. L. (1997). Hexavalent chromium-contaminated soils: options for risk assessment and risk management. *Regul Toxicol Pharmacol* **25**, 43–59.

Ferris, F. G., Schultze-Lam, S., Witten, T. C., Fyfe, W. S. & Beveridge, T. J. (1989). Metal interactions with microbial biofilms in acidic and neutral pH environments. *Appl Environ Microbiol* **55**, 1249–1257.

Ferris, F. G., Konhauser, K. O., Lyven, B. & Pedersen, K. (1999). Accumulation of metals by bacteriogenic iron oxides in a subterranean environment. *Geomicrobiol J* **16**, 181–192.

Filgueiras, A. V., Lavilla, I. & Bendicho, C. (2002). Chemical sequential extraction for metal partitioning in environmental solid samples. *J Environ Monit* **4**, 823–857.

Finlay, B. J. (2002). Global dispersal of free-living microbial eukaryote species. *Science* **296**, 1061–1063.

Griffiths, R. I., Manefield, M., Ostle, N., McNamara, N., O'Donnell, A. G., Bailey, M. J. & Whiteley, A. S. (2004). $^{13}CO_2$ pulse labelling of plants in tandem with stable isotope probing: methodological considerations for examining microbial function in the rhizosphere. *J Microbiol Methods* **58**, 119–129.

Gustafsson, J. P., Pechova, P. & Berggren, D. (2003). Modelling metal binding to soils: the role of natural organic matter. *Environ Sci Technol* **37**, 2767–2774.

Haack, E. A. & Warren, L. A. (2003). Biofilm hydrous manganese oxyhydroxides and metal dynamics in acid rock drainage. *Environ Sci Technol* **37**, 4138–4147.

Hare, L. (1992). Aquatic insects and trace metals: bioavailability, bioaccumulation, and toxicity. *Crit Rev Toxicol* **22**, 327–369.

Hassler, C. S., Slaveykova, V. I. & Wilkinson, K. J. (2004). Some fundamental (and often overlooked) considerations underlying the free ion activity and biotic ligand models. *Environ Toxicol Chem* **23**, 283–291.

Holden, J. F. & Adams, M. W. W. (2003). Microbe-metal interactions in marine hydrothermal environments. *Curr Opin Chem Biol* **7**, 160–165.

Holt, S. C. & Leadbetter, E. R. (1992). Structure-function relationships in prokaryotic cells. In *Prokaryotic Structure and Function: a New Perspective* (Society for General Microbiology Symposium no. 47), pp. 11–44. Edited by S. Mohan, C. Dow & J. A. Coles. Cambridge: Cambridge University Press.

Hunt, A. P., Hamilton-Taylor, J. & Parry, J. D. (2001). Trace metal interactions with epilithic biofilms in small acidic mountain streams. *Arch Hydrobiol* **153**, 155–176.

Islam, F. S., Gault, A. G., Boothman, C., Polya, D. A., Charnock, J. M., Chatterjee, D. & Lloyd, J. R. (2004). Role of metal-reducing bacteria in arsenic release from Bengal delta sediments. *Nature* **430**, 68–71.

Jacob, D. L. & Otte, M. L. (2004). Long-term effects of submergence and wetland vegetation on metals in a 90-year old abandoned Pb-Zn mine tailings pond. *Environ Pollut* **130**, 337–345.

Jenne, E. A. (1968). Controls on Mn, Fe, Co, Ni, Cu and Zn concentrations in soils and water: the significant role of hydrous Mn and Fe oxides. In *Trace Inorganics in Water*, American Chemical Society Publication no. 73, pp. 337–387. Edited by R. A. Baker. Washington, DC: American Chemical Society.

Johnson, B. B. (1990). Effect of pH, temperature, and concentration on the adsorption of cadmium on goethite. *Environ Sci Technol* **24**, 112–118.

Kaeberlein, T., Lewis, K. & Epstein, S. S. (2002). Isolating "uncultivable" microorganisms in pure culture in a simulated natural environment. *Science* **296**, 1127–1129.

Konhauser, K. O., Fyfe, W. S., Ferris, F. G. & Beveridge, T. J. (1993). Metal sorption and mineral precipitation by bacteria in two Amazonian river systems: Rio Solimões and Rio Negro, Brazil. *Geology* **21**, 1103–1106.

Labrenz, M. & Banfield, J. K. (2004). Sulfate-reducing bacteria-dominated biofilms that precipitate ZnS in a subsurface circumneutral-pH mine drainage system. *Microb Ecol* **47**, 205–217.

Lawrence, J. R., Korber, D. R., Wolfaardt, G. M. & Caldwell, D. E. (1995). Behavioural strategies of surface-colonizing bacteria. *Adv Microb Ecol* **14**, 1–75..

Lawrence, J. R., Swerhone, G. D. W. & Kwong, Y. T. J. (1998). Natural attenuation of aqueous metal contamination by an algal mat. *Can J Microbiol* **44**, 825–832.

Ledin, M. & Pedersen, K. (1996). The environmental impact of mine wastes – roles of microorganisms and their significance in treatment of mine wastes. *Earth Sci Rev* **41**, 67–108.

Lee, D. Y., Fortin, C. & Campbell, P. G. C. (2004). Influences of chloride on silver uptake by two green algae, *Pseudokirchneriella subcapitata* and *Chlorella pyrenoidosa*. *Environ Toxicol Chem* **23**, 1012–1018.

Leppard, G. G., Droppo, I. G., West, M. M. & Liss, S. N. (2003). Compartmentalization of metals within the diverse colloidal matrices comprising activated sludge microbial flocs. *J Environ Qual* **32**, 2100–2108.

Leppard, G. G., Mavrocordatos, D. & Perret, D. (2004). Electron-optical characterization of nano- and macro-particles in raw and treated waters: an overview. *Water Sci Technol* **50** (12), 1–8.

Leveille, S. A., Leduc, L. G., Ferroni, G. D., Telang, A. J. & Voordouw, G. (2001). Monitoring of bacteria in acid mine environments by reverse sample genome probing. *Can J Microbiol* **47**, 431–442.

Little, B. J., Wagner, P. A. & Lewandowski, Z. (1997). Spatial relationships between bacteria and mineral surfaces. In *Geomicrobiology: Interactions between Microbes and Minerals*, Reviews in Mineralogy no. 35, pp. 123–159. Edited by J. F. Banfield & K. H. Nealson. Washington, DC: American Mineralogical Society.

Liu, C. X., Gorby, Y. A., Zachara, J. M., Fredrickson, J. K. & Brown, C. F. (2002). Reduction kinetics of Fe(III), Co(III), U(VI), Cr(VI), and Tc(VII) in cultures of dissimilatory metal-reducing bacteria. *Biotechnol Bioeng* **80**, 637–649.

Lores, E. M. & Pennock, J. R. (1998). The effect of salinity on binding of Cd, Cr, Cu and Zn to dissolved organic matter. *Chemosphere* **37**, 861–874.

Lovley, D. R. & Anderson, R. T. (2000). Influence of dissimilatory metal reduction on fate of organic and metal contaminants in the subsurface. *Hydrogeol J* **8**, 77–88.

Lu, Y. H., Lueders, T., Friedrich, M. W. & Conrad, R. (2005). Detecting active methanogenic populations on rice roots using stable isotope probing. *Environ Microbiol* **7**, 326–336.

Lueders, T., Pommerenke, B. & Friedrich, M. W. (2004). Stable-isotope probing of microorganisms thriving at thermodynamic limits: syntrophic propionate oxidation in flooded soil. *Appl Environ Microbiol* **70**, 5778–5786.

Lünsdorf, H., Brümmer, I., Timmis, K. N. & Wagner-Döbler, I. (1997). Metal selectivity of in situ microcolonies in biofilms of the Elbe River. *J Bacteriol* **179**, 31–40.

Manefield, M., Whiteley, A. S. & Bailey, M. J. (2004). What can stable isotope probing do for bioremediation? *Int Biodeterior Biodegrad* **54**, 163–166.

Martell, A. E., Motekaitis, R. J. & Smith, R. M. (1988). Structure-stability relationships of metal complexes and metal speciation in environmental aqueous solutions. *Environ Toxicol Chem* **7**, 414–434.

Martínez, C. E. & McBride, M. B. (1998). Solubility of Cd^{2+}, Cu^{2+}, Pb^{2+}, and Zn^{2+} in aged coprecipitates with amorphous iron hydroxides. *Environ Sci Technol* **32**, 743–748.

Martino, M., Turner, A. & Nimmo, M. (2004). Distribution, speciation and particle-water interactions of nickel in the Mersey Estuary, UK. *Mar Chem* **88**, 161–177.

Meylan, S., Behra, R. & Sigg, L. (2004). Influence of metal speciation in natural freshwater on bioaccumulation of copper and zinc in periphyton: a microcosm study. *Environ Sci Technol* **38**, 3104–3111.

Moreau, J. W., Webb, R. I. & Banfield, J. K. (2004). Ultrastructure, aggregation-state, and crystal growth of biogenic nanocrystalline sphalerite and wurtzite. *Am Miner* **89**, 950–960.

Morel, F. M. M. & Hering, J. G. (1993). *Principles and Applications of Aquatic Chemistry*. London: Wiley Interscience.

Nealson, K. H. (2003). Harnessing microbial appetites for remediation. *Nat Biotechnol* **21**, 243–244.

Nealson, K. H. & Stahl, D. H. (1997). Micro-organisms and biogeochemical cycles: what can we learn from stratified communities? In *Geomicrobiology: Interactions between Microbes and Minerals*, Reviews in Mineralogy no. 35, pp. 5–34. Edited by J. F. Banfield & K. H. Nealson. Washington, DC: American Mineralogical Society.

Nelson, Y. M., Lo, W., Lion, L. W., Shuler, M. L. & Ghiorse, W. C. (1995). Lead distribution in a simulated aquatic environment: effects of bacterial biofilms and iron oxide. *Water Res* **29**, 1934–1944.

Nelson, Y. M., Lion, L. W., Ghiorse, W. C. & Shuler, M. L. (1999). Production of biogenic Mn oxides by *Leptothrix discophora* SS-1 in a chemically defined growth medium and evaluation of their Pb adsorption characteristics. *Appl Environ Microbiol* **65**, 175–180.

Neu, T. R. & Lawrence, J. R. (1997). Development and structure of microbial biofilms in river water studied by confocal laser scanning microscopy. *FEMS Microbiol Ecol* **24**, 11–25.

Newman, D. K. & Banfield, J. F. (2002). Geomicrobiology: how molecular-scale interactions underpin biogeochemical systems. *Science* **296**, 1071–1077.

Nordstrom, D. K. & Southam, G. (1997). Geomicrobiology of sulfide mineral oxidation. In *Geomicrobiology: Interactions between Microbes and Minerals*, Reviews in Mineralogy no. 35, pp. 361–390. Edited by J. F. Banfield & K. H. Nealson. Washington, DC: American Mineralogical Society.

O'Day, P. A. (1999). Molecular environmental geochemistry. *Rev Geophys* **37**, 249–274.

O'Day, P. A., Vlassopoulos, D., Root, R. & Rivera, N. (2004). The influence of sulfur and iron on dissolved arsenic concentrations in the shallow subsurface under changing redox conditions. *Proc Natl Acad Sci U S A* **101**, 13703–13708.

Pace, N. R. (1997). A molecular view of microbial diversity and the biosphere. *Science* **276**, 734–740.

Parmar, N., Warren, L. A., Roden, E. E. & Ferris, F. G. (2000). Solid phase capture of strontium by the iron reducing bacteria *Shewanella alga* strain BrY. *Chem Geol* **169**, 281–288.

Pearson, A., Sessions, A. L., Edwards, K. J. & Hayes, J. M. (2004). Phylogenetically specific separation of rRNA from prokaryotes for isotopic analysis. *Mar Chem* **92**, 295–306.

Philip, L., Iyengar, L. & Venkobachar, C. (2000). Site of interaction of copper on *Bacillus polymyxa*. *Water Air Soil Pollut* **119**, 11–21.

Puls, R. W., Powell, R. M., Clark, D. & Eldred, C. J. (1991). Effects of pH, solid/solution ratio, ionic strength, and organic acids on Pb and Cd sorption on kaolinite. *Water Air Soil Pollut* **57–58**, 423–430.

Reysenbach, A.-L. & Shock, E. (2002). Merging genomes with geochemistry in hydrothermal ecosystems. *Science* **296**, 1077–1082.

Richard, F. C. & Bourg, A C. M. (1991). Aqueous geochemistry of chromium: a review. *Water Res* **25**, 807–816.

Roberts, K. A., Santschi, P. H., Leppard, G. G. & West, M. M. (2004). Characterization of organic-rich colloids from surface and ground waters at the actinide-contaminated Rocky Flats Environmental Technology Site (RFETS), Colorado, USA. *Colloids Surfaces A Physicochem Eng Aspects* **244**, 105–111.

Schorer, M. & Eisele, M. (1997). Accumulation of inorganic and organic pollutants by biofilms in the aquatic environment. *Water Air Soil Pollut* **99**, 651–659.

Singer, P. C. & Stumm, W. (1970). Acid mine drainage: the rate limiting step. *Science* **167**, 1121–1123.

Smiejan, A., Wilkinson, K. J. & Rossier, C. (2003). Cd bioaccumulation by a freshwater bacterium, *Rhodospirillum rubrum*. *Environ Sci Technol* **37**, 701–706.

Southam, G. & Beveridge, T. J. (1992). Enumeration of thiobacilli from pH-neutral and acidic mine tailings and their role in the development of secondary mineral soil. *Appl Environ Microbiol* **58**, 1904–1912.

Spear, J. R., Walker, J. J., McCollom, T. M. & Pace, N. R. (2005). Hydrogen and bioenergetics in the Yellowstone geothermal ecosystem. *Proc Natl Acad Sci U S A* **102**, 2555–2560.

Stumm, W. (1992). *Chemistry of the Solid-Water Interface: Processes at the Mineral-Water and Particle-Water Interface in Natural Systems*. New York: Wiley.

Stumm, W. & Morgan, J. J. (1996). *Aquatic Chemistry. Chemical Equilibria and Rates in Natural Waters*, 3rd edn. New York: Wiley.

Takai, K., Gamo, T., Tsunogai, U., Nakayama, N., Hirayama, H., Nealson, K. H. & Horikoshi, K. (2004). Geochemical and microbiological evidence for a hydrogen-based, hyperthermophilic subsurface lithoautotrophic microbial ecosystem (HyperSLiME) beneath an active deep-sea hydrothermal field. *Extremophiles* **8**, 269–282.

Tani, Y., Ohasi, M., Miyata, N., Seyama, H., Iwahori, K. & Soma, M. (2004a). Sorption of Co(II), Ni(II), and Zn(II) on biogenic manganese oxides produced by a Mn-oxidizing fungus, strain KR21-2. *J Environ Sci Health Part A Toxic Hazard Subst Environ Eng* **39**, 2641–2660.

Tani, Y., Miyata, N., Osahi, M., Ohnuki, T., Seyama, H., Iwahori, K. & Soma, M. (2004b). Interaction of inorganic arsenic with biogenic manganese oxide produced by a Mn-oxidizing fungus, strain KR21-2. *Environ Sci Technol* **38**, 6618–6624.

Templeton, A. S., Trainor, T. P., Traina, S. J., Spormann, A. M. & Brown, G. E., Jr (2001). Pb(II) distributions at biofilm-metal oxide surfaces. *Proc Natl Acad Sci U S A* **98**, 11897–11902.

Templeton, A. S., Trainor, T. P., Spormann, A. M. & Brown, G. E., Jr (2003a). Selenium speciation and partitioning within *Burkholderia cepacea* biofilms formed on α-Al_2O_3 surfaces. *Geochim Cosmochim Acta* **67**, 3547–3557.

Templeton, A. S., Trainor, T. P., Spormann, A. M., Newville, M., Sutton, S. R., Dohnalkova, A., Gorby, Y. & Brown, G. E., Jr (2003b). Sorption versus biomineralization of Pb(II) within *Burkholderia cepacea* biofilms. *Environ Sci Technol* **37**, 300–307.

Tessier, A. & Campbell, P. G. C. (1988). Comments on the testing of the accuracy of an extraction procedure for determining the partitioning of trace metals in sediments. *Anal Chem* **60**, 1475–1476.

Tessier, A., Campbell, P. G. C. & Bisson, M. (1979). Sequential extraction procedure for the speciation of particulate trace metals. *Anal Chem* **51**, 844–851.

Unz, R. F. & Shuttleworth, K. L. (1996). Microbial mobilization and immobilization of heavy metals. *Curr Opin Biotechnol* **7**, 307–310.

Urrutia, M. M. (1997). General bacterial sorption processes. In *Biosorbents for Metal Ions*, pp. 39–66. Edited by J. Wase & C. Forster. London: Taylor and Francis.

Vigneault, B., Campbell, P. G. C., Tessier, A. & De Vitre, R. (2001). Geochemical changes in sulfidic mine tailings stored under a shallow water cover. *Water Res* **35**, 1066–1076.

Warren, L. A. (2004). A special issue dedicated to microbial geochemistry. *Geochim Cosmochim Acta* **68**, 3139.

Warren, L. A. & Ferris, F. G. (1998). Continuum between sorption and precipitation of Fe(III) on microbial surfaces. *Environ Sci Technol* **32**, 2331–2337.

Warren, L. A. & Haack, E. A. (2001). Biogeochemical controls on metal behaviour in freshwater environments. *Earth Sci Rev* **54**, 261–320.

Warren, L. A. & Kauffman, M. E. (2003). Microbial geoengineers. *Science* **299**, 1027–1029.

Warren, L. A. & Zimmerman, A. P. (1994). The influence of temperature and NaCl on cadmium, copper and zinc partitioning among suspended particulate and dissolved phases in an urban river. *Water Res* **28**, 1921–1931.

Warren, L. A., Tessier, A. & Hare, L. (1998). Modelling cadmium accumulation by benthic invertebrates in situ: the relative contributions of sediment and overlying water reservoirs to organism cadmium concentrations. *Limnol Oceanogr* **43**, 1442–1454.

Zachara, J. M., Ainsworth, C. C., Cowan, C. E. & Resch, C. T. (1989). Adsorption of chromate by subsurface soil horizons. *Soil Sci Soc Am J* **53**, 418–428.

Minerals, mats, pearls and veils: themes and variations in giant sulfur bacteria

Neil D. Gray and Ian M. Head

School of Civil Engineering and Geosciences, Institute for Research on the Environment and Sustainability and Centre for Molecular Ecology, University of Newcastle, Newcastle upon Tyne, NE1 7RU, UK

MINERALS, MATS, PEARLS AND VEILS: A PANOPLY OF GIANT SULFUR BACTERIA

The biogeochemical cycling of sulfur has been at the heart of microbial ecology since the mid-19th century. This is due, at least in part, to the striking forms of many of the organisms involved in the transformation of reduced sulfur species. Giant sulfur bacteria were among the earliest micro-organisms to capture the interest of micro-biologists exploring the links between geochemical cycling of the elements and the microbiota responsible. Consequently, giant sulfur bacteria were among the first bacteria described. Organisms resembling *Beggiatoa* ('*Oscillatoria alba*') were described as early as 1803 (Vaucher, 1803), but were included in the genus *Beggiatoa* some time later (Trevisan, 1842). *Thiothrix* (Rabenhorst, 1865; Winogradsky, 1888), *Achromatium* (Schewiakoff, 1893) and *Thioploca* (Lauterborn, 1907) were all described by the early 20th century and Winogradsky (1887, 1888) had already formulated the principles of lithotrophic growth based on sulfide oxidation, from his work on *Beggiatoa* species. Surprisingly for such conspicuous organisms, novel giant sulfur bacteria are still being described (Guerrero *et al.*, 1999; Schulz *et al.*, 1999).

Achromatium

Bacteria of the genus *Achromatium* are remarkable. Cells of up to 125 µm in length have been reported (Babenzien *et al.*, 2005; Head *et al.*, 2000a) and, in addition to characteristic sulfur globules that become visible on treatment of the cells with dilute acid, their large oval cells are typically filled with enormous inclusions of calcium carbonate (Fig. 1). As with many giant sulfur bacteria, they have as yet eluded

Fig. 1. DAPI (4,6-diamidino-2-phenylindole)-stained preparation of a crudely purified suspension of *Achromatium* cells. The diatom cells illustrate the large size of the *Achromatium* cells.

cultivation in the laboratory. Nevertheless, the ability to physically enrich cells from environmental samples (De Boer *et al.*, 1971; Head *et al.*, 2000a) has permitted the inference of some physiological properties (Gray *et al.*, 1997, 1999a, b, 2000, 2004). *Achromatium* species are capable of oxidizing reduced sulfur species completely to sulfate (Gray *et al.*, 1997). This is a feature of bacteria capable of conserving energy from sulfur oxidation, in contrast to chemoheterotrophic sulfur bacteria that appear to utilize sulfide oxidation as a means to detoxify sulfide or neutralize reactive oxygen species produced during aerobic metabolism (Burton & Morita, 1964), although considerable ambiguity and debate still surround the precise role of sulfide oxidation in these bacteria (Strohl, 2005). Recent evidence also suggests that, although oxygen is probably the preferred electron acceptor for *Achromatium* species, nitrate is also probably used as an alternative electron acceptor (Gray *et al.*, 2004). The use of nitrate as a terminal electron acceptor is a trait shared with several other giant sulfur bacteria. Although the name of only one species of *Achromatium* has been validly published (*Achromatium oxaliferum*; Schewiakoff, 1893), it is clear from comparative analysis of 16S rRNA gene sequences recovered from *Achromatium* cells physically enriched from sediment samples that several distinct species exist and, typically, several species coexist in a single sediment (Babenzien *et al.*, 2005; Gray *et al.*, 1999b; Head *et al.*, 1996). All *Achromatium* species investigated to date form a distinct monophyletic group within

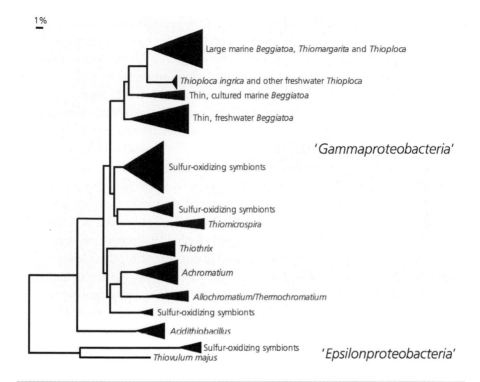

1%

Large marine *Beggiatoa*, *Thiomargarita* and *Thioploca*

Thioploca ingrica and other freshwater *Thioploca*
Thin, cultured marine *Beggiatoa*

Thin, freshwater *Beggiatoa*

'*Gammaproteobacteria*'

Sulfur-oxidizing symbionts

Sulfur-oxidizing symbionts
Thiomicrospira

Thiothrix

Achromatium

Allochromatium/Thermochromatium
Sulfur-oxidizing symbionts

Acidithiobacillus

Sulfur-oxidizing symbionts   '*Epsilonproteobacteria*'
Thiovulum majus

Fig. 2. Phylogenetic tree of representatives of the sulfur-oxidizing bacteria from the '*Gammaproteobacteria*' and '*Epsilonproteobacteria*', including the giant sulfur bacteria from the genera *Achromatium*, *Beggiatoa*, *Thiomargarita*, *Thioploca*, *Thiothrix* and *Thiovulum*.

the '*Gammaproteobacteria*' (Fig. 2) and, in most analyses, appear to be related to anoxygenic, photosynthetic bacteria of the family *Chromatiaceae*. However, the fifth release of the taxonomic outline of prokaryotes from the second edition of *Bergey's Manual of Systematic Bacteriology* lists *Achromatium* in the order '*Thiotrichales*' as genus II within the family '*Thiotrichaceae*' (Garrity *et al.*, 2005).

Beggiatoa

In contrast to *Achromatium* species, which are found as large populations of free-living unicells, *Beggiatoa* species often form extensive, conspicuous mats on the surface of sediments or associated with sulfur springs, hydrothermal vents or cold-seep environments (Strohl, 2005; Teske & Nelson, 2004). The mats can be extremely thin and compact (hundreds of micrometres) or much more diffuse and thick (centimetres), depending on the prevailing hydrodynamic conditions, which dictate whether substrate transport to the cells is diffusion-limited (Gunderson *et al.*, 1992; Jørgensen & Revsbech, 1983). Indeed, there have been reports of *Beggiatoa*-dominated mats as thick as 30–60 cm in some marine hydrothermal systems associated with populations of tube worms (McHatton *et al.*, 1996; Nelson *et al.*, 1989). Several species have been obtained in pure culture, but the largest mat-forming *Beggiatoa* species discovered have yet to be grown in pure culture. *Beggiatoa* species are physiologically varied: some strains have been shown to be obligate chemolithoautotrophs, whilst others are facultative chemo-lithoautotrophs or mixotrophs (Hagen & Nelson, 1996; Nelson & Jannasch, 1983). Strains capable of autotrophic growth have been obtained from marine environments, whereas evidence to date suggests that freshwater isolates are primarily heterotrophs and evidence for lithoautotrophic growth of freshwater *Beggiatoa* isolates has been contentious (Strohl, 2005). Nevertheless, lithoautotrophic growth has been shown for more recently isolated freshwater *Beggiatoa* strains (Grabovich *et al.*, 2001; Patritskaya *et al.*, 2001) and evidence for biochemical control of the switch from heterotrophic growth to mixotrophic or lithoautotrophic growth has been provided (Eprintsev *et al.*, 2003, 2004; Stepanova *et al.*, 2002). The heterotrophic freshwater strains produce characteristically thin filaments, as do cultivated marine strains capable of autotrophic growth. Neither of these thin-filament types produces large vacuoles and both have been cultivated in the laboratory. *Beggiatoa* cells with wide filaments (up to 200 μm in diameter) have yet to be grown in laboratory culture and are characterized by large vacuoles that are believed to play a role in storing nitrate, which is used as an oxidant in these enormous bacteria. However, recently studied vacuolated filaments of giant bacteria that resemble *Thiothrix* species morphologically, but are related most closely to *Beggiatoa* species, do not appear to store nitrate, suggesting that the role of large intracellular vacuoles may be more complex than simply providing a repository for large amounts of oxidant (Kalanetra *et al.*, 2004). Most large-vacuolated *Beggiatoa* species accumulate high concentrations of nitrate (Kalanetra *et al.*, 2004; McHatton *et al.*, 1996; Mußmann *et al.*, 2003), which is used as an oxidant either by reduction to ammonia or through denitrification, thus providing a link between the biogeochemical cycles of nitrogen and sulfur (Teske & Nelson, 2004). Like *Achromatium* species, *Beggiatoa* species belong to the '*Gammaproteobacteria*', where they form a mono-phyletic group with *Thioploca* and *Thiomargarita* species (Fig. 2; Kalanetra *et al.*, 2004; Kojima *et al.*, 2003; Mußmann *et al.*, 2003; Teske *et al.*, 1996). The uncultured vacuolate bacteria (*Beggiatoa*, *Thioploca* and *Thiomargarita* species) themselves form a distinct phylogenetic group within this, with freshwater and marine thin-filament *Beggiatoa* species each occupying distinct lineages (Teske & Nelson, 2004). It is also clear that there is considerable diversity within communities of *Beggiatoa*, which often exhibit filaments with different diameters (Nelson *et al.*, 1989).

Thioploca

In many respects, *Thioploca* species resemble *Beggiatoa* species, especially the large-vacuolated *Beggiatoa* found in marine environments. Members of the genus *Thioploca* are differentiated from those of the genus *Beggiatoa* on the basis that trichomes of *Thioploca* species are usually enclosed in a thick polysaccharide sheath. Whilst some *Beggiatoa* strains have been isolated in pure culture, no *Thioploca* strains have been grown axenically in the laboratory. Like *Beggiatoa* strains, they form multicellular filaments that can be several centimetres long. Several morphotypes can be differentiated on the basis of filament diameter (Schulz *et al.*, 1996) and these correspond to different species, defined on the basis of 16S rRNA gene sequence identity (Teske *et al.*, 1996). Moreover, most *Thioploca* species described to date, in common with large *Beggiatoa* species, harbour vacuoles that are presumed to store nitrate, which is used as an oxidant by both *Beggiatoa* and *Thioploca* species (McHatton *et al.*, 1996). These occur even in *Thioploca* species with relatively thin (2–5 µm diameter) filaments (Zemskaya *et al.*, 2001), whereas thin *Beggiatoa* species apparently do not have large vacuoles (Teske & Nelson, 2004). rRNA gene sequence-based analyses of *Thioploca* isolates recovered from naturally occurring mats have shown that they are related closely to, and form a monophyletic group with, the large-vacuolated *Beggiatoa* species and members of the genus *Thiomargarita*, which also exhibit a large internal vacuole implicated in nitrate storage (Mußmann *et al.*, 2003; Teske *et al.*, 1996, 1999). Given the similarity with *Beggiatoa* and the fact that, under certain circumstances, *Thioploca* trichomes leave their sheaths and are then morphologically indistinguishable from *Beggiatoa* filaments, it is likely that the classification of these two very similar genera will be reconciled as more phylogenetic and physiological information accumulates. Extensive mats of *Thioploca* cells have been discovered worldwide in both marine and freshwater environments (Teske & Nelson, 2004) and Jørgensen & Gallardo (1999) have suggested that the discontinuous occurrence of *Thioploca* mats that extend for a distance of over 3000 km along the west coast of South America probably represent the largest communities of sulfur bacteria on the planet.

Mats of *Thioploca* cells differ in structure from those composed of *Beggiatoa* cells. This is principally a consequence of the presence of multiple *Thioploca* trichomes within an extensive sheath that can penetrate several centimetres below the sediment surface (Schulz *et al.*, 1996). This innovation, coupled with the intracellular storage of large amounts of nitrate, has suggested a novel ecological strategy whereby *Thioploca* species exploit pools of oxidant and reductant that are separated in space. *Thioploca* trichomes exhibit a tactic response towards nitrate and, when nitrate concentrations are high in the overlying water, the trichomes extend into the water column, where it is believed that they accumulate nitrate within their extensive system of vacuoles. They then retreat back into their sheaths, carrying their store of oxidant to greater depth in the sediment, where electron donor in the form of sulfide is freely available either from sulfate reduction or geothermal sources (Huettel *et al.*, 1996).

Fig. 3. Mats and streamers of filamentous sulfur bacteria from springs on Sulfur Mountain, CA, USA.

Thiothrix

Conspicuous streamers of *Thiothrix* cells are often noticeable in flowing waters from sulfur springs (Fig. 3). In this respect, they differ from the sulfur bacteria considered above. *Beggiatoa* species, in particular, rely on molecular diffusion through a stagnant boundary layer to provide electron donor and acceptor to the cells (Schulz & Jørgensen, 2001). By virtue of specialized holdfast cells, members of the genus *Thiothrix* can maintain themselves even in vigorously flowing waters that would wash away filamentous sulfur bacteria lacking this adaptation. The rapid fluid flow decreases the

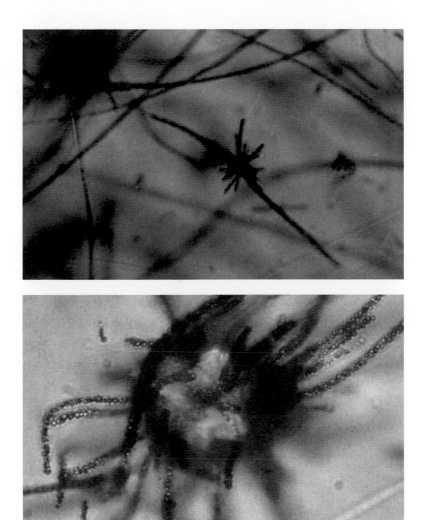

Fig. 4. Rosettes of *Thiothrix* sp. from a sulfur spring, Gilsland Spa, Northumberland, UK.

unmixed boundary layer, freeing *Thiothrix* cells from the limitations of molecular diffusion. It has long been considered that defining features of members of the genus *Thiothrix* are their ability to deposit intracellular sulfur and their growth in character-istic rosette structures that emanate long filaments (Fig. 4; Larkin, 1989). This has, however, been called into question by a number of observations. The genus *Thiothrix* includes organisms with highly variable morphology, such as *Thiothrix eikelboomii* and *Thiothrix defluvii*, that do not necessarily produce rosettes (Howarth *et al.*, 1999; Unz & Head, 2005). Furthermore, there is evidence that *Leucothrix* species, which also produce rosettes and were until recently considered as heterotrophs, may deposit intracellular sulfur when growing lithoheterotrophically (Grabovich *et al.*, 2002). In addition, recently characterized bacteria that are morphologically reminiscent of *Thiothrix* species have been shown to be related more closely to vacuolated *Beggiatoa* and *Thioploca* strains (Kalanetra *et al.*, 2004). Spectacular accumulations of *Thiothrix* cells have been reported from sulfur springs (Brigmon *et al.*, 2003; Larkin & Strohl, 1983; McGlannan & Makemson, 1990) and in a range of cave environments (Brigmon *et al.*, 1994; Engel *et al.*, 2004), associated with methane seeps (Pimenov *et al.*, 2000) and, more recently, they have been reported as an important component of novel prokaryotic communities with a string-of-pearls morphology (Moissl *et al.*, 2002).

Although not related closely to the genera *Beggiatoa* or *Thioploca*, *Thiothrix* species also belong to the '*Gammaproteobacteria*' and this group is replete with morphologically distinctive sulfur bacteria (Fig. 2).

Thiomargarita

Thiomargarita cells are enormous. The bacterium produces large, spherical cells, and cells in excess of 700 µm in diameter have been reported (Schulz *et al.*, 1999). The large, spherical cells reside in a mucus sheath that links the otherwise unconnected cells in a chain. Despite the huge size of *Thiomargarita* cells, the bacterium was discovered only recently (Schulz & Jorgensen, 2005; Schulz *et al.*, 1999). This may reflect the habitat occupied by the bacterium. *Thiomargarita* isolates were first reported from seafloor sediments at a depth of approximately 100 m, off the coast of south-western Africa. Like *Thioploca*-dominated sediments off the western seaboard of South America, the overlying water has low oxygen tension and relatively high nitrate concentrations, which are a consequence of the very high productivity in these regions that are characterized by upwelling of nutrient-rich deep waters. Like the large, marine *Beggiatoa* and *Thioploca* species, *Thiomargarita* cells have large, central vacuoles that constitute approximately 98 % of the biovolume of the cells, and accumulate nitrate to levels of hundreds of millimolar (Schulz *et al.*, 1999). *Thiomargarita* species do not exhibit the gliding motility that is characteristic of *Beggiatoa* and *Thioploca* species and rely instead on being transported passively into the water column by processes such as gas venting, storms and wave action. This permits them to access their sources of oxidant (oxygen and nitrate), which can be used to oxidize stored elemental sulfur or sulfide when the cells are deposited back into the organic-rich sulfidic sediment that they normally inhabit. The kinds of event that transport *Thiomargarita* cells passively into the water column are innately sporadic and this may explain their huge size, which allows them to store large quantities of oxidant that may be required for long-term survival between resuspension events. In addition, they are tolerant of high oxygen concentrations, whereas *Beggiatoa* and *Thioploca* species are microaerophilic and have a phobic response to high oxygen concentrations (Huettel *et al.*, 1996). Empirical observations suggest that *Thiomargarita* cells may survive in the absence of nitrate for over 700 days (Schulz & Jørgensen, 2001).

This large-vacuolated bacterium is phylogenetically related most closely to the vacuolated *Beggiatoa* and *Thioploca* species and, based on comparative 16S rRNA gene sequence analysis, forms a monophyletic group with them in the '*Gammaproteobacteria*' (Fig. 2; Schulz *et al.*, 1999).

Thiovulum

In 1913, Hinze described two species of *Thiovulum* (Hinze, 1913; Starr & Skerman, 1965): *Thiovulum majus* (the type species) and '*Thiovulum minus*'. These were identified in marine environments and all isolated *Thiovulum* species to date come from marine habitats. Although perhaps not in the same realm as the microbial giants described above (*Thiovulum* cells range from 5 to 25 μm in diameter; Fenchel, 1994), conspicuous 'veils' of *Thiovulum* cells are produced, associated with gradients of sulfide and oxygen (Jørgensen & Revsbech, 1983). Perhaps their most remarkable feature is their rapid motility, which has been measured at over 600 μm (24–60 cell lengths) s^{-1}. This is equivalent to a 2 m fish swimming at up to >400 km h^{-1} (60 body lengths s^{-1}). Their rapid swimming is facilitated by extensive flagellation over the surface of the cells and they use this either to propel the multicellular floating veils that they form into the optimum position in opposed gradients of sulfide and oxygen (Jørgensen & Revsbech, 1983) or to direct water flow around tethered veils to reduce diffusional limitation of substrate transport to the cells (Fenchel & Glud, 1998). It is now clear that a number of other bacteria have adopted similar strategies to enhance transport of substrates to the cells (Thar & Kühl, 2002).

Apart from aspects of their chemosensory behaviour, motility and ultrastructure, there are few studies on the physiology of *Thiovulum* species and only one partial 16S rRNA gene sequence from authenticated *Thiovulum* cells has been deposited in the public databases. The one *Thiovulum* species that has been characterized phylogenetically to date belongs to the '*Epsilonproteobacteria*' (Fig. 2; Lane *et al.*, 1992; Romaniuk *et al.*, 1987), although it seems likely that, with more extensive sampling, the picture of the taxonomy of the genus *Thiovulum* will change.

GIANT SULFUR BACTERIA: ENERGY AND GEOCHEMICAL SIGNIFICANCE

Bacteria involved in the oxidative side of the sulfur cycle couple the oxidation of reduced sulfur species to the reduction of dissolved oxygen, nitrate and possibly oxidized metals. Many generate energy from these processes and reduce CO_2 for biosynthesis. Some also have the capacity to fix dinitrogen gas and *Beggiatoa* species have been best studied in this respect (Nelson *et al.*, 1982; Polman & Larkin, 1988). Sulfur bacteria therefore play a pivotal role in the biogeochemical cycling of sulfur, carbon, nitrogen and possibly metals. Their activities also affect the precipitation and dissolution of minerals and can have consequences for higher trophic levels in a variety of ecosystems.

The reduced sulfur that provides the energy to drive the biogeochemical activities of sulfur bacteria can come from a variety of sources. Reduced sulfur can be liberated from decaying organic matter, be produced from dissimilatory sulfate reduction or released as a result of volcanic activity or from other geological sources. Environments characterized by different sources of reduced sulfur present different geochemical constraints on electron-donor and -acceptor availability, which have led to the numerous ecological adaptations observed in sulfur-oxidizing bacteria. One major ecological adaptation has been the evolution of sulfur bacteria with a large cell size.

The size of giant sulfur bacteria not only represents a remarkable evolutionary adaptation to the lifestyles that they have adopted, but also permits them to be manipulated experimentally under *in situ* conditions and used as models to establish links between environment, physiology and evolutionary diversification that may be extrapolated more widely. Some of the adaptations in the energetics and biochemistry of giant sulfur bacteria have consequences for ecosystem-scale processes. These are discussed in the context of the geochemical features of habitats dominated by conspicuous sulfur bacteria.

Energetic considerations of sulfur oxidation

The transformation of S^{2-} to SO_4^{2-} by bacteria requires the transfer of eight electrons to an appropriate electron acceptor. What is known or assumed about electron transport in giant sulfur bacteria has been gleaned both directly (Cannon *et al.*, 1979; Schmidt & DiSpirito, 1990; Strohl *et al.*, 1986) and by analogy with mechanisms observed in other sulfur bacteria. What therefore follows, by necessity, does not represent a comprehensive picture of sulfur-oxidation pathways, but focuses on those areas where direct or circumstantial evidence has been obtained from studies of giant sulfur bacteria. Readers seeking a comprehensive review of sulfur-oxidation pathways are directed to some excellent reviews (Friedrich *et al.*, 2001; Kappler & Dahl, 2001; Kelly, 1999; Kelly *et al.*, 1997). What these studies indicate is that there is no universal pathway of sulfur oxidation and the giant sulfur bacteria, like all sulfur oxidizers, display considerable variation in pathways of dissimilatory sulfur metabolism, even between closely related species. For instance, the oxidation of hydrogen sulfide to sulfur, the first step in sulfur oxidation, is linked either to the reduction of a *c*-type cytochrome, mediated by flavocytochrome *c*–sulfide dehydrogenase, or reduction of a quinone, mediated by sulfide–quinone reductase, in phototrophic and lithotrophic bacteria (Friedrich *et al.*, 2001). However, very little information is available on how these processes are mediated in the giant sulfur bacteria. Some studies have shown that the conversion of sulfide to sulfur in freshwater *Beggiatoa alba* cells was not prevented by dibromothymoquinone, an inhibitor of ubiquinone reduction, whereas inhibitors of flavoproteins (e.g. thenoyl-trifluoroacetone) did suppress sulfide oxidation (Schmidt *et al.*, 1987). This suggests that electrons from sulfide enter the electron-transport chain via a flavocytochrome in *Beggiatoa alba* (Schmidt *et al.*, 1987). How typical this is of other giant sulfur bacteria is not clear and *Beggiatoa alba* is not considered to be capable of chemolithotrophic growth. This first step in the oxidation process is critical, because the sulfur produced can then be either oxidized further or stored by precipitation as intracellular inclusions. Although a characteristic of all giant sulfur bacteria is the deposition of intracellular sulfur, a central problem still to be resolved in sulfur oxidation by chemolithotrophs is the means by which zero-valent sulfur is transformed to SO_3^{2-} and whether energy is conserved from this step (Friedrich *et al.*, 2001; Kelly *et al.*, 1997). It has been proposed that the process may involve oxidation of S^0 to sulfite by a sulfur oxygenase; sulfite generated may then react with elemental sulfur to generate thiosulfate, which is observed as a product *in vitro* (Kelly, 1999). The requirement for molecular oxygen by a sulfur oxygenase means that this mechanism would not allow energy conservation and would not be feasible during anaerobic sulfur oxidation by nitrate-reducing *Thioploca*, *Beggiatoa* or *Thiomargarita* species, for example. Electrons from *Thiobacillus denitrificans* growing with nitrate as an electron acceptor in the absence of oxygen have been shown to be linked to a respiratory chain (Kelly, 1999) and there is evidence that, even in aerobic sulfur oxidizers, energy is conserved during elemental sulfur oxidation and does not involve a non-energy-conserving oxygenase. It is clear that several mechanisms may be involved in elemental sulfur oxidation and the details may be different in different bacterial taxa (Kelly, 1999). The final step in the sulfide-oxidation process is the transformation of sulfite to sulfate and it is this section of the sulfur-oxidation pathway for which most data are available in the giant sulfur bacteria.

Sulfite oxidation to sulfate is known to occur by three different pathways: one direct and two indirect (Fig. 5; Kappler & Dahl, 2001). The first of the indirect pathways is in essence a reversal of sulfate reduction: sulfite is oxidized to sulfate via the formation of adenosine 5′-phosphosulfate (APS), mediated by APS reductase, and subsequent decomposition by reaction with pyrophosphate to form ATP and sulfate (catalysed by ATP sulfurylase; Fig. 5, pathway IIa). Alternatively, APS and orthophosphate may be converted to ADP by ADP sulfurylase (adenylylsulfate : phosphate andenylyl-transferase), with the ADP being converted to ATP by adenylate kinase (Fig. 5, pathway IIb). Both indirect pathways therefore produce a molecule of ATP by substrate-level

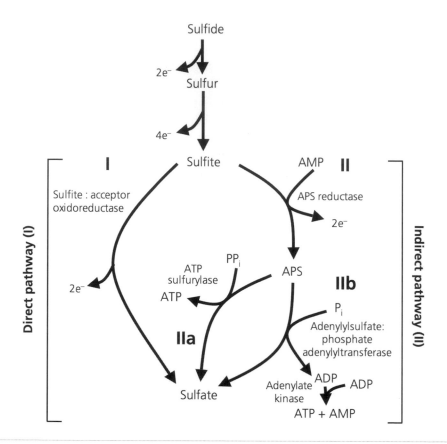

Fig. 5. Some of the alternative pathways of sulfur oxidation thought to be predominant in giant sulfur bacteria. The pathways for oxidation of sulfide through sulfur to sulfite and the involvement of intermediates, such as thiosulfate and tetrathionate, have not been elucidated in giant sulfur bacteria and a minimal representation of oxidation of sulfide to sulfite is provided. This shows the number of electrons generated at each step with potential to contribute to oxidative phosphorylation, but does not represent the full range of reactions that may contribute to sulfide and sulfur oxidation.

phosphorylation. The direct mechanism of sulfite oxidation involves the enzyme sulfite:acceptor oxidoreductase, which oxidizes sulfite to sulfate without the need for AMP (Fig. 5, pathway I). This pathway only generates ATP by oxidative phosphorylation. Many sulfur bacteria carry both enzyme systems. However, whilst sulfite:acceptor oxidoreductase is widespread among sulfur-oxidizing lithoautotrophs, the APS pathway is less prevalent (Kappler & Dahl, 2001). Nevertheless, some *Beggiatoa* strains exhibit both a direct and an indirect pathway (Hagen & Nelson, 1997) and the APS reductase pathway occurs in sulfur-oxidizing symbionts of invertebrates and APS reductase genes have been detected in *Achromatium* species (Head *et al.*, 2000b; Nelson & Hagen, 1995). In contrast, some *Beggiatoa* isolates do not exhibit APS reductase activity (Grabovich *et al.*, 1998, 2001; Hagen & Nelson, 1997).

However, there is potential for substrate-level phosphorylation in some sulfur-oxidizing pathways and, indeed, in some *Thiothrix* species, substrate-level phosphorylation is believed to be the sole mechanism of energy conservation (Grabovich *et al.*, 1999; Odintsova & Dubinina, 1993).

The thermodynamics and, hence, the energy yield of these oxidative processes are dependent upon the particular electron acceptor used, the chemistry of the reduced sulfur compound being oxidized and the reduced and oxidized products of the reaction. Thermodynamic and kinetic considerations of sulfur oxidation have considerable bearing on the ecology and physiology of the sulfur bacteria. For instance, the energy yield from the oxidation of 1 mol HS^- by 2 mol O_2 to sulfate ostensibly yields $-732\cdot6$ kJ (mol sulfur)$^{-1}$ (Kelly, 1999). Whilst this may be important for giant sulfur bacteria, many that are adapted to low oxygen concentrations may use nitrate as an alternative electron acceptor. There is good evidence that the large-vacuolated *Beggiatoa* and *Thioploca* species reduce nitrate to ammonium rather than dinitrogen gas, which only accounted for approximately 15 % of the nitrate reduced (Otte *et al.*, 1999), and there has been some debate regarding the validity of some earlier suggestions that *Beggiatoa* species are capable of denitrification (Fossing *et al.*, 1995; McHatton *et al.*, 1996; Sweerts *et al.*, 1990). Based on the free energy of formation values used by Kelly (1999), it is possible to calculate the energy yield from the oxidation of sulfide coupled to nitrate reduction to N_2 according to the equation

$$5HS^- + 8NO_3^- + 3H^+ \rightarrow 5SO_4^{2-} + 4N_2 + 4H_2O$$

This yields approximately the same amount of energy [$-744\cdot76$ kJ (mol sulfur)$^{-1}$] as sulfide oxidation with oxygen [$-732\cdot6$ kJ (mol sulfur)$^{-1}$]. Note that the energy yield from reduction of nitrate to dinitrogen gas or ammonium is considerably greater than that from reduction of nitrate to nitrite. It should also be noted that these values were calculated under standard conditions and are useful for comparative purposes, but probably do not reflect the energy yield under *in situ* conditions.

In contrast, the reaction

$$HS^- + NO_3^- + H^+ + H_2O \rightarrow NH_4^+ + SO_4^{2-}$$

yields $-463\cdot8$ kJ (mol sulfur)$^{-1}$. On the face of it, one might conclude that these free-energy values make the dissimilatory reduction of nitrate to ammonium (nitrate ammonification or DNRA) less favourable, but several giant sulfur bacteria adopt this way of life and reduction of NO_3^- to NH_4^+ is quantitatively important in many organic-rich coastal marine sediments (Sørensen, 1978). In these sediments, whilst reduction of nitrate to N_2 and consumption of O_2 are restricted to the upper few millimetres of sediment, the reduction of NO_3^- to NH_4^+ is significant at greater depth. The enzymology of the two pathways of nitrate reduction is, however, quite different, as shown by acetylene treatment of sediment microcosms. Acetylene prevents denitrification, but DNRA is unaffected (Bonin *et al.*, 1998). Interestingly, the relative balance of these two pathways has a significant effect on the sedimentary nitrogen cycle. This is because denitrification produces nitrogen gas, which is biologically and chemically inert; in contrast, DNRA produces ammonium that, under suitable conditions, can be reoxidized to nitrate. Thus, biologically available nitrogen is more likely to be retained in sediments where the majority of nitrate is reduced to ammonium (Sørensen, 1978). Whilst this is a useful outcome with respect to the bacteria that utilize nitrate, it does not in itself explain why, given the higher energy yield, nitrate is not always reduced preferentially to N_2. A number of studies have indicated that, in high-sulfide environments, DNRA coupled to sulfide oxidation is favoured because of the inhibition of denitrification at high sulfide concentrations (e.g. Brunet & Garcia-Gil, 1996). However, it is also noteworthy that, in reduced environments, because of the stoichiometry of the two reactions, DNRA is probably no less thermodynamically favourable than denitrification. In reducing marine sediments, sulfide is unlikely to be limiting, whereas nitrate is likely to be less abundant. Only 1 mol nitrate is required to oxidize 1 mol sulfide to sulfate when nitrate is reduced to ammonium (an eight-electron transfer), whereas 1·6 mol nitrate is required to oxidize 1 mol sulfide to sulfate when the nitrate is reduced to N_2 (a five-electron transfer). Despite a more favourable energy yield for denitrification than for DNRA when the values are expressed as kJ (mol sulfur oxidized)$^{-1}$, the energy yield in kJ (mol nitrate reduced)$^{-1}$ is broadly the same for the two reactions, i.e. $-465\cdot5$ kJ (mol nitrate)$^{-1}$ for denitrification and $-463\cdot8$ kJ (mol nitrate)$^{-1}$ for DNRA. From this different perspective, if nitrate is limiting compared with the availability of electron donor, it would appear that the two processes are equally energetically favourable.

However, the energy yield of electron transport is not dictated solely by thermodynamics. The biology of the system may also have significance regarding the ability to realize the maximal thermodynamic yield. To generate ATP by oxidative phosphorylation, it is necessary to generate a proton-motive force. This depends on the ability of each step in the electron-transport pathway to export protons efficiently across the cell membrane. The efficiency is dependent on the mode of respiration employed. The aerobic electron-transport chain, because of its topography, is more efficient at converting the chemical energy of electron donors into ATP (Berks *et al.*, 1995). This is because cytochrome aa_3 is not only a proton pump, but it also delivers electrons to the cytoplasmic side of the membrane, where they combine with oxygen and protons and hence create a greater electrochemical gradient. In contrast, in denitrification, at least in some organisms, the enzymes nitrite reductase, nitric oxide reductase and nitrous oxide reductase are located on the periplasmic side of the membrane, the consumption of protons occurs externally and the net gain in translocated protons is lower than might be suggested solely by the reduction potentials of the respective electron acceptors (Berks *et al.*, 1995). Thus, despite similar thermodynamics, different respiratory processes can have very different energetic efficiency in terms of ATP formation. In the process of DNRA, after the reduction of nitrate to nitrite, nitrite is reduced to ammonium without the release of intermediates (Simon, 2002). In this process, nitrite is reduced by a cytochrome *c*–nitrite reductase complex, which mediates the multi-electron and multi-proton reduction of nitrite to ammonium. It is clear that this process, when coupled to sulfide oxidation, produces ATP (Simon, 2002). In the small number of cases studied to date, nitrite reduction to ammonium occurs on the outside of the cell membrane where protons are consumed, reducing the effective proton-motive force as in denitrification (Simon, 2002).

Geochemical significance of giant sulfur bacteria

The metabolic diversity of giant sulfur bacteria and the huge abundances that they can achieve in appropriate environments mean that they play a significant role in a range of biogeochemical processes. A further consequence of the activity of sulfur-oxidizing bacteria is that they may have pronounced effects at higher trophic levels. The various giant sulfur bacteria oxidize reduced sulfur species, deposit elemental sulfur, fix inorganic carbon, utilize organic compounds as electron donors or carbon sources, consume oxygen and reduce nitrate and nitrite to ammonium and perhaps, to a lesser extent, gaseous products. Several are capable of fixing dinitrogen gas and some may mobilize insoluble metals and other minerals. Giant sulfur bacteria can therefore have important implications for the cycling of many elements.

Giant sulfur bacteria and the sulfur cycle. The oxidation of sulfide to elemental sulfur and ultimately sulfate by giant sulfur bacteria clearly has considerable consequences for sulfur cycling, especially where extensive mats of bacteria occur. The sediments off the Pacific coast of South America, which harbour extensive mats of *Thioploca* cells, exhibit high rates of sulfate reduction (up to 1500 nmol cm^{-3} day^{-1}); however, sulfide rarely accumulates above a few tens of micromolar (Ferdelman *et al.*, 1997; Thamdrup & Canfield, 1996). The low concentrations of sulfide can be explained to some extent by reaction of sulfide with reactive iron, which is considered an important sink for sulfide in the Pacific coast sediments (Thamdrup & Canfield, 1996). Nevertheless, it has been estimated that anywhere between 16 and 91 % of sulfide reoxidation can be attributed to mats of *Thioploca* cells (Ferdelman *et al.*, 1997; Otte *et al.*, 1999). This has the dual role of replenishing an important oxidant for anaerobic carbon mineralization (sulfate) and reducing the steady-state concentration of toxic sulfide. Sediments off the Namibian coast also exhibit low levels of dissolved sulfide (Brüchert *et al.*, 2003). This is particularly unexpected, because of the relatively low levels of reactive iron species in Namibian coastal sediments (Brüchert *et al.*, 2000;

Morse & Emeis, 1990). The effective removal of sulfide in this environment is therefore due, in large part, to the effective reoxidation of sulfide by the mats of *Thiomargarita* cells. In this case, the giant sulfur bacteria could account for up to 55 % of the reoxidation of sulfide and a positive correlation was found between the measured sulfide flux from the sediment and the size of the population of *Thiomargarita* cells (Brüchert *et al.*, 2003). Furthermore, considering that *Thiomargarita* cells may contain up to 1·7 M sulfur in their cytoplasm and that biomass densities as high as 176 g m^{-2} may occur off the coast of south-west Africa (Brüchert *et al.*, 2003), they represent a major repository of elemental sulfur in these sediments. For example, the biovolume of *Thiomargarita* cells in the top few centimetres of Namibian coastal sediments was approximately 4 μl mm^{-3} (Schulz *et al.*, 1999). Given that only 2 % of this volume is cytoplasm, this gives a value of 0·08 μl cytoplasm per cubic millimetre of sediment, which will contain up to 4·35 μg sulfur. Converting this to an areal basis, this equates to approximately 4·35 g sulfur m^{-2} contributed by the *Thiomargarita* cells. Interestingly, measurements of intracellular and extracellular elemental sulfur in *Thioploca*-dominated sediments off the coast of Chile showed that extracellular sulfur was abundant and had a different depth distribution from intracellular sulfur, suggesting that chemical oxidation also played an important role in sulfide oxidation in these sediments (Zopfi *et al.*, 2001). This is in agreement with the conclusions drawn based upon the relative availability of reactive iron in Namibian and Chilean coastal sediments (Brüchert *et al.*, 2003).

In freshwater sediments, the concentration of sulfate is orders of magnitude lower than that in marine systems and it is rarely possible to measure dissolved sulfide in these environments. Consequently, fewer studies of sulfur cycling have been conducted in freshwater sediments. In many circumstances where there are extensive accumulations of sulfur bacteria in freshwater environments, this may be due to locally high levels of sulfide associated with decaying organic matter, rather than from sulfate reduction. Nevertheless, even when no dissolved sulfide is detectable, large populations of giant sulfur bacteria may be supported (Head *et al.*, 1996, 2000b). These may have greater significance for biogeochemical cycling in freshwater systems than the sulfate concentration would suggest. *Achromatium* species are extremely effective at reoxidizing reduced sulfur generated by sulfate reduction (Gray *et al.*, 1997). This is the case even in environments with extremely high levels of reactive iron, which would be expected to effectively outcompete *Achromatium* species for the low levels of sulfide produced by sulfate reduction (Head *et al.*, 2000b). This not only has consequences for the sulfur bacteria themselves, but also suggests that a greater fraction of organic carbon mineralization in *Achromatium*-bearing sediments is channelled through sulfate reduction than would be predicted from measurements of sulfate reduction in incubations performed under anoxic conditions (Gray *et al.*, 1997; Head *et al.*, 2000a, b). It has been suggested that, for sulfur bacteria to thrive under conditions of low sulfide and high reactive-iron concentrations, they must compete effectively with the rapid chemical reaction between sulfide and iron (Thamdrup & Canfield, 1996). In the case of *Thioploca* cells in marine sediments, this may be achieved through close physical association with the sulfate-reducing bacteria that generate the sulfide (Thamdrup & Canfield, 1996). However, this does not seem to be the case for *Achromatium* species, where other bacterial cells have not been found associated with *Achromatium* cells. *Achromatium* may therefore possess high-affinity sulfide-uptake systems that are capable of competing effectively with chemical reactions with iron to permit the utilization of low levels of dissolved sulfide. The kinetics of sulfide oxidation by *Achromatium* species have yet to be studied in detail.

On the other hand, *Achromatium* cells may not use dissolved sulfide directly, but may instead use solid-phase iron sulfides as their principal source of electron donor (Gray *et al.*, 1997; Head *et al.*, 2000a, b). If this is the case, the precipitation of intracellular calcium carbonate by these bacteria may provide a means by which they could dissolve solid-phase iron sulfides and thus drive reoxidation of sulfur at greater rates than would be predicted from very low free-sulfide concentrations. Several explanations for calcite deposition in *Achromatium* species, i.e. neutralization of acid formed by sulfur oxidation, regulation of cell buoyancy or maintenance of high CO_2 partial pressures to facilitate autotrophic growth, have been proposed (Head *et al.*, 2000b). However, this physiological feature, which is unique to *Achromatium* species in the bacterial domain, suggests that it represents an adaptation to an ecophysiological challenge not faced by other organisms.

Calcite precipitation linked to sulfide dissolution may therefore be an ability unique to members of the genus *Achromatium*. Protons exported during chemiosmotic energy generation are consumed in the mineral dissolution reaction

$$FeS + H^+ \rightarrow Fe^{2+} + HS^-$$

and the sulfide generated can then be oxidized to sulfate:

$$HS^- + 2O_2 \rightarrow HSO_4^-$$

As some of the exported protons will be consumed by their reaction with metal sulfides, they will be unavailable for transport across the cell membrane by ATP synthase. An inevitable consequence is that a surplus of hydroxyl ions will occur on the cytoplasmic side of the cell membrane. This is exacerbated if nitrate is used as an electron acceptor, as sulfide oxidation with nitrate consumes protons:

$$HS^- + NO_3^- + H^+ + H_2O \rightarrow NH_4^+ + SO_4^{2-}$$

$$5HS^- + 8NO_3^- + 3H^+ \rightarrow 5SO_4^{2-} + 4N_2 + H_2O$$

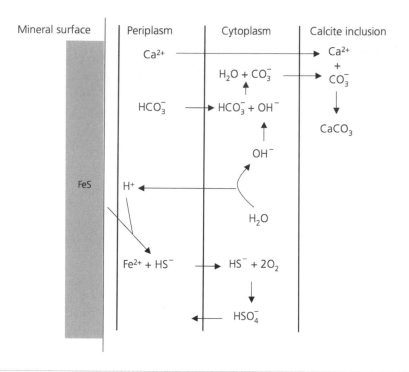

Fig. 6. Putative mechanism for the coupling of intracellular calcite precipitation and sulfide mineral dissolution and oxidation in *Achromatium* species.

A potential sink for excess hydroxyl ions is through buffering with bicarbonate and the formation and precipitation of calcite:

$$OH^- + HCO_3^- \rightarrow CO_3^{2-} + H_2O$$

$$Ca^{2+} + CO_3^{2-} \rightarrow CaCO_3 \,(s)$$

The coccolithophorid algae have previously been proposed to buffer increases in cytoplasmic pH via intracellular precipitation of calcite (Borowitzka, 1982). These photosynthetic organisms deposit intracellular calcite to maintain an high internal partial pressure of CO_2 to facilitate carbon fixation by ribulose-1,5-bisphosphate carboxylase/oxygenase (RuBisCO) (Borowitzka, 1982). Under conditions of low dissolved CO_2 and high dissolved oxygen, bicarbonate is converted to CO_2 and hydroxyl ions by carbonic anhydrase (Borowitzka, 1982). The subsequent fixation of CO_2, however, raises cytoplasmic pH, due to the residual hydroxyl ions. To neutralize the cytoplasmic pH, a second molecule of bicarbonate is precipitated as calcium carbonate within an intracellular membrane-bound structure, the coccolith-containing vesicle (Fig. 6). This reaction consumes hydroxyl ions with the generation of carbonate and water:

$$OH^- + HCO_3^- \rightarrow CO_3^{2-} + H_2O$$

$$Ca^{2+} + CO_3^{2-} \rightarrow CaCO_3 \text{ (s)}$$

An alternative mechanism by which *Achromatium* species might maintain a neutral pH in the cytoplasm would be to liberate protons directly from the reaction of bicarbonate ions with calcium. This is unlikely, as calcite is precipitated not at the cell membrane, but within membrane-bound intracellular inclusions (Fig. 6).

$$Ca^{2+} + HCO_3^- \rightarrow CaCO_3 + H^+$$

Despite the fact that this mechanism of carbonate precipitation is encountered widely in the literature, the direct precipitation of calcium carbonate in this manner does not occur in nature. This is because calcite is ionic and can only be formed from its component ions. As a result, CO_3^{2-} must be formed before calcite formation can take place (Wright & Oren, 2005).

This proposed basis for calcite precipitation in members of the genus *Achromatium* means that there should be a direct link between calcite precipitation and energy generation. For this reason, the energetics of the whole process need to be evaluated. Interestingly, the stoichiometry for iron sulfide oxidation linked to calcium carbonate precipitation shows that there is no net gain or loss of protons and, thus, the process should be chemiosmotically neutral:

$$FeS + Ca^{2+} + 2O_2 + HCO_3^- \rightarrow HSO_4^- + CaCO_3 + Fe^{2+}$$

However, there are other factors that should be considered: for instance, *Achromatium* cells contain large amounts of calcite, even when calcium is below detection limits in its environment. This implies a massive accumulation of calcium against a considerable concentration gradient. There is thus likely to be an energetic cost of precipitating calcite that must be balanced against the enhanced energy generation supported by the consumption of intracellular hydroxyl ions. In environments with high levels of dissolved sulfide, it is unlikely that this mechanism to access solid-phase sulfides would be competitive, because of the cost of calcium carbonate precipitation. Interestingly, the only *Achromatium* species reported routinely from sulfidic marine systems is '*Achromatium volutans*', which does not precipitate calcite.

Giant sulfur bacteria and the nitrogen cycle. Some giant sulfur bacteria have an important role to play in the biogeochemical cycling of nitrogen. There is evidence that some *Beggiatoa* species are capable of fixing dinitrogen gas (Polman & Larkin, 1988), but more attention has been diverted toward the ability of a number of giant sulfur bacteria to use nitrate as a respiratory oxidant. Initially, this was presumed to be linked to denitrification (Fossing *et al.*, 1995; Sweerts *et al.*, 1990), but more recent evidence strongly supports the notion that the principal route for respiratory nitrate reduction is through DNRA (Otte *et al.*, 1999). This has prompted a considerable reappraisal of the significance of mats of giant sulfur bacteria in the nitrogen dynamics of sedimentary environments.

In many coastal marine environments, nitrogen is the most important limiting nutrient (Nixon, 1981; Ryther & Dunstan, 1971) and removal of nitrogen via denitrification can be an important control on eutrophication (Seitzinger, 1988).

As organic carbon inputs to sediments increase in marine environments in particular, denitrification becomes less important and DNRA more so. This has been attributed to increased sulfate reduction and the accumulation of sulfide to concentrations that inhibit key enzymes in the denitrification pathway (Brunet & Garcia-Gil, 1996), but, as discussed above, energetic considerations may also play a role. Nevertheless, it is clear that a shift in nitrogen-reduction pathways from denitrification to a conservative route of nitrate reduction, such as DNRA, will have important consequences for the overall budget of nitrogen and the productivity of a system. In this context, the ability of giant sulfur bacteria, such as members of the genera *Thioploca* and *Beggiatoa*, which accumulate high levels of nitrate, to reduce nitrate to ammonium coupled with sulfide oxidation (McHatton *et al.*, 1996; Otte *et al.*, 1999) clearly has significance.

Studies of sediments off the coast of Chile have shown that nitrate fluxes into sediments with large populations of *Thioploca* cells are > 50 % higher than in sediments with a low abundance of *Thioploca* cells (Zopfi *et al.*, 2001). It has been suggested that, because *Thioploca* cells can extend into nitrate-containing waters overlying the sediments that they inhabit, they may monopolize the available nitrate before it can be transported to other sediment-dwelling bacteria. This would imply that other forms of nitrate respiration might be of minimal significance in nitrogen cycling in sediments harbouring large populations of *Thioploca* cells. Experimental evidence refutes this suggestion and denitrification rates in *Thioploca*-rich sediments were actually slightly higher $(4 \cdot 5-9 \, \text{mmol m}^{-2} \text{day}^{-1})$ than typical values for coastal sediments (Herbert, 1999). Thus, there was still sufficient nitrate reaching the sediment to support significant denitrification rates, despite considerable assimilation of water-column nitrate by *Thioploca* cells. Furthermore, nitrous oxide was also consumed rapidly, suggesting that complete denitrification occurred in the sediments (Zopfi *et al.*, 2001). Although the sole route of nitrate reduction in these sediments is not through DNRA linked to sulfide oxidation and not all sulfide oxidation is mediated by *Thioploca* species, it has been estimated that if 25 % of sulfide oxidation is mediated by nitrate-reducing *Thioploca* cells [toward the lower range of reported estimates (16–91 %)], this would result in an increase in the net flux of ammonium from the sediments of 83 %.

This represents a significant increase in the amount of nitrogen retained in the system and is likely to have a significant effect on the overall primary productivity in such coastal ecosystems, which are typically nitrogen-limited. Indeed, although Graco *et al.* (2001) found that the main source of ammonium in Chilean coastal sediments was in fact mineralization of organic matter and only 17 % of the total recycled ammonium resulted from the DNRA activity of giant sulfur bacteria, the activity of the sulfur bacteria still made a significant contribution to the recycled nitrogen and could account for > 10 % of the primary productivity supported by recycled nitrogen. Similar conclusions have been drawn from studies of sediments in Tokyo Bay, which have also found that nitrate-accumulating, vacuolated sulfur bacteria have an important role in nitrogen retention in this coastal ecosystem. Interestingly, conditions of high productivity are likely to exacerbate the situation, as higher inputs of organic matter stimulate sulfate reduction, leading to higher sulfide concentrations that may repress denitrification and favour DNRA (Brunet & Garcia-Gil, 1996), resulting in greater nitrogen retention in the system with potential effects on long-term eutrophication. It is interesting to speculate about the wider implications of nitrogen retention on the biogeochemical cycling of sulfur and carbon. Effective recycling of nitrogen within the system will ultimately lead to greater sulfate regeneration and, consequently, the importance of carbon mineralization through sulfate reduction may also be increased. This would be further enhanced by the greater stoichiometric efficiency of DNRA relative to denitrification.

Giant sulfur bacteria and the carbon cycle. The vast accumulations of biomass in mats of giant sulfur-oxidizing bacteria represent a significant pool of organic carbon. It has been shown that sedimentary carbon dynamics are affected by the presence of mats of *Thioploca* and *Beggiatoa* cells, and sediments characterized by sulfur-oxidizing bacterial mats have been shown to promote higher rates of carbon burial (Graco *et al.*, 2001). It has been suggested that this may be a consequence of sheaths and cells of *Thioploca* and *Beggiatoa* containing refractory carbon and limited grazing on the giant bacterial cells, although no direct measurements of the resistance of sulfur bacterial sheaths have been made (Graco *et al.*, 2001).

Most studies of the effect of sulfur bacterial mats on carbon cycling have focused on cold-seep environments, where methane and hydrocarbons are important carbon sources driving the microbial ecosystem. Mats of different colour have often been reported from such sites and recent studies have shown that the colour of the mats, which is believed to reflect high cytochrome content of the orange-mat bacteria, correlates with the type of carbon metabolism exhibited by the sulfur bacteria (Nikolaus *et al.*, 2003). However, in this case, it was indicated that the water-soluble pigments, which had an absorbance maximum at 390 nm, were unlike typical cytochromes (Nikolaus *et al.*, 2003). The non-pigmented mats expressed high but variable levels of RuBisCO activity, suggesting that the non-pigmented sulfur bacteria had chemoautotrophic potential. RuBisCO activity in the pigmented filaments was several orders of magnitude lower, suggesting that they were more likely to be heterotrophic, perhaps utilizing hydrocarbons or hydrocarbon-oxidation products prevalent at the seep site (Nikolaus *et al.*, 2003).

Sulfur bacterial mats are characteristic of many methane-rich environments and it has even been suggested that the mats may form a relatively impermeable barrier that restricts the movement of methane out of the sediment (Orphan *et al.*, 2004). This may help to promote anaerobic oxidation of methane by holding methane in surface sediments, close to the large pool of sulfate in the overlying water and also to a source of sulfate from reoxidation of sulfide by the sulfur bacteria themselves. Consequently, high levels of anaerobic methane oxidation appear to be associated with conspicuous sulfur bacterial mats (Joye *et al.*, 2004; Treude *et al.*, 2003). The high levels of anaerobic methane oxidation, which are typically linked to sulfate reduction in marine systems, appear to be supported by the effective reoxidation of sulfide by the bacterial mats (Joye *et al.*, 2004). Isotopic data, obtained from a combination of fluorescent *in situ* hybridization and secondary-ion mass spectrometry (FISH–SIMS; Orphan *et al.*, 2001), suggest that the filamentous mat bacteria also incorporate methane-derived light carbon (Orphan *et al.*, 2004). It is clear from these data and the precipitation of isotopically light carbonates in seep environments that much of the carbon from active seep sites associated with sulfur bacterial mats is retained in the system by the combined activities of the anaerobic methane-oxidizing archaeal–sulfate-reducer consortia and mats of giant sulfur bacteria (Boetius & Suess, 2004). It is not only the microbial component of the ecosystem that is affected by the carbon dynamics in seep environments; higher trophic levels are also influenced by the presence of sulfur bacterial mats. One study has shown that both the density and diversity of fauna are greater at seep sites with extensive sulfur bacterial mats, compared with similar settings that lack the bacterial mats. Mats of *Thioploca* cells in the Gulf of Mexico in particular appeared to support a more diverse foraminiferan community, compared with mats of *Beggiatoa* cells and off-mat sites (Robinson *et al.*, 2004).

EVOLUTIONARY AND ECOLOGICAL DIVERSITY IN THE GIANT SULFUR BACTERIA

Giant sulfur bacteria have been shown by comparative 16S rRNA gene sequence analysis to be distributed between two classes of the phylum *Proteobacteria* (Teske *et al.*, 1996). The majority (*Beggiatoa–Thioploca–Thiomargarita*, *Achromatium* and *Thiothrix*) form well-defined clades within the '*Gammaproteobacteria*'. However, the genus *Thiovolum* clusters with the '*Epsilonproteobacteria*' (Fig. 2).

The different genera of giant sulfur bacteria have each evolved major morphological and physiological adaptations that allow them to exploit sulfur oxidation in different geochemical settings. Within each of these genera, however, there is evidence for more recent diversification and adaptation. It has been possible to correlate fine-scale patterns of genetic diversity with ecological function in giant sulfur bacteria and, by extrapolation, this is providing insights into the ecological mechanisms that underpin diversification in the microbial world more generally.

Within-genus diversity in morphology and physiology

On the basis of comparative 16S rRNA gene sequence analysis, the genera *Beggiatoa*, *Thioploca* and *Thiomargarita* occupy a monophyletic group within the *Proteobacteria*. Within this, four distinct lineages are evident (Teske & Nelson, 2004). The large-vacuolated, marine *Beggiatoa* and *Thioploca* species, *Thiomargarita namibiensis* and a recently described, vacuolated, rosette-forming sulfur bacterium (Kalanetra *et al.*, 2004) occupy one group, whilst freshwater *Thioploca* species form a second group of closely related organisms. The non-vacuolated, marine *Beggiatoa* species form the third group, with the fourth comprising freshwater *Beggiatoa* isolates (Teske & Nelson, 2004). Until relatively recently, our knowledge of the diversity of these bacteria has largely been defined by a few recognized species, identified on the basis of filament diameter, the presence of nitrate-storing vacuoles and the presence or absence of sheath material (Teske & Nelson, 2004). Some of these morphological features map onto the 16S rRNA gene sequence-based phylogeny (e.g. filament diameter and presence of vacuoles), whereas others (sheath formation) do not. As more 16S rRNA gene sequence data become available for giant sulfur bacteria, it is becoming apparent that there is considerably greater genetic diversity within each of the described genera than the original, morphologically based descriptions would suggest. For instance, a recent study of the diversity and population structure of filamentous sulfur bacteria in the Danish Limfjorden and the German Wadden Sea showed high diversity in nitrate-storing, vacuolated *Beggiatoa* species coexisting within the same sediments (Mußmann *et al.*, 2003). These organisms represented novel phylogenetic clusters distinct from those characterized previously. It was predicted from the wide spectrum of filament diameters encountered in these environments and the relatively restricted diameter of individual *Beggiatoa* species identified by using FISH probes that the entire diversity that was present within the genus *Beggiatoa* had not been surveyed completely. Interestingly, the coexisting *Beggiatoa* species identified by FISH showed depth-related differences in their distribution (Mußmann *et al.*, 2003). This was attributed either to the ability of species with a wider average filament diameter to store nitrate in greater quantities and hence reside at greater depths for longer, or the ability of species with a smaller average diameter to scavenge more efficiently for dissolved sulfide, which is in limited supply in the near-surface environment. This is echoed by the findings that the depth distribution of *Thioploca chileae*, *Thioploca araucae* and a third *Thioploca* species, designated SCM (short-cell morphotype), in South American coastal sediments was different, and the nitrate and sulfur content of the different *Thioploca* species correlated with the different location of the bacteria in the sediment (Zopfi *et al.*, 2001). As in most other locations, nitrate-accumulating *Beggiatoa* cells from organic-rich sediments in Tokyo Bay also had varying widths of trichome (Kojima & Fukui, 2003). The morphotype with wider trichomes was related most closely to uncultured *Beggiatoa* species from other geographical localities, identified from 16S rRNA gene sequence analysis, and distinct from the Tokyo Bay morphotype with narrow trichomes. Among the narrower types, which formed a new branch within the *Beggiatoa–Thopmargarita–Thioploca* clade, a sample of cells from a tidal flat harboured a narrow *Beggiatoa*-like morphotype that was genetically distinct from those identified in samples taken in water depths of 10 and 20 m. This suggests that there may be some ecological differentiation between closely related *Beggiatoa* genotypes. Analysis of 16S rRNA gene sequences from Chilean *Thioploca* isolates revealed several coexisting, genetically distinct species that can be differentiated by filament diameter. Not only were these found in the same sediment sample, but they also occurred commonly within the same sheath (Teske *et al.*, 1996). In contrast, a more recent study of freshwater *Thioploca* isolates from Lake Biwa, Japan, and Lake Constance, Germany, indicated that two freshwater *Thioploca* isolates from geographically separated environments were barely distinguishable on the basis of trichome diameter and had almost identical 16S rRNA gene sequences (Kojima *et al.*, 2003).

As a relatively large number of *Beggiatoa* species have been isolated in axenic culture, a great deal more is known about the physiology of members of the genus *Beggiatoa* than other large sulfur bacteria. Members of the genus *Beggiatoa* have a diverse carbon metabolism, ranging from obligate chemolithoautotrophy, facultative chemolithoautotrophy and mixotrophy through to lithoheterotrophy. This phenotypic diversity is observed at the species level, so it appears that carbon metabolism has been a key driver of ecological and evolutionary diversification. For instance, although it appears that the marine *Beggiatoa* species (vacuolated and unvacuolated) display predominantly chemolithoautotrophic nutrition (Teske & Nelson, 2004), it also appears that they can be either obligate or facultative autotrophs, as demonstrated by the related marine strains MS-81-1c and MS-81-6 (Hagen & Nelson, 1996). For instance, in the presence of acetate, carbon fixation by the obligately chemolithotrophic strain MS-81-1c was not reduced significantly (Hagen & Nelson, 1996). In addition, the use of radiolabelled substrates demonstrated that this organism was unable to respire acetate and that 2-oxoglutarate reductase, an enzyme necessary for the respiration of organic substrates in the tricarboxylic cycle, was absent. Even in the presence of organic compounds, 80 %

of cell carbon was obtained from CO_2. Strain MS-81-1c also obtained more energy from the oxidation of reduced sulfur species than the facultatively chemolitho-autotrophic marine strain MS-81-6 (Hagen & Nelson, 1997). MS-81-6, although able to grow chemolithoautotrophically, was found to be metabolically more versatile and could utilize acetate freely for energy or biosynthesis (Hagen & Nelson, 1996). *Beggiatoa* sp. strains MS-81-1c and MS-81-6 were originally isolated from the same environment (Great Sippewissett salt marsh, Woods Hole, MA, USA) and the physio-logical differences between the strains may be a factor in the maintenance of the diversity of *Beggiatoa* species in this single geographical location.

In contrast to marine organisms, there has been considerable debate as to whether genetically differentiated freshwater *Beggiatoa* strains are capable of genuine autotrophic or lithoheterotrophic growth (Hagen & Nelson, 1997; Strohl, 2005; Teske & Nelson, 2004). Whilst mixotrophic nutrition has been claimed for a large number of freshwater strains (Hagen & Nelson, 1996), it has only recently been shown that a freshwater isolate can grow lithoautotrophically (Grabovich *et al.*, 2001). Nevertheless, it is clear that the genetic differentiation between the marine and freshwater *Beggiatoa* species is correlated with fundamental differences in carbon metabolism, given the predominance of heterotrophic/mixotrophic metabolism in the freshwater strains and autotrophic metabolism in the marine species. The ecological sense of these differences may be explained by the availability of sulfide in marine and freshwater environments. The metabolic versatility in the freshwater strains is consistent with the potentially low and variable supply of sulfide in these environments and hence the requirement to supplement growth by the utilization of organic carbon.

At present, very little is known about the genetic diversity of the genus *Thiomargarita* and 16S rRNA gene sequence data are only available for *Thiomargarita namibiensis* from Walvis Bay, Namibia, south-west Africa (Schulz *et al.*, 1999). Unlike most com-munities of *Achromatium*, *Beggiatoa* and *Thioploca* species, samples of *Thiomargarita namibiensis* from Walvis Bay only produced a single 16S rRNA gene sequence, suggest-ing either that *Thiomargarita namibiensis* is genetically much more homogeneous than other giant sulfur bacteria or that any genetic differentiation of *Thiomargarita* species cannot be resolved on the basis of 16S rRNA gene sequence analysis. It may be of significance that the habitats where several distinct species of giant sulfur bacteria coexist tend to be spatially structured, relatively undisturbed sediments. The different genotypes observed in these systems may therefore reflect species adapted to different conditions in the sediment environment (Gray *et al.*, 1999b). Members of the genus *Thiomargarita* are believed to have quite a different lifestyle from sulfur bacteria that exhibit genetic diversity at a single location. It is thought that *Thiomargarita* cells survive by living in an environment that is subject to periodic disturbance and mixing.

The mixing brings sessile cells in sulfide-rich sediment into contact with water containing nitrate and oxygen, the oxidants that it uses for sulfide oxidation (Schulz & Jørgensen, 2001). It is often considered that mixed environments sustain less diversity than structured environments, as they provide a narrower range of environmental conditions that can be exploited by ecologically distinct organisms. This may in part explain the apparent homogeneity of communities of *Thiomargarita* cells. Nonetheless, chains of *Thiomargarita* cells also fall into distinct cell-size classes that, by analogy with other giant sulfur bacteria, may correlate with different species (Schulz *et al.*, 1999). Alternatively, the failure to identify different *Thiomargarita* species may be due to limited sampling of the communities.

The genus *Thiothrix* forms a deep branch within the '*Gammaproteobacteria*'; however, there is considerable physiological variation within the genus (Aruga *et al.*, 2002; Howarth *et al.*, 1999; Rossetti *et al.*, 2003). *Thiothrix* species, like *Beggiatoa* species, are varied in their carbon metabolism and heterotrophs, mixotrophs and chemolitho-autotrophs are all represented by different *Thiothrix* species. At present, no studies have established how this genetic and physiological diversity relates to the distribution and abundance of *Thiothrix* species in engineered or natural ecosystems. However, a number of studies have shown that bacteria related closely to *Thiothrix unzii*, which can grow with organic carbon sources but has an obligate requirement for sulfide or thiosulfate (Unz & Head, 2005), appear to be prevalent in some sulfidic cave environments (Brigmon *et al.*, 2003; Engel *et al.*, 2004).

Genetic and ecological diversity in the genus *Achromatium*

The tantalizing link between fine-scale patterns of genetic diversity and ecological differentiation in giant sulfur bacteria has been explored most thoroughly in the genus *Achromatium*. Natural communities of *Achromatium* cells that were originally thought to be genetically homogeneous in fact comprise a number of phylogenetically distinct subpopulations, distinguishable by FISH (Glöckner *et al.*, 1999; Gray *et al.*, 1999b). The degree of identity observed in *Achromatium*-derived 16S rRNA gene sequences (<97·5 %) from individual samples of freshwater sediment indicates that different species of *Achromatium* exist both in geographically separated locations and within a single sediment (Gray *et al.*, 1999b). As with the marine *Beggiatoa* species described above, it has been shown that individual, coexisting *Achromatium* species exhibit physiological and ecological differentiation. Coexisting *Achromatium* species in sediment from the margins of Rydal Water in the English Lake District, like *Beggiatoa* and *Thioploca* species, fall into distinct size classes (Gray *et al.*, 1999b). Furthermore, the composition of the *Achromatium* community was different in oxidizing and reducing zones within the sediment (Gray *et al.*, 1999b). On this basis, it was hypothesized that genetic diversity in coexisting *Achromatium* communities reflected exploitation of different redox-related niches. This was based on the competitive-exclusion principle, which states that if two similar species coexist in a stable environment, they do so as a result of niche differentiation. If, however, there is no niche differentiation, the species will compete and one will eliminate or exclude the other (Begon *et al.*, 1996; Gause, 1934; Hardin, 1960). Redox-related niche differentiation in *Achromatium* communities was demonstrated experimentally (Gray *et al.*, 2004). It was reasoned that, in sediment microcosms harbouring *Achromatium* communities, growth or maintenance of a particular *Achromatium* species exposed to particular redox conditions, accompanied by neutral or antagonistic effects on other coexisting *Achromatium* species, would be indicative of niche differentiation. In anoxic microcosms, *Achromatium* sp. RY8 decreased and *Achromatium* spp. RY5 and RYKS increased over time. Addition of increasing concentrations of nitrate, however, maintained the population of *Achromatium* sp. RY8. When high levels of nitrate were maintained throughout the incubation, the composition of the *Achromatium* community remained stable over time (Fig. 7). This suggested that all of the coexisting *Achromatium* species are obligate or facultative anaerobes that can utilize nitrate as an electron acceptor, but it was not possible to establish whether *Achromatium* species utilized DNRA or denitrification. *Achromatium* sp. RY8 was clearly more sensitive to sediment redox conditions than the other *Achromatium* species. These results give a much clearer picture of the mechanism that supports coexistence of several *Achromatium* species adapted to different redox conditions (Gray *et al.*, 2004). Redox conditions in sediments vary with depth, and coexisting *Achromatium* species are adapted to different redox conditions and thus avoid direct competition. Clearly, then, the genetic diversity observed in *Achromatium* communities correlates to functional diversity. This in turn may have an impact on geochemical cycling. The inherently low levels of sulfate that typify sediments inhabited by freshwater *Achromatium* species suggest that regeneration of sulfate is an important process and may lead to a greater flow of electrons from organic carbon through sulfate reduction than would be suggested by the steady-state sulfate concentration. Consequently, *Achromatium* community composition may have a direct effect on the efficiency of the ecosystem process by allowing efficient reoxidation of sulfide under a range of different redox conditions.

GENOMICS MEETS ECOLOGY: UNDERSTANDING ECOLOGICAL DIVERSITY BY USING METAGENOMICS

Many giant sulfur bacteria exist as communities comprising several related species that are ecologically and physiologically distinct. Although techniques such as combined microautoradiography and FISH can provide some information on physiological differences between coexisting species in natural communities (Gray *et al.*, 2000), this provides at best a shuttered view of the properties that distinguish the coexisting

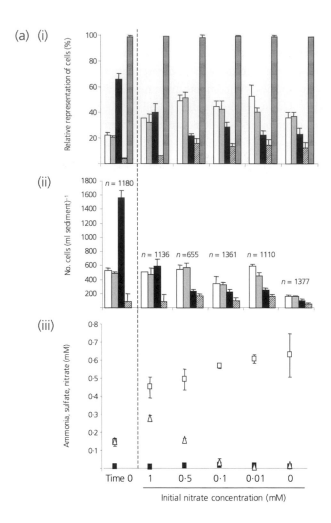

Achromatium or *Thioploca* species. With the advent of genomic technologies and the emergence of metagenomic analysis (DeLong, 2002), the potential now exists to recover large amounts of genomic information directly from natural microbial communities. Recent forays into metagenomics have helped to frame some of the limitations of the approach. The largest-scale metagenomic analysis conducted to date examined the microbial diversity of the Sargasso Sea (Venter *et al.*, 2004). This breathtaking work resulted in the recovery of 1 Gb non-redundant sequence, yet still barely scratched the surface of the diversity present. Even with extreme brute force, the metagenome of marine ecosystems is a powerful adversary. In contrast, a more comprehensive sampling with less intensive sequencing effort was obtained by focusing on an ecosystem known to have very limited diversity (acid minewaters from Iron Mountain, CA, USA; Tyson *et al.*, 2004). It proved possible to reconstruct the genomes from the principal players

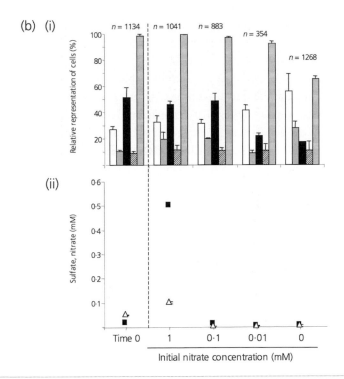

Fig. 7. The effect of nitrate availability on different *Achromatium* species in a freshwater sediment. (a) Relative abundance (i) and absolute numbers (ii) of different *Achromatium* species within sediment microcosms incubated for 7 days under anoxic conditions with different initial nitrate concentrations. Empty bars, *Achromatium* sp. RY5; shaded bars, *Achromatium* sp. RYKS; filled bars, *Achromatium* sp. RY8; diagonally hatched bars, *Achromatium* sp. RY1; horizontally hatched bars, Eub338-positive cells. Ammonium (□), sulfate (△) and nitrate (■) in sediment microcosms incubated for 7 days under different redox conditions (iii). (b) Relative abundance of different *Achromatium* species in sediment microcosms incubated for 13 days and supplied repeatedly with nitrate at different concentrations (i). Empty bars, *Achromatium* sp. RY5, shaded bars, *Achromatium* sp. RYKS; filled bars, *Achromatium* sp. RY8; diagonally hatched bars, *Achromatium* sp. RY1; horizontally hatched bars, Eub338-positive cells. Sulfate (△) and nitrate (■) in sediment microcosms after 13 days incubation (48 h after last addition of nitrate-containing overlying water) (ii). Time-zero analyses were conducted on replicate microcosms sampled sacrificially at the beginning of the experiments. Error bars on all data indicate SEM of triplicate incubations. Where error bars are not shown, the error bars were smaller than the symbols. *n*, No. cells counted for each replicate set of microcosms. Reproduced from Gray *et al.* (2004) with permission from Blackwell Publishing.

in the highly acidic environment and, consequently, to link the genomic information with the specific metabolic and ecological role of the most abundant organisms present. The analyses of the Sargasso Sea and the acid minewater were both essentially exploratory in nature. Whilst this is a valid way to proceed and leads to many novel and exciting discoveries, different opportunities are offered by hypothesis-driven research. Testing of hypotheses is only possible within the framework of a rigorous experimental design. The poor sampling afforded by metagenomic analysis of many environments precludes such an experimental design. The work of Tyson *et al.* (2004), however, illustrates that it is possible to obtain meaningful sampling of a microbial community that would lend itself to a defensible comparative analysis. Communities of giant sulfur bacteria offer an ideal system for the pursuit of hypothesis-driven metagenomics. Giant sulfur bacteria occur naturally as highly enriched, high-biomass communities. The communities typically comprise several coexisting species that can be distinguished genetically at the level of comparative 16S rRNA gene sequence analysis and the different genotypes can be shown to be physiologically and ecologically distinct. Extraction of total nucleic acids and generation of high-coverage clone libraries from such communities have a high chance of reconstructing the genomes of the most abundant community members. Comparative genome analysis would then provide the opportunity to identify the genetic basis for the ecological differentiation observed in natural communities of giant sulfur bacteria.

REFERENCES

Aruga, S., Kamagata, Y., Kohno, T., Hanada, S., Nakamura, K. & Kanagawa, T. (2002). Characterization of filamentous Eikelboom type 021N bacteria and description of *Thiothrix disciformis* sp. nov. and *Thiothrix flexilis* sp. nov. *Int J Syst Evol Microbiol* **52**, 1309–1316.

Babenzien, H.-D., Glöckner, F.-O. & Head, I. M. (2005). Genus *Achromatium*. In *Bergey's Manual of Systematic Bacteriology*, 2nd edn, vol. 2, part B: *Gammaproteobacteria*, pp. 142–147. Edited by G. Garrity. New York: Springer.

Begon, M., Harper, J. L. & Townsend, C. R. (1996). *Ecology: Individuals, Populations and Communities*, 3rd edn. Oxford: Blackwell Science.

Berks, B. C., Ferguson, S. J., Moir, J. W. B. & Richardson, D. J. (1995). Enzymes and associated electron transport systems that catalyse the respiratory reduction of nitrogen oxides and oxyanions. *Biochim Biophys Acta* **1232**, 97–173.

Boetius, A. & Suess, E. (2004). Hydrate Ridge: a natural laboratory for the study of microbial life fueled by methane from near-surface gas hydrates. *Chem Geol* **205**, 291–310.

Bonin, P., Omnes, P. & Chalamet, A. (1998). Simultaneous occurrence of denitrification and nitrate ammonification in sediments of the French Mediterranean Coast. *Hydrobiologia* **389**, 169–182.

Borowitzka, M. A. (1982). Morphological and cytological aspects of algal calcification. *Int Rev Cytol* **74**, 127–162.

Brigmon, R. L., Martin, H. W., Morris, T. L., Bitton, G. & Zam, S. G. (1994). Biogeochemical ecology of *Thiothrix* spp. in underwater limestone caves. *Geomicrobiol J* **12**, 141–159.

Brigmon, R. L., Furlong, M. & Whitman, W. B. (2003). Identification of *Thiothrix unzii* in two distinct ecosystems. *Lett Appl Microbiol* **36**, 88–91.

Brüchert, V., Pérez, M. E. & Lange, C. B. (2000). Coupled primary production, benthic foraminiferal assemblage, and sulfur diagenesis in organic-rich sediments of the Benguela upwelling system. *Mar Geol* **163**, 27–40.

Brüchert, V., Jørgensen, B. B., Neumann, K., Riechmann, D., Schlösser, M. & Schulz, H. (2003). Regulation of bacterial sulfate reduction and hydrogen sulfide fluxes in the central Namibian coastal upwelling zone. *Geochim Cosmochim Acta* **67**, 4505–4518.

Brunet, R. C. & Garcia-Gil, L. J. (1996). Sulfide-induced dissimilatory nitrate reduction to ammonia in anaerobic freshwater sediments. *FEMS Microbiol Ecol* 21, 131–138.

Burton, S. D. & Morita, R. Y. (1964). Effect of catalase and cultural conditions on growth of *Beggiatoa*. *J Bacteriol* 88, 1755–1761.

Cannon, G. C., Strohl, W. R., Larkin, J. M. & Shively, J. M. (1979). Cytochromes in *Beggiatoa alba*. *Curr Microbiol* 2, 263–266.

De Boer, W. E., La Rivière, J. W. M. & Schmidt, K. (1971). Some properties of *Achromatium oxaliferum*. *Antonie van Leeuwenhoek* 37, 553–563.

DeLong, E. F. (2002). Microbial population genomics and ecology. *Curr Opin Microbiol* 5, 520–524.

Engel, A. S., Porter, M. L., Stern, L. A., Quinlan, S. & Bennett, P. C. (2004). Bacterial diversity and ecosystem function of filamentous microbial mats from aphotic (cave) sulfidic springs dominated by chemolithoautotrophic "*Epsilonproteobacteria*". *FEMS Microbiol Ecol* 51, 31–53.

Eprintsev, A. T., Falaleeva, M. I., Stepanova, I. Y. & Parfenova, N. V. (2003). Purification and physicochemical properties of malate dehydrogenase from bacteria of the genus *Beggiatoa*. *Biochemistry (Mosc)* 68, 172–176.

Eprintsev, A. T., Falaleeva, M. I., Grabovich, M. Yu., Parfenova, N. V., Kashirskaya, N. N. & Dubinina, G. A. (2004). The role of malate dehydrogenase isoforms in the regulation of anabolic and catabolic processes in the colorless sulfur bacterium *Beggiatoa leptomitiformis* D-402. *Microbiology (English translation of Mikrobiologiia)* 73, 367–371.

Fenchel, T. (1994). Motility and chemosensory behaviour of the sulphur bacterium *Thiovulum majus*. *Microbiology* 140, 3109–3116.

Fenchel, T. & Glud, R. N. (1998). Veil architecture in a sulphide-oxidizing bacterium enhances countercurrent flux. *Nature* 394, 367–369.

Ferdelman, T. G., Lee, C., Pantoja, S., Harder, J., Bebout, B. M. & Fossing, H. (1997). Sulfate reduction and methanogenesis in a *Thioploca*-dominated sediment off the coast of Chile. *Geochim Cosmochim Acta* 61, 3065–3079.

Fossing, H., Gallardo, V. A., Jørgensen, B. B. & 12 other authors (1995). Concentration and transport of nitrate by the mat-forming sulphur bacterium *Thioploca*. *Nature* 374, 713–715.

Friedrich, C. G., Rother, D., Bardischewsky, F., Quentmeier, A. & Fischer, J. (2001). Oxidation of reduced inorganic sulfur compounds by bacteria: emergence of a common mechanism? *Appl Environ Microbiol* 67, 2873–2882.

Garrity, G. M., Bell, J. A. & Lilburn, T. G. (2005). Taxonomic outline of the prokaryotes. In *Bergey's Manual of Systematic Bacteriology*, 2nd edn, vol. 2. Edited by G. Garrity. New York: Springer (http://dx.doi.org/10.1007/bergeysoutline).

Gause, G. F. (1934). *The Struggle for Existence*. Baltimore, MD: Williams & Wilkins.

Glöckner, F. O., Babenzien, H.-D., Wulf, J. & Amann, R. (1999). Phylogeny and diversity of *Achromatium oxaliferum*. *Syst Appl Microbiol* 22, 28–38.

Grabovich, M. Yu., Dubinina, G. A., Lebedeva, V. Yu. & Churikova, V. V. (1998). Mixotrophic and lithoheterotrophic growth of the freshwater filamentous sulfur bacterium *Beggiatoa leptomitiformis* D-402. *Microbiology (English translation of Mikrobiologiia)* 67, 383–388.

Grabovich, M. Yu., Muntyan, M. S., Lebedeva, V. Yu., Ustiyan, V. S. & Dubinina, G. A. (1999). Lithoheterotrophic growth and electron transfer chain components of the filamentous gliding bacterium *Leucothrix mucor* DSM 2157 during oxidation of sulfur compounds. *FEMS Microbiol Lett* 178, 155–161.

Grabovich, M. Yu., Patritskaya, V. Yu., Muntyan, M. S. & Dubinina, G. A. (2001). Lithoautotrophic growth of the freshwater strain *Beggiatoa* D-402 and energy conservation in a homogeneous culture under microoxic conditions. *FEMS Microbiol Lett* **204**, 341–345.

Grabovich, M. Yu., Dul'tseva, N. M. & Dubinina, G. A. (2002). Carbon and sulfur metabolism in representatives of two clusters of bacteria of the genus *Leucothrix*: a comparative study. *Microbiology (English translation of Mikrobiologiia)* **71**, 255–261.

Graco, M., Farías, L., Molina, V., Gutiérrez, D. & Nielsen, L. P. (2001). Massive developments of microbial mats following phytoplankton blooms in a naturally eutrophic bay: implications for nitrogen cycling. *Limnol Oceanogr* **46**, 821–832.

Gray, N. D., Pickup, R. W., Jones, J. G. & Head, I. M. (1997). Ecophysiological evidence that *Achromatium oxaliferum* is responsible for the oxidation of reduced sulfur species to sulfate in a freshwater sediment. *Appl Environ Microbiol* **63**, 1905–1910.

Gray, N. D., Howarth, R., Pickup, R. W., Jones, J. G. & Head, I. M. (1999a). Substrate uptake by uncultured bacteria from the genus *Achromatium* determined by microautoradiography. *Appl Environ Microbiol* **65**, 5100–5106.

Gray, N. D., Howarth, R., Rowan, A., Pickup, R. W., Jones, J. G. & Head, I. M. (1999b). Natural communities of *Achromatium oxaliferum* comprise genetically, morphologically, and ecologically distinct subpopulations. *Appl Environ Microbiol* **65**, 5089–5099.

Gray, N. D., Howarth, R., Pickup, R. W., Jones, J. G. & Head, I. M. (2000). Use of combined microautoradiography and fluorescence in situ hybridization to determine carbon metabolism in mixed natural communities of uncultured bacteria from the genus *Achromatium*. *Appl Environ Microbiol* **66**, 4518–4522.

Gray, N. D., Comaskey, D., Miskin, I. P., Pickup, R. W., Suzuki, K. & Head, I. M. (2004). Adaption of sympatric *Achromatium* spp. to different redox conditions as a mechanism for coexistence of functionally similar sulphur bacteria. *Environ Microbiol* **6**, 669–677.

Guerrero, R., Haselton, A., Solé, M., Wier, A. & Margulis, L. (1999). *Titanospirillum velox*: a huge, speedy, sulfur-storing spirillum from Ebro Delta microbial mats. *Proc Natl Acad Sci U S A* **96**, 11584–11588.

Gunderson, J. K., Jørgensen, B. B., Larsen, E. & Jannasch, H. W. (1992). Mats of giant sulphur bacteria on deep-sea sediments due to fluctuating hydrothermal flow. *Nature* **360**, 454–456.

Hardin, G. (1960). The competitive exclusion principle. *Science* **131**, 1292–1297.

Hagen, K. D. & Nelson, D. C. (1996). Organic carbon utilization by obligately and facultatively autotrophic *Beggiatoa* strains in homogeneous and gradient cultures. *Appl Environ Microbiol* **62**, 947–953.

Hagen, K. D. & Nelson, D. C. (1997). Use of reduced sulfur compounds by *Beggiatoa* spp. enzymology and physiology of marine and freshwater strains in homogeneous and gradient cultures. *Appl Environ Microbiol* **63**, 3957–3964.

Head, I. M., Gray, N. D., Clarke, K. J., Pickup, R. W. & Jones, J. G. (1996). The phylogenetic position and ultrastructure of the uncultured bacterium *Achromatium oxaliferum*. *Microbiology* **142**, 2341–2354.

Head, I. M., Gray, N. D., Babenzien, H.-D. & Glöckner, F. O. (2000a). Uncultured giant sulfur bacteria of the genus *Achromatium*. *FEMS Microbiol Ecol* **33**, 171–180.

Head, I. M., Gray, N. D., Howarth, R., Pickup, R. W., Clarke, K. J. & Jones, J. G. (2000b). *Achromatium oxaliferum* – understanding the unmistakable. *Adv Microbiol Ecol* **16**, 1–40.

Herbert, R. A. (1999). Nitrogen cycling in coastal marine ecosystems. *FEMS Microbiol Rev* **23**, 563–590.

Hinze, G. (1913). Beitrage zur Kenntnis der farblosen Schwefelbakterien. *Ber Dtsch Bot Ges* **31**, 189–202 (in German).

Howarth, R., Unz, R. F., Seviour, E. M., Seviour, R. J., Blackall, L. L., Pickup, R. W., Jones, J. G., Yaguchi, J. & Head, I. M. (1999). Phylogenetic relationships of filamentous sulfur bacteria (*Thiothrix* spp. and Eikelboom type 021N bacteria) isolated from wastewater-treatment plants and description of *Thiothrix eikelboomii* sp. nov., *Thiothrix unzii* sp. nov., *Thiothrix fructosivorans* sp. nov. and *Thiothrix defluvii* sp. nov. *Int J Syst Bacteriol* **49**, 1817–1827.

Huettel, M., Forster, S., Klöser, S. & Fossing, H. (1996). Vertical migration in the sediment-dwelling sulfur bacteria *Thioploca* spp. in overcoming diffusion limitations. *Appl Environ Microbiol* **62**, 1863–1872.

Jørgensen, B. B. & Gallardo, V. A. (1999). *Thioploca* spp.: filamentous sulfur bacteria with nitrate vacuoles. *FEMS Microbiol Ecol* **28**, 301–313.

Jørgensen, B. B. & Revsbech, N. P. (1983). Colorless sulfur bacteria, *Beggiatoa* spp. and *Thiovulum* spp., in O_2 and H_2S microgradients. *Appl Environ Microbiol* **45**, 1261–1270.

Joye, S. B., Boetius, A., Orcutt, B. N., Montoya, J. P., Schulz, H. N., Erickson, M. J. & Lugo, S. K. (2004). The anaerobic oxidation of methane and sulfate reduction in sediments from Gulf of Mexico cold seeps. *Chem Geol* **205**, 219–238.

Kalanetra, K. M., Huston, S. L. & Nelson, D. C. (2004). Novel, attached, sulfur-oxidizing bacteria at shallow hydrothermal vents possess vacuoles not involved in respiratory nitrate accumulation. *Appl Environ Microbiol* **70**, 7487–7496.

Kappler, U. & Dahl, C. (2001). Enzymology and molecular biology of prokaryotic sulfite oxidation. *FEMS Microbiol Lett* **203**, 1–9.

Kelly, D. P. (1999). Thermodynamic aspects of energy conservation by chemolithotrophic sulfur bacteria in relation to the sulfur oxidation pathways. *Arch Microbiol* **171**, 219–229.

Kelly, D. P., Shergill, J. K., Lu, W.-P. & Wood, A. P. (1997). Oxidative metabolism of inorganic sulfur compounds by bacteria. *Antonie van Leeuwenhoek* **71**, 95–107.

Kojima, H. & Fukui, M. (2003). Phylogenetic analysis of *Beggiatoa* spp. from organic rich sediment of Tokyo Bay, Japan. *Water Res* **37**, 3216–3223.

Kojima, H., Teske, A. & Fukui, M. (2003). Morphological and phylogenetic character-izations of freshwater *Thioploca* species from Lake Biwa, Japan, and Lake Constance, Germany. *Appl Environ Microbiol* **69**, 390–398.

Lane, D. J., Harrison, A. P., Jr, Stahl, D., Pace, B., Giovannoni, S. J., Olsen, G. J. & Pace, N. R. (1992). Evolutionary relationships among sulfur- and iron-oxidizing eubacteria. *J Bacteriol* **174**, 269–278.

Larkin, J. M. (1989). Genus II. *Thiothrix* Winogradsky 1888. In *Bergey's Manual of Systematic Bacteriology*, vol. 3, pp. 2098–2101. Edited by J. T. Staley, M. P. Bryant, N. Pfennig & J. G. Holt. Baltimore, MD: Williams & Wilkins.

Larkin, J. M. & Strohl, W. R. (1983). *Beggiatoa*, *Thiothrix*, and *Thioploca*. *Annu Rev Microbiol* **37**, 341–367.

Lauterborn, R. (1907). Eine neue Gattung der Schwefelbakterien (*Thioploca schmidlei* nov. gen. nov. spec.). *Ber Dtsch Bot Ges* **25**, 238–242 (in German).

McGlannan, M. F. & Makemson, J. C. (1990). HCO_3^- fixation by naturally occurring tufts and pure cultures of *Thiothrix nivea*. *Appl Environ Microbiol* **56**, 730–738.

McHatton, S. C., Barry, J. P., Jannasch, H. W. & Nelson, D. C. (1996). High nitrate concentrations in vacuolate, autotrophic marine *Beggiatoa* spp. *Appl Environ Microbiol* **62**, 954–958.

Moissl, C., Rudolph, C. & Huber, R. (2002). Natural communities of novel archaea and bacteria with a string-of-pearls-like morphology: molecular analysis of the bacterial partners. *Appl Environ Microbiol* **68**, 933–937.

Morse, J. W. & Emeis, K. C. (1990). Controls on C/S ratios in hemipelagic sediments. *Am J Sci* **290**, 1117–1135.

Mußmann, M., Schulz, H. N., Strotmann, B., Kjær, T., Nielsen, L. P., Rosselló-Mora, R. A., Amann, R. I. & Jørgensen, B. B. (2003). Phylogeny and distribution of nitrate-storing *Beggiatoa* spp. in coastal marine sediments. *Environ Microbiol* **5**, 523–533.

Nelson, D. C. & Hagen, K. D. (1995). Physiology and biochemistry of symbiotic and free-living chemoautotrophic sulfur bacteria. *Am Zool* **35**, 91–101.

Nelson, D. C. & Jannasch, H. W. (1983). Chemoautotrophic growth of a marine *Beggiatoa* in sulfide-gradient cultures. *Arch Microbiol* **136**, 262–269.

Nelson, D. C., Waterbury, J. B. & Jannasch, H. W. (1982). Nitrogen fixation and nitrate utilization by marine and freshwater *Beggiatoa*. *Arch Microbiol* **133**, 172–177.

Nelson, D. C., Wirsen, C. O. & Jannasch, H. W. (1989). Characterization of large, autotrophic *Beggiatoa* spp. abundant at hydrothermal vents of the Guaymas Basin. *Appl Environ Microbiol* **55**, 2909–2917.

Nikolaus, R., Ammerman, J. W. & MacDonald, I. R. (2003). Distinct pigmentation and trophic modes in *Beggiatoa* from hydrocarbon seeps in the Gulf of Mexico. *Aquat Microb Ecol* **32**, 85–93.

Nixon, S. W. (1981). Remineralization and nutrient cycling in coastal marine ecosystems. In *Estuaries and Nutrients*, pp. 111–138. Edited by B. J. Nelson & L. E. Cronin. Totowa, NJ: Humana Press.

Odintsova, E. V. & Dubinina, G. A. (1993). Role of reduced sulfur compounds in *Thiothrix ramosa* metabolism. *Microbiology (English translation of Mikrobiologiia)* **62**, 139–146.

Orphan, V. J., House, C. H., Hinrichs, K.-U., McKeegan, K. D. & DeLong, E. F. (2001). Methane-consuming archaea revealed by directly coupled isotopic and phylogenetic analysis. *Science* **293**, 484– 487.

Orphan, V. J., Ussler, W., III, Naehr, T. H., House, C. H., Hinrichs, K.-U. & Paull, C. K. (2004). Geological, geochemical, and microbiological heterogeneity of the seafloor around methane vents in the Eel River Basin, offshore California. *Chem Geol* **205**, 265–289.

Otte, S., Kuenen, J. G., Nielsen, L. P. & 7 other authors (1999). Nitrogen, carbon, and sulfur metabolism in natural *Thioploca* samples. *Appl Environ Microbiol* **65**, 3148–3157.

Patritskaya, V. Yu., Grabovich, M. Yu., Muntyan, M. S. & Dubinina, G. A. (2001). Lithoautotrophic growth of the freshwater colorless sulfur bacterium *Beggiatoa "leptomitiformis"* D-402. *Microbiology (English translation of Mikrobiologiia)* **70**, 145–150.

Pimenov, N. V., Savvichev, A. S., Rusanov, I. I., Lein, A. Yu. & Ivanov, M. V. (2000). Microbiological processes of the carbon and sulfur cycles at cold methane seeps of the North Atlantic. *Microbiology (English translation of Mikrobiologiia)* **69**, 709–720.

Polman, J. K. & Larkin, J. M. (1988). Properties of in vivo nitrogenase activity in *Beggiatoa alba*. *Arch Microbiol* **150**, 126–130.

Rabenhorst, L. (1865). *Flora Europaea Algarum aquae dulcis et submarinae. Sectio II: Algas phycochromaceas complectens*, pp. 1–319. Leipzig, Germany: Eduard Kummer (in Latin).

Robinson, C. A., Bernhard, J. M., Levin, L. A., Mendoza, G. F. & Blanks, J. K. (2004). Surficial hydrocarbon seep infauna from the Blake Ridge (Atlantic Ocean, 2150 m) and the Gulf of Mexico (690–2240 m). *Mar Ecol* **25**, 313–336.

Romaniuk, P. J., Zoltowska, B., Trust, T. J., Lane, D. J., Olsen, G. J., Pace, N. R. & Stahl, D. A. (1987). *Campylobacter pylori*, the spiral bacterium associated with human gastritis, is not a true *Campylobacter* sp. *J Bacteriol* **169**, 2137–2141.

Rossetti, S., Blackall, L. L., Levantesi, C., Uccelletti, D. & Tandoi, V. (2003). Phylogenetic and physiological characterization of a heterotrophic, chemolithoautotrophic *Thiothrix* strain isolated from activated sludge. *Int J Syst Evol Microbiol* **53**, 1271–1276.

Ryther, J. H. & Dunstan, W. M. (1971). Nitrogen, phosphorus, and eutrophication in the coastal marine environment. *Science* **171**, 1008–1013.

Schewiakoff, W. (1893). *Über einen neuen bacterienähnlichen Organismus des Susswassers*. Habilitationsschrift, C. Winter, pp. 1–36. Heidelberg: Universität Heidelberg (in German).

Schmidt, T. M. & DiSpirito, A. A. (1990). Spectral characterization of c-type cytochromes purified from *Beggiatoa alba*. *Arch Microbiol* **154**, 453–458.

Schmidt, T. M., Arieli, B., Cohen, Y., Padan, E. & Strohl, W. R. (1987). Sulfur metabolism in *Beggiatoa alba*. *J Bacteriol* **169**, 5466–5472.

Schulz, H. N. & Jørgensen, B. B. (2001). Big bacteria. *Annu Rev Microbiol* **55**, 105–137.

Schulz, H. N. & Jørgensen, B. B. (2005). Genus *Thiomargarita*. In *Bergey's Manual of Systematic Bacteriology*, 2nd edn, vol. 2, part B: *Gammaproteobacteria*, pp. 169–171. Edited by G. Garrity. New York: Springer.

Schulz, H. N., Jørgensen, B. B., Fossing, H. A. & Ramsing, N. B. (1996). Community structure of filamentous, sheath-building sulfur bacteria, *Thioploca* spp., off the coast of Chile. *Appl Environ Microbiol* **62**, 1855–1862.

Schulz, H. N., Brinkhoff, T., Ferdelman, T. G., Mariné, M. H., Teske, A. & Jørgensen, B. B. (1999). Dense populations of a giant sulfur bacterium in Namibian shelf sediments. *Science* **16**, 493–495.

Seitzinger, S. P. (1988). Denitrification in freshwater and coastal marine ecosystems: ecological and geochemical significance. *Limnol Oceanogr* **33**, 702–724.

Simon, J. (2002). Enzymology and bioenergetics of respiratory nitrite ammonification. *FEMS Microbiol Rev* **26**, 285–309.

Sørensen, J. (1978). Capacity for denitrification and reduction of nitrate to ammonia in a coastal marine sediment. *Appl Environ Microbiol* **35**, 301–305.

Starr, M. P. & Skerman, V. B. D. (1965). Bacterial diversity: the natural history of selected morphologically unusual bacteria. *Annu Rev Microbiol* **19**, 407–454.

Stepanova, I. Yu., Eprintsev, A. T., Falaleeva, M. I., Parfenova, N. V., Grabovich, M. Yu., Patritskaya, V. Yu. & Dubinina, G. A. (2002). Dependence of malate dehydrogenase structure on the type of metabolism in freshwater filamentous colorless sulfur bacteria of the genus *Beggiatoa*. *Microbiology (English translation of Mikrobiologiia)* **71**, 377–382.

Strohl, W. R. (2005). Genus *Beggiatoa*. In *Bergey's Manual of Systematic Bacteriology*, 2nd edn, vol. 2, part B: *Gammaproteobacteria*, pp. 148–161. Edited by G. Garrity. New York: Springer.

Strohl, W. R., Schmidt, T. M., Vinci, V. A. & Larkin, J. M. (1986). Electron transport and respiration in *Beggiatoa* and *Vitreoscilla*. *Arch Microbiol* **145**, 71–75.

Sweerts, J.-P. R. A., De Beer, D., Nielsen, L. P., Verdouw, H., Van den Heuvel, J. C., Cohen, Y. & Cappenberg, T. E. (1990). Denitrification by sulphur oxidizing *Beggiatoa* spp. mats on freshwater sediments. *Nature* **344**, 762–763.

Teske, A. & Nelson, D. C. (2004). The genera *Beggiatoa* and *Thioploca*. In *The Prokaryotes: an Evolving Electronic Resource for the Microbiological Community*, 3rd edn, release 3.17. Edited by M. Dworkin, S. Falkow, E. Rosenberg, K.-H. Schleifer & E. Stackebrandt. New York: Springer (http://link.springer-ny.com/link/service/books/10125/).

Teske, A., Ramsing, N. B., Küver, J. & Fossing, H. (1996). Phylogeny of *Thioploca* and related filamentous sulfide-oxidizing bacteria. *Syst Appl Microbiol* **18**, 517–526.

Teske, A., Sogin, M. L., Nielsen, L. P. & Jannasch, H. W. (1999). Phylogenetic relationships of a large marine *Beggiatoa*. *Syst Appl Microbiol* **22**, 39–44.

Thamdrup, B. & Canfield, D. E. (1996). Pathways of carbon oxidation in continental margin sediments off central Chile. *Limnol Oceanogr* **41**, 1629–1650.

Thar, R. & Kühl, M. (2002). Conspicuous veils formed by vibrioid bacteria on sulfidic marine sediment. *Appl Environ Microbiol* **68**, 6310–6320.

Trevisan, V. (1842). *Prospetto della flora Euganea*, pp. 1–68. Padova, Italy: Coi Tipi Del Seminario (in Italian).

Treude, T., Boetius, A., Knittel, K., Wallmann, K. & Jørgensen, B. B. (2003). Anaerobic oxidation of methane above gas hydrates at Hydrate Ridge, NE Pacific Ocean. *Mar Ecol Prog Ser* **264**, 1–14.

Tyson, G. W., Chapman, J., Hugenholtz, P. & 7 other authors (2004). Community structure and metabolism through reconstruction of microbial genomes from the environment. *Nature* **428**, 37–43.

Unz, R. F. & Head, I. M. (2005). Genus *Thiothrix*. In *Bergey's Manual of Systematic Bacteriology*, 2nd edn, vol. 2, part B: *Gammaproteobacteria*, pp. 131–142. Edited by G. Garrity. New York: Springer.

Vaucher, J. P. (1803). *Histoire des conferves d'eau douce, contenant leurs different modes de reproduction, et la description de leurs principales espèces*, pp. 1–285. Geneva: J. Paschoud (in French).

Venter, J. C., Remington, K., Heidelberg, J. F. & 20 other authors (2004). Environmental genome shotgun sequencing of the Sargasso Sea. *Science* **304**, 66–74.

Winogradsky, S. (1887). Über Schwefelbakterien. *Bot Ztg* **45**, 489–610 (in German).

Winogradsky, S. (1888). *Beiträge zur Morphologie und Physiologie der Bacterien. Heft 1. Zur Morphologie und Physiologie der Schwefelbacterien*, pp. 1–120. Leipzig: Arthur Felix (in German).

Wright, D. & Oren, A. (2005). Nonphotosynthetic bacteria and the formation of carbonates and evaporites through time. *Geomicrobiol J* **22**, 27–53.

Zemskaya, T. I., Namsaraev, B. B., Dul'tseva, N. M., Khanaeva, T. A., Golobokova, L. P., Dubinina, G. A., Dulov, L. E. & Wada, E. (2001). Ecophysiological characteristics of the mat-forming bacterium *Thioploca* in bottom sediments of the Frolikha Bay, northern Baikal. *Microbiology (English translation of Mikrobiologiia)* **70**, 335–341.

Zopfi, J., Kjær, T., Nielsen, L. P. & Jørgensen, B. B. (2001). Ecology of *Thioploca* spp.: nitrate and sulfur storage in relation to chemical microgradients and influence of *Thioploca* spp. on the sedimentary nitrogen cycle. *Appl Environ Microbiol* **67**, 5530–5537.

Soil micro-organisms in Antarctic dry valleys: resource supply and utilization

D. W. Hopkins,[1] B. Elberling,[2] L. G. Greenfield,[3]
E. G. Gregorich,[4] P. Novis,[5] A. G. O'Donnell[6] and
A. D. Sparrow[3,7]

[1]School of Biological and Environmental Sciences, University of Stirling, Stirling FK9 4LA, Scotland, UK

[2]Institute of Geography, University of Copenhagen, Øster Voldgade 10, DK-1350, Copenhagen K., Denmark

[3]School of Biological Sciences, University of Canterbury, Private Bag 4800, Christchurch, New Zealand

[4]Agriculture Canada, Central Experimental Farm, Ottawa, Canada K1A 0C6

[5]Manaaki Whenua – Landcare Research, PO Box 69, Lincoln 8152, New Zealand

[6]Institute for Research on Environment and Sustainability, University of Newcastle upon Tyne, Newcastle upon Tyne NE1 7RU, UK

[7]Department of Natural Resources and Environmental Sciences, University of Nevada, 1000 Valley Rd, Reno, NV 89512, USA

INTRODUCTION

In 1903, the explorer Robert Scott was one of the first humans ever to see the dry valleys of Antarctica. He called them 'valley(s) of the dead' in which 'we have seen no sign of life, … not even a moss or lichen'. A century later, we know that the soils and rocks are home to many microscopic organisms that Scott could not have seen.

The dry valleys are part of the small percentage of the land surface of the Antarctic continent that is ice-free, amounting to about 4000 km², and thus have rock and soil surfaces that can be colonized by terrestrial organisms. They are an ancient polar desert, perhaps as much as 2 million years old, located in Victoria Land between about 77 and 79° south (Fig. 1). The valleys are in a precipitation shadow caused by the Transantarctic Mountains, which rise over 4000 m. The Antarctic dry valleys are now recognized as one of the harshest terrestrial environments on Earth, characterized by summer maximum temperatures that rarely exceed 0 °C and only a few tens of millimetres of precipitation, most of which falls as snow and is ablated by strong winds carrying dry air from the polar plateau – potential evaporation far exceeds precipitation (Fig. 1). The long periods of winter darkness are punctuated by a short summer, when 24 h daylight is reached for a few weeks either side of the summer solstice and when the ground surface temperature may rise to a few degrees above zero. This harsh environment has led to the Antarctic dry valleys being considered as analogues of

SGM symposium 65: Micro-organisms and Earth systems – advances in geomicrobiology.
Editors G. M. Gadd, K. T. Semple & H. M. Lappin-Scott. Cambridge University Press. ISBN 0 521 86222 1 ©SGM 2005

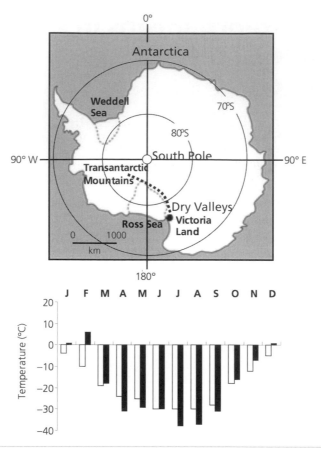

Fig. 1. Map of Antarctica showing the location of the dry valleys (●) and the Transantarctic Mountains (dotted line), and summary temperatures for the dry valleys (filled bars) and McMurdo Sound (empty bars). McMurdo Sound data were measured at Scott Base on the Ross Sea coast at 78° S; mean annual temperature, –20 °C; mean annual precipitation, 180 mm. Dry valley data were measured at Vanda Station in the Wright Valley at 78° S; mean annual temperature, –20 °C; mean annual precipitation, 45 mm.

extraterrestrial habitats by the 'astrobiology/exobiology' community (Mahaney *et al.*, 2001; Onofri *et al.*, 2004).

The dry valleys are highly significant sites for studies of ecosystem processes because of their relative biological simplicity (low diversity and small terrestrial biomass), and for monitoring the effects of environmental changes because they operate under extreme (both coldness and dryness) climatic conditions; hence, the Taylor Valley is included in the US National Science Foundation network of Long-Term Ecological Research sites (e.g. Virginia & Wall, 1999; Wall & Virginia, 1999; Greenland & Kittel, 2002; Hobbie, 2003; Turner *et al.*, 2003). The terrestrial ecosystem comprises micro-organisms, mosses, lichens and a restricted invertebrate community. Given the absence of a conspicuous community of terrestrial autotrophs in these resource-poor ecosystems, understanding the carbon and energy supply to terrestrial organisms is an important geomicrobiological question.

ORGANISMS IN THE DRY VALLEYS

Although there have been no systematic surveys of the diversity of organisms in the dry valleys, it is known that the soils of the dry valleys support sparse moss and lichen communities (Bargagli *et al.*, 1999), a low diversity of invertebrates (e.g. Treonis *et al.*, 1999; Stevens & Hogg, 2002) and small microbial communities, the diversity of which is largely uncharacterized (Wynn-Williams, 1996; Cowan & Tow, 2004). Although there is little information about the soil micro-organisms, there is evidence of microbial activity or potential activity in the soil (Burkins *et al.*, 2002; Treonis *et al.*, 2002; Parsons *et al.*, 2004; B. Elberling, E. G. Gregorich, D. W. Hopkins, A. D. Sparrow, P. Novis & L. G. Greenfield, unpublished results). Probably the most specialized terrestrial microbial community in the dry valleys is the endo/cryptoendolithic community of lichens, algae and fungi, living a few millimetres inside relatively coarse-grained rocks. The temperature and moisture regimes inside the rocks are less hostile and the organisms are sustained by light penetrating the translucent mineral grains (Friedmann, 1982; Friedmann *et al.*, 1993; Nienow & Friedmann, 1993). However, the most researched group of terrestrial organisms is the invertebrate animals – the largest residents of the dry valleys. One hundred years after Scott proclaimed the dry valleys sterile, Wilson (2002) referred to the soil invertebrates as 'McMurdo's equivalent of elephants and tigers'. The invertebrate community includes rotifers, tardigrades, acari, collembola and, most notably, nematodes, which are the largest terrestrial consumers in the dry valleys (Treonis *et al.*, 1999; Virginia & Wall, 1999; Courtright *et al.*, 2001; Doran *et al.*, 2002; Stevens & Hogg, 2002, 2003). The most abundant nematode is usually *Scottnema lindsayae*, which feeds on bacteria and algae and typically numbers between about 500 and 2000 (kg soil)$^{-1}$, whilst at particularly favoured sites, its number may rise to several thousand kg^{-1} (Table 1). The enduring presence of relatively large consumers indicates active energy processing and nutrient cycling.

MICROBIAL ACTIVITY IN DRY VALLEY SOILS

The soils in the dry valleys are characterized by the absence of structure, cohesion and moisture, typically alkaline pH, the absence of leaching and localized salt accumulations (Campbell & Claridge, 2000). The soils contain only small reserves of organic matter and nitrogen (Beyer *et al.*, 1999; Burkins *et al.*, 2002; Moorhead *et al.*, 2003; Barrett *et al.*, 2005; B. Elberling and others, unpublished results). With the exception of moist soils at the lake margins and in transient stream beds, the soils generally contain less (typically two orders of magnitude) carbon and nitrogen than the normal range for temperate soils (Table 2; Barrett *et al.*, 2002; B. Elberling and others, unpublished

Table 1. Numbers of soil invertebrates in four samples of polygon soils from the Garwood Valley, Antarctica

Data are means ± SD (D. H. Wall & B. Adams, personal communication).

Invertebrate	No. invertebrates (kg dry soil)$^{-1}$								
	Males		Females		Juveniles		Total alive	Total dead	Total alive+dead
	Alive	Dead	Alive	Dead	Alive	Dead			
Nematodes									
Scottnema lindsayae	506 ± 310	101 ± 76	714 ± 479	148 ± 158	2097 ± 1576	336 ± 175	3317 ± 2362	585 ± 393	3902 ± 2746
Eudorylaimus antarcticus	36 ± 51	0	30 ± 29	0	120 ± 160	11 ± 16	185 ± 231	11 ± 16	196 ± 246
Plectus sp.	0	0	0	0	8 ± 17	0	8 ± 17	0	8 ± 17
Tardigrades									5 ± 6
Rotifers									49 ± 51

Table 2. Selected properties of surface (0–1 cm) soils from different landscape units in the Garwood Valley, Antarctica

Data (means ± SD) are by B. Elberling and others (unpublished results).

Soil	Water content [g H$_2$O (g soil)$^{-1}$]	Organic carbon [mg C (g soil)$^{-1}$]	Total nitrogen [µg N (g soil)$^{-1}$]	NH$_4^+$ nitrogen [µg N (g soil)$^{-1}$]	NO$_3^-$ nitrogen [µg N (g soil)$^{-1}$]	pH
Hill-slope	1·9 ± 0·9	1·38 ± 0·30	58·6 ± 36·3	0·41 ± 0·14	1·00 ± 0·49	8·6
Polygon	5·7 ± 3·6	1·63 ± 0·61	114 ± 48	1·27 ± 0·46	1·46 ± 1·23	8·6
Lake margin	62·7 ± 27·9	29·9 ± 26·0	3259 ± 2807	86·4 ± 67·8	7·25 ± 2·97	7·0

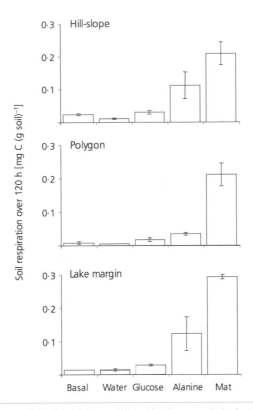

Fig. 2. Soil respiration over 120 h for soils from different landscape units in the Garwood Valley, incubated at 5 °C in the laboratory in unamended soil (basal) and in response to addition of water, glucose, alanine and samples of the microbial mat from the lake margin. Each bar is the mean of three replicates and the bars are ± SEM.

results). Nevertheless, *in situ* microbial respiration is consistently measured as CO_2 efflux at the soil surface (Parsons *et al.*, 2004; B. Elberling and others, unpublished results) and Barrett *et al.* (2005) reported the presence of a significant amount of labile soil organic matter. Assuming a steady state, the estimated turnover times for organic matter (mass/flux) in the dry valley soils is remarkably fast: approximately 23 years in the Taylor Valley (Burkins *et al.*, 2002) and 30–123 years in the Garwood Valley, depending on soil type (B. Elberling and others, unpublished results). By contrast, turnover times in most temperate region soils are generally several centuries or even millennia (Hopkins & Gregorich, 2005). In the absence of a conspicuous community of terrestrial autotrophs in the dry valleys, the rapid turnover of organic carbon suggests either that the soil organic carbon is not in a steady state or the presence of modern additions of labile organic matter (Barrett *et al.*, 2005). The presence of labile organic matter is supported by *in situ* measurements of soil respiration (B. Elberling and others, unpublished results). In the Garwood Valley, soil respiration at the lake margin was greater than that in the drier soils from elsewhere in the valley (Fig. 2). This difference

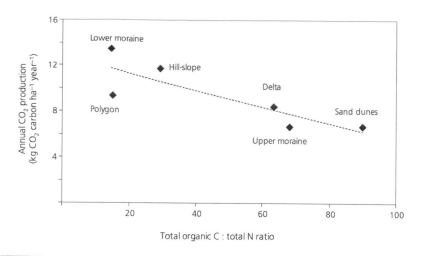

Fig. 3. Relationship of *in situ* soil respiration and organic C : N ratio for different soils in the Garwood Valley (B. Elberling and others, unpublished results).

was not related solely to the water content (Fig. 2) and experimental additions of glucose and alanine and of detritus from the lake margin microbial mats (C : N ratio of 10–12) indicated that soil respiration is limited by both carbon and nitrogen (Fig. 2). In the case of the addition of microbial-mat material, the soil amendment may also have contained viable organisms that may have contributed to the respiratory response. The influence of nitrogen on soil respiration is consistent with *in situ* measurements of soil CO_2 respiration rates (Fig. 3), which indicate a correlation between the ratio of the total organic C : total N and soil CO_2 effluxes, with the largest effluxes associated with the smallest ratios of organic C : total N (B. Elberling and others, unpublished results).

SOURCES OF RESOURCES IN DRY VALLEY SOILS

For most terrestrial habitats, there is little debate over the source of organic resources for soil micro-organisms. In the dry valleys, however, the sparse and discontinuous presence of large photoautotrophs means that alternative sources of organic resources need to be considered. The possible sources of fixed carbon are: (i) modern autotrophic activity *in situ*, which would include the cryptoendolithic communities, mosses, cyanobacteria and heterotrophic algae in the soils, and autotrophic bacteria such as nitrifiers; (ii) ancient *in situ*, or 'legacy', organic deposits from a time when the climate was warmer and conditions were wetter, and organic lake sediments accumulated on the surfaces that developed into the modern dry valleys; (iii) spatial subsidies from the coastal regions, where there are abundant marine and ornithogenic deposits carried into the dry valleys by aeolian dispersal. This would represent relatively long-range transport of up to about 50 km inland; and (iv) spatial subsidies from the margins of modern lakes, where microbial mats accumulate under favourable conditions. This

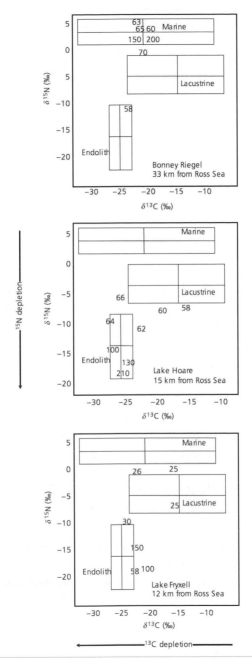

Fig. 4. Isotopic signals for soils from different locations at three sites in the Taylor Valley, shown with boxes covering the range of isotopic signals for materials from the marine and lacustrine environments and for cryptoendolithic materials. The sites are each shown by a number indicating the altitude in m above sea level. Data extracted and redrawn from Burkins *et al.* (2000) with permission from The Ecological Society of America.

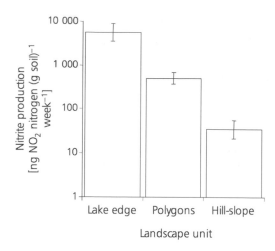

Fig. 5. Potential nitrification activities for soils from different landscape units in the Garwood Valley. The soils were incubated in a shaking incubator at 5 °C in a suspension containing 0·25 mM $(NH_4)_2SO_4$ as described by Hopkins et al. (1988).

material may subsequently be dispersed onto the surrounding soils, representing relatively short-range aeolian dispersal.

These sources are not necessarily mutually exclusive and, depending on particular site details, they may all contribute to the overall carbon cycling, albeit in different proportions. There have been few detailed investigations of the provenance of organic matter in the dry valley soils (Burkins et al., 2000; B. Elberling and others, unpublished results) and it is not therefore possible to partition soil biological activity between different resource drivers. Nevertheless, from the evidence available, it is possible to evaluate the likelihood of some of the different sources of organic resources being important contributors to carbon cycling.

Burkins et al. (2000) used the carbon and nitrogen isotopic signatures of soils and organic materials from within and around the dry valleys to investigate the provenance of organic matter in the Taylor Valley (Fig. 4). These studies did not provide evidence of substantial ornithogenic sources of soil organic matter. They did show that, depending on altitude, organic matter with cryptoendolithic and lacustrine signals was present, with the cryptoendolithic material occurring at greater altitude, i.e. at drier and colder sites further from the valley floor and lakes. However, the conclusions were not clear-cut because, counter-intuitively, the strongest evidence for marine-derived organic matter occurred at the most inland site (33 km from the coast) examined, rather than at sites closer to the coast.

The data shown in Fig. 5 provide evidence for a chemoautotrophic contribution to organic matter input that is supported indirectly by the soil NO_3^- concentrations (Table 2). However, given the low productivity (mol NH_4^+ oxidized by autotrophic nitrifying bacteria)$^{-1}$ (Wood, 1986) and the fact that the measurements in Fig. 5 were obtained under laboratory conditions where NH_4^+ nitrogen was not limiting, it is not likely that nitrification is a substantial source of soil organic carbon.

The 'legacy' model for the Taylor Valley proposes that the origin of much of the soil organic carbon is material laid down in ancient lake beds. It is supported indirectly by geomorphological data on landscape origin (e.g. Burkins *et al.*, 2000; Higgins *et al.*, 2000), in particular the presence of the palaeolake Washburn in the Taylor Valley between 22 800 and 8500 years ago, and by some isotopic signatures in soils (Burkins *et al.*, 2000). However, it is difficult to reconcile the relatively rapid estimated rates of carbon turnover with the persistence of an ancient but active reserve of carbon in the soil, and Barrett *et al.* (2005) now discuss multiple organic matter sources.

The 'legacy' model may be most relevant in large expanses of dry ground remote from lakes, in valleys with relatively small areas covered by lakes and where the geology is not conducive to endoliths. However, in smaller valleys with a larger proportion covered by lakes, at lake and stream margins and in ephemeral stream beds in both small and large valleys, which are recognized as hot-spots of biological activity (Greenfield, 1998; McKnight *et al.*, 1999; Moorhead *et al.*, 2003), modern aquatic-derived detritus may assume greater importance (Moorhead *et al.*, 2003; B. Elberling and others, unpublished results). The lakes and ponds in the dry valleys are the focus of productivity by carbon- and nitrogen-fixing cyanobacteria and eukaryotic algae (Olson *et al.*, 1998; Hawes & Schwarz, 1999), representing draw-down and concentration of resources from the atmosphere. Periodicity in lake and stream level may then cause exposure of resource-rich cyanobacterial and algal detritus and wash-up of nitrogen-rich foams originating from decomposition of algal biomass within the lake. This conceptual model for nutrient cycling in such dry valleys is summarized in Fig. 6. Wilson (1965), Parker *et al.* (1982) and Greenfield (1998) suggested a linkage between aquatic productivity and nitrogen fixation by cyanobacteria, with terrestrial nutrient cycling consistent with the spatial-subsidy model (Fig. 6). Whilst this model remains largely unparameterized, such redistribution has been noted in the field (Wilson, 1965; Parker *et al.*, 1982; Nienow & Friedmann, 1993; Moorhead *et al.*, 2003) and is likely to produce a gradient in soil carbon stocks, declining with distance from lakes, and vertical concentration gradients in soil profiles (B. Elberling and others, unpublished results).

The redistribution of lacustrine microbial mat from lake shores to soils may, as suggested above, also lead to redistribution of viable organisms to the soils, although

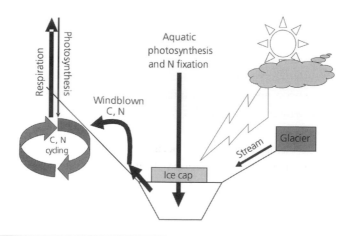

Fig. 6. Conceptual model for resource transfer in one of the small Antarctic dry valleys.

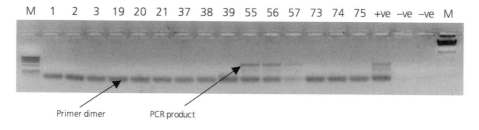

Fig. 7. Gel showing amplification of methanogenesis-specific methyl co-enzyme reductase (MCR) in soils from the Garwood Valley by using the ME1/ME2 primer set (Hales *et al.*, 1996). Lanes 37–39 and 55–57 correspond to the hill-slope and polygon soils in Table 2, respectively.

few studies provide direct evidence for this. However, we have preliminary indirect evidence from the incidence of methanogenesis and methanogens in the Garwood Valley. We have detected substantial *in situ* methanogenesis as CH_4 efflux at the surface in soils within 1 m of lake margins in the range of $1\cdot3$–69 mg m^{-2} day^{-1} during the 2002–2003 austral summer (E. G. Gregorich, D. W. Hopkins, B. Elberling, A. D. Sparrow, P. Novis & L. G. Greenfield, unpublished results). These rates are, incidentally, comparable to CH_4 emissions from temperate and sub-Arctic wetlands (Moore & Knowles, 1989, 1990). By using PCR, we have detected the gene for the methanogenesis-specific methyl co-enzyme reductase (MCR) using ME1/ME2 methanogenic primers (Hales *et al.*, 1996) in soils between 20 and 50 m from lake margin (Fig. 7). DNA amplified by using this primer set suggests the presence of methanogenic bacteria in the dry valley soils that may have arisen from the wet, organic-rich lake margins (Table 2). These data provide circumstantial evidence for transfer of methanogens from the lake margin to the surrounding dry soils, where they were most unlikely to be active under the dry and well-aerated surface conditions.

CONCLUDING REMARKS

Understanding the provenance of resources for soil micro-organisms in nutrient- and energy-poor ecosystems, such as polar deserts, requires consideration of a wider range of possible sources than in most other ecosystems. When the indigenous stocks of soil organic carbon and nitrogen are as low as occurs in the dry valleys, even very modest inputs from other sources may represent a significant subsidy. Whilst there is a parallel in temperate ecosystems at the early stages of primary successions (Hodkinson *et al.*, 2002), for most ecosystems, *in situ* primary production is usually, and correctly, assumed to be the energy source for the below-ground microbial community (Hopkins & Gregorich, 2005). The Antarctic dry valleys may represent ecosystems dependent to a greater extent on spatial and/or temporal subsidies than other terrestrial ecosystems. Clearly, to better understand the relative contributions of resources from the different sources, parameterization of the models is necessary. Our priority is the spatial-subsidy model (Fig. 6) and this will require measurements of the dispersal of lake-derived organic matter and organisms and investigation of the periodicity of production and dispersal in relation to lake level changes.

The dry valleys may also provide an opportunity to examine whether the biological diversity of the soil community is causally related to function. This question has emerged recently as a highly topical ecological issue. However, for most soils, the biological diversity is so large, particularly amongst the heterotrophs, that it is difficult to quantify and relate to the key carbon-cycling processes. In resource-poor systems, where the diversity is likely to be lower, there may be a realistic opportunity to examine soil biodiversity and function.

ACKNOWLEDGEMENTS

We are grateful to Antarctica New Zealand, the UK Natural Environment Research Council, the Carnegie Trust for the Universities of Scotland, the Royal Society of London, the Transantarctic Association, the Danish Natural Science Research Council, Agriculture and Agri-Food Canada and the University of Canterbury (Christchurch, NZ) for support for different parts of the research outlined here. In addition, we express our gratitude to David Wardle (Lincoln, NZ) and collaborators in the US Antarctic Program supported by the US National Science Foundation, in particular Diana Wall, Ross Virginia, Jeb Barrett and Byron Adams, for access to some of their data and stimulating discussions. We are grateful to Lorna English, Patrick St Georges and M. R. Nielsen for technical assistance, and the numerous staff of Antarctica New Zealand for logistic support. Finally, D. W. H. wishes to acknowledge the influential role of the late David Wynn-Williams for an introduction to biological research in Antarctica.

REFERENCES

Bargagli, R., Smith, R. I. L., Martella, L., Monaci, F., Sanchez-Hernandez, J. C. & Ugolini, F. C. (1999). Solution geochemistry and behaviour of major and trace elements during summer in a moss community at Edmonson Point, Victoria Land, Antarctica. *Antarct Sci* **11**, 3–12.

Barrett, J. E., Virginia, R. A. & Wall, D. H. (2002). Trends in resin and KCl-extractable soil nitrogen across landscape gradients in Taylor Valley, Antarctica. *Ecosystems* **5**, 289–299.

Barrett, J. E., Virginia, R. A., Parsons, A. N. & Wall, D. H. (2005). Potential soil organic matter turnover in Taylor Valley, Antarctica. *Arct Antarct Alp Res* **37**, 108–117.

Beyer, L., Bockheim, J. G., Campbell, I. B. & Claridge, G. G. C. (1999). Genesis, properties and sensitivity of Antarctic Gelisols. *Antarct Sci* **11**, 387–398.

Burkins, M. B., Virginia, R. A., Chamberlain, C. P. & Wall, D. H. (2000). Origin and distribution of soil organic matter in Taylor Valley, Antarctica. *Ecology* **81**, 2377–2391.

Burkins, M. B., Virginia, R. A. & Wall, D. H. (2002). Organic carbon cycling in Taylor Valley, Antarctica: quantifying soil reservoirs and soil respiration. *Global Change Biol* **7**, 113–125.

Campbell, I. B. & Claridge, G. G. C. (2000). Soil temperature, moisture and salinity patterns in Transantarctic Mountain cold desert ecosystems. In *Antarctic Ecosystems: Models for Wider Ecological Understanding*, pp. 233–240. Edited by W. Davidson, C. Howard-Williams & P. Broady. Christchurch, New Zealand: Caxton Press.

Courtright, E. M., Wall, D. H. & Virginia, D. A. (2001). Determining habitat suitability for soil invertebrates in an extreme environment: the McMurdo Dry Valleys, Antarctica. *Antarct Sci* **13**, 9–17.

Cowan, D. A. & Tow, L. A. (2004). Endangered Antarctic environments. *Annu Rev Microbiol* **58**, 649–690.

Doran, P. T., Priscu, J. C., Lyons, W. B. & 10 other authors (2002). Antarctic climate cooling and terrestrial ecosystem response. *Nature* **415**, 517–520.

Friedmann, E. I. (1982). Endolithic microorganisms in the Antarctic cold desert. *Science* **215**, 1045–1053.

Friedmann, E. I., Kappen, L., Meyer, M. A. & Nienow, J. A. (1993). Long-term productivity in the cryptoendolithic microbial community of the Ross Desert, Antarctica. *Microb Ecol* **25**, 51–69.

Greenfield, L. G. (1998). Nitrogen in soil *Nostoc* mats: foams, release and implications for nutrient cycling in Antarctica. *N Z J Nat Sci* **23**, 101–107.

Greenland, D. & Kittel, T. G. F. (2002). Temporal variability of climate at the US Long-Term Ecological Research (LTER) sites. *Clim Res* **19**, 213–231.

Hales, B. A., Edwards, C., Ritchie, D. A., Hall, G., Pickup, R. W. & Saunders, J. R. (1996). Isolation and identification of methanogen-specific DNA from blanket bog peat by PCR amplification and sequence analysis. *Appl Environ Microbiol* **66**, 668–675.

Hawes, I. & Schwarz, A. M. (1999). Photosynthesis in an extreme shade environment: benthic microbial mats from Lake Hoare, a permanently ice-covered Antarctic lake. *J Phycol* **35**, 448–459.

Higgins, S. M., Denton, G. H. & Hendy, C. H. (2000). Glacial geomorphology of Bonney Drift, Taylor Valley, Antarctica. *Geogr Ann Ser A Phys Geogr* **82**, 365–389.

Hobbie, J. E. (2003). Scientific accomplishments of the long term ecological research program: an introduction. *Bioscience* **53**, 17–20.

Hodkinson, I. D., Webb, N. R. & Coulson, S. J. (2002). Primary community assembly on land – missing stages: why are the heterotrophic organisms always there first? *J Ecol* **90**, 569–577.

Hopkins, D. W. & Gregorich, E. G. (2005). Carbon as a substrate for soil organisms. In *Biodiversity and Function in Soils*. Edited by R. D. Bardgett, M. B. Usher & D. W. Hopkins. Cambridge: Cambridge University Press (in press).

Hopkins, D. W., O'Donnell, A. G. & Shiel, R. S. (1988). The effect of fertilisation on soil nitrifier activity in experimental grassland plots. *Biol Fertil Soils* **5**, 344–349.

Mahaney, W. C., Dohm, J. M., Baker, V. R., Newsom, H. E., Malloch, D., Hancock, R. G. V., Campbell, I., Sheppard, D. & Milner, M. W. (2001). Morphogenesis of Antarctic paleosols: Martian analogue. *Icarus* **154**, 113–130.

McKnight, D. M., Niyogi, D. K., Alger, A. S., Bomblies, A., Conovitz, P. A. & Tate, C. M. (1999). Dry valley streams in Antarctica: ecosystems waiting for water. *Bioscience* **49**, 985–995.

Moore, T. R. & Knowles, R. (1989). The influence of water table levels on methane and carbon dioxide emissions from peatland soils. *Can J Soil Sci* **69**, 33–38.

Moore, T. R. & Knowles, R. (1990). Methane emissions from fen, bog and swamp peatlands in Quebec. *Biogeochemistry* **11**, 45–61.

Moorhead, D. L., Barrett, J. E., Virginia, R. A., Wall, D. H. & Porazinska, D. (2003). Organic matter and soil biota of upland wetlands in Taylor Valley, Antarctica. *Polar Biol* **26**, 567–576.

Nienow, J. A. & Friedmann, E. I. (1993). Terrestrial lithophytic (rock) communities. In *Antarctic Microbiology*, pp. 343–412. Edited by E. I. Friedmann. New York: Wiley.

Olson, J. B., Steppe, T. F., Litaker, R. W. & Paerl, H. W. (1998). N_2-fixing microbial consortia associated with the ice cover of Lake Bonney, Antarctica. *Microb Ecol* **36**, 231–238.

Onofri, S., Selbmann, L., Zucconi, L. & Pagano, S. (2004). Antarctic microfungi as models for exobiology. *Planet Space Sci* **52**, 229–237.

Parker, B. C., Simmons, G. M., Jr, Wharton, R. A., Jr, Seaburg, K. G. & Love, F. G. (1982). Removal of organic and inorganic matter from Antarctic lakes by aerial escape of bluegreen algal mats. *J Phycol* **18**, 72–78.

Parsons, A. N., Barrett, J. E., Wall, D. H. & Virginia, R. A. (2004). Soil carbon dioxide flux in Antarctic dry valley ecosystems. *Ecosystems* **7**, 286–295.

Stevens, M. I. & Hogg, I. D. (2002). Expanded distributional records of Collembola and Acari in southern Victoria Land, Antarctica. *Pedobiologia* **46**, 485–495.

Stevens, M. I. & Hogg, I. D. (2003). Long-term isolation and recent range expansion from glacial refugia revealed for the endemic springtail *Gomphiocephalus hodgsoni* from Victoria Land, Antarctica. *Mol Ecol* **12**, 2357–2369.

Treonis, A. M., Wall, D. H. & Virginia, R. A. (1999). Invertebrate biodiversity in Antarctic dry valley soils and sediments. *Ecosystems* **2**, 482–492.

Treonis, A. M., Wall, D. H. & Virginia, R. A. (2002). Field and microcosm studies of decomposition and soil biota in a cold desert soil. *Ecosystems* **5**, 159–170.

Turner, M. G., Collins, S. L., Lugo, A. L., Magnuson, J. J., Rupp, T. S. & Swanson, F. J. (2003). Disturbance dynamics and ecological response: the contribution of long-term ecological research. *Bioscience* **53**, 46–56.

Virginia, R. A. & Wall, D. H. (1999). How soils structure communities in the Antarctic dry valleys. *Bioscience* **49**, 973–983.

Wall, D. H. & Virginia, R. A. (1999). Controls on soil biodiversity: insights from extreme environments. *Appl Soil Ecol* **13**, 137–150.

Wilson, A. T. (1965). Escape of algae from frozen lakes and ponds. *Ecology* **46**, 376.

Wilson, E. O. (2002). *The Future of Life.* New York: Alfred A. Knopf.

Wood, P. M. (1986). Nitrification as a bacterial energy source. In *Nitrification* (Special Publication of the Society for General Microbiology no. 10), pp. 39–62. Edited by J. I. Prosser. Oxford: IRL Press.

Wynn-Williams, D. D. (1996). Antarctic microbial biodiversity: the basis of polar ecosystem processes. *Biodivers Conserv* **5**, 1271–1293.

New insights into bacterial cell-wall structure and physico-chemistry: implications for interactions with metal ions and minerals

V. R. Phoenix, A. A. Korenevsky, V. R. F. Matias and
T. J. Beveridge

Department of Molecular and Cellular Biology, College of Biological Science,
University of Guelph, Guelph, Ontario, Canada N1G 2W1

INTRODUCTION

Prokaryotes are the Earth's smallest life form and, yet, have the largest surface area : volume ratio of all cells (Beveridge, 1988, 1989a). They are also the most ancient form of life and have persisted on Earth for at least 3.6×10^9 years, even in some of the most extreme environments imaginable, such as the deep subsurface. Most of these early primitive (and today's modern) natural environments possess reasonably high amounts of metal ions that are capable of precipitation under suitable pH or redox conditions. Deep-seated in such geochemical situations is the likelihood of suitable interfaces that lower the local free energy, so that interfacial metal precipitation is promoted. Bacteria, being minute and having highly reactive surfaces (interfaces), are exquisitely efficient environmental particles for metal-ion adsorption and mineral nucleation. Metal ions interact with available reactive groups (or ligands) on the bacterial surface and precipitates grow as environmental counter-ions interact with more and more metal at the site (Beveridge & Murray, 1976, 1980; Beveridge et al., 1982; Ferris & Beveridge, 1986; Fortin et al., 1998). Once formed, these precipitates are under the influence of natural geochemical and additional microbially mediated conditions (Lee & Beveridge, 2001) that instigate the development of fine-grain minerals, usually via dehydration, so that crystalline phases are eventually developed (Beveridge et al., 1983). These minerals commence as so-called 'nano-mineral phases' and grow with time to become larger and larger. This bacterially induced mineral-ization is probably the natural phenomenon that so encases some cells in fine-grain minerals that they die and become bona fide 'microfossils' (Ferris et al., 1988). In ancient times, these mineral-encased prokaryotes, enduring low-temperature

SGM symposium 65: Micro-organisms and Earth systems – advances in geomicrobiology.
Editors G. M. Gadd, K. T. Semple & H. M. Lappin-Scott. Cambridge University Press. ISBN 0 521 86222 1 ©SGM 2005

metamorphic geological conditions, survived as microfossils that are still existent in such very old Precambrian formations as the $\sim 2 \cdot 0 \times 10^9$ years Gun Flint Chert, north of Lake Superior in Canada.

It is certain that bacterial surfaces interact with environmental metal ions and can provide nucleation sites for mineral precipitation, but it has been extremely difficult to study such systems with high precision on a cell-to-cell basis, even though a wide base of techniques exists (Beveridge *et al.*, 1997). This is because the cells are extremely small and the interactive structures even smaller, and because the reactive sites responsible for adsorbing metals retain their reactivity only over a certain range of pH and E_h, which is difficult to monitor over microscale distances. This chapter will outline our new advances in elucidating the structure of bacterial surfaces, as well as the determination of reactive sites via potentiometric examination. Additional techniques, such as zeta potentials and hydrophobic/hydrophilic determinations, will also be given, so as to provide a more general picture of how surface structure and surface reactivity affect the natural physico-chemical traits of bacteria.

NEW OBSERVATIONS OF BACTERIAL SURFACES BY USING CRYO-TRANSMISSION ELECTRON MICROSCOPY (cryoTEM)

By using conventional fixation, embedding and thin-section techniques, the TEM observation of bacteria has been a most powerful method for elucidating the internal cytoplasmic organization and the juxtaposition of encompassing envelope layers of cells (Beveridge, 1989b; Koval & Beveridge, 1999). Yet, there are many drawbacks to such conventional techniques, as the cells are fixed chemically by using harsh fixatives (such as glutaraldehyde and osmium tetroxide) and dehydrated before embedding in a plastic resin for thin-sectioning (Beveridge *et al.*, 2005). Essential lipids are extracted, proteins are denatured and nucleic acids are condensed artificially ... the cells are a spectre of their former selves. Clearly, the images of such cells have been beneficial to our initial perception of the structural organization of prokaryotic cells (Beveridge, 1989b), but hydration (which these embedded cells no longer have) is a necessary prerequisite for the maintenance of native structure. With proper expertise, care and equipment, it is now possible through the use of cryoTEM to obtain a better and more natural view of bacteria (Dubochet *et al.*, 1983; Umeda *et al.*, 1987; Beveridge & Graham, 1991; Paul *et al.*, 1993).

Freeze-substitution

One cryoTEM technique that became popular during the 1980s and 1990s was freeze-substitution (Hobot *et al.*, 1984; Umeda *et al.*, 1987; Graham & Beveridge, 1990, 1994). Here, cells are frozen rapidly at approximately $-196\,^{\circ}C$ so as to vitrify them in amorphous ice, which is not crystalline and is a kind of glass (Koval & Beveridge, 1999;

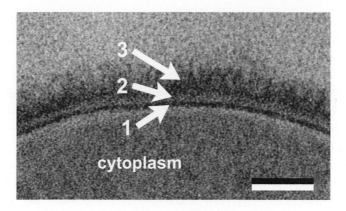

Fig. 1. Thin-section image of a freeze-substituted wall of a Gram-positive *B. subtilis* cell, showing three distinct regions in the cell wall: the inner (1), middle (2) and outer (3) regions that correspond to cell-wall turnover and to the available reactive groups within the cell-wall network. Bar, 50 nm.

Beveridge *et al.*, 2005). Hence, the cells are physically 'fixed', as there is no time during freezing for structure to degrade; in fact, freezing occurs within milli- to microseconds and all molecular motion is stopped. If the bacteria are thawed, they come back to the living state and continue to grow and divide. This is a clear-cut measure of how well the cells are preserved! Once vitrification is accomplished and cellular structure is preserved, the temperature is raised from −196 to −80 °C and the cells are put into a freeze-substitution mixture. This consists of a cryogenic fluid (such as acetone) containing a chemical fixative (osmium tetroxide), a heavy-metal stain (uranyl acetate) and a molecular sieve (for trapping water) (Graham & Beveridge, 1990, 1994). At this low temperature, the cells do not melt; instead they remain vitrified and structure is preserved. The cellular and surrounding ice (water) sublimes and is trapped in the molecular sieve. Eventually, the specimen is dehydrated thoroughly and fixed in this freeze-substitution mixture, with the structure maintaining many of its native features. Now, it can be embedded in plastic resin, cured and thin-sectioned for viewing by TEM.

Freeze-substitution and Gram-positive (*Bacillus subtilis*) cell walls

The results of freeze-substitution are breathtaking, especially when examining the surface structures (Fig. 1). The fabric of Gram-positive cell walls is no longer featureless (as seen in conventional embeddings; Beveridge, 2000), but is a tripartite structure (Fig. 1). The region associated with the plasma membrane (immediately above the bilayer) is highly contrasted, due to the acquisition of large quantities of heavy-metal stain. The middle region is lightly contrasted, as the wall polymers (mainly peptidoglycan) are stretched almost to breaking point, so that the mass : volume ratio is much reduced compared with other wall regions. The outermost region consists of thin fibres that extend into the external milieu. This tripartite cell-wall structure is compatible with current models of cell-wall turnover (Graham & Beveridge, 1994; Beveridge, 2000).

More important for this chapter is the fact that the polymeric structure of the wall has been preserved by freeze-substitution and the available reactive sites within the wall have been 'decorated' with the heavy-metal stain, so that we obtain a clear picture of where the reactive sites reside. Many sites are in the region immediately apposed to the membrane as, here, new wall polymers are being extruded and compacted via penicillin-binding proteins in the membrane. As an area of de novo wall assembly, where new polymers are being cross-linked into the pre-existing wall fabric, many reactive groups are available for decoration in this region, as the mass is great and the availability of reactive groups is high. However, the middle region is different. It is the area in the cell wall that resists turgor pressure and maintains the integrity of the cell. It has to be highly cross-linked to hold the cell together under such a pressure load and is considerably stretched. This region, then, has few reactive groups left available for interaction with the heavy-metal stain, as most have been used to cross-link the network together for strength. The outermost region is an area where the cell wall is being broken down by the wall's constituent autolysins. Covalent bonds are being broken and new reactive groups are being made. This region probably has little mass (as it is being shed during cell-wall turnover), but an excess of reactive sites that are decorated readily by the stain.

Frozen hydrated thin sections

A more difficult and therefore less used cryoTEM technique is the use of frozen hydrated sections (Dubochet *et al.*, 1983, 1988). This technique requires skill and perseverance. As in freeze-substitution, cells are vitrified, but now, instead of processing the cells so that conventional, plastic thin sections are obtained, this frozen material is put immediately into a 'cryo'-ultramicrotome and thin-sectioned. The cells are sectioned whilst vitrified and the frozen sections are mounted immediately into a 'cryo'-specimen holder and inserted into the 'cryo'-chamber of a 'cryo'TEM. We emphasize 'cryo' because the temperature must be maintained at between −196 and −140 °C during all manipulations, otherwise the amorphous ice embedding the cells will become crystalline and ruin the native structure to be observed. No chemical fixatives or heavy-metal stains are used during the entire process and, as the sections remain vitrified, all cellular macromolecules and polymers remain in a hydrous state.

Frozen hydrated Gram-positive (*B. subtilis*) cell walls

One of the advantages of the cryo-sectioning technique is that no artificial chemical fixatives or heavy-metal stains need to be used. However, implicit in the use of TEM is that the specimen must possess enough density to efficiently scatter high-voltage electrons from the electron gun of the microscope (i.e. the electron potential is typically approx. 100 000–200 000 eV). Biomaterials, once thin-sectioned, rarely have enough density to effectively scatter such high-powered electrons, as the thin sections are only ~50 nm thick and the biomatter possesses only low-atomic-number elements (such as H, C, O, N etc.). This is the primary reason that conventional and freeze-substitution thin sections use stained material; the heavy-metal stains increase the density (or the mass : volume ratio) of the specimen so that the contrast becomes great enough for the cells to be visualized easily (Beveridge *et al.*, 2005). Frozen hydrated thin sections of unstained bacteria do not have this luxury, as they cannot be stained once sectioned; the staining fluids would immediately freeze over the sample and obliterate the structure of the cells.

Clearly, then, these frozen hydrated sections of bacteria are difficult to see, as their contrast is close to that of the surrounding vitrified ice. For this reason, we rely on the inherent phase function of the lenses of the cryoTEM and use phase contrast to help imaging by underfocusing to see the bacteria. Certain microscopes (e.g. those with energy filters) can also derive more additional contrast for the specimen. However, even then, there are additional problems in visualizing frozen sections. The energy of the electron beam is often high enough to locally increase the specimen's temperature, resulting in the formation of crystalline ice (from amorphous ice) and, eventually, in ice sublimation. As bacteria are excellent nucleation particles (remember how efficient they are at forming fine-grain minerals), the amorphous-to-crystalline phase transition of ice frequently occurs on the bacteria and their structure is obscured. Furthermore, as the specimen is kept so cold, the frozen sections act as 'cold traps' for the condensation of extraneous molecules within the high vacuum of the microscope column, thereby often contaminating the structure of the specimen. With all these associated problems, it is a wonder that frozen hydrated sections of bacteria can be imaged, but they can and they are extraordinary (Fig. 2).

These images differ from what we see in freeze-substitution images. Remember, now we have no heavy-metal stains to assist contrast and must rely on the inherent density imparted on the cell by the constituent atoms within its molecules. Proteins will be discerned more readily from the surrounding ice than, say, carbohydrates, because they are usually larger, contain nitrogen and (sometimes) sulfur and tend to fold more tightly. Most importantly, all cellular constituents remain hydrated and, therefore, not artificially condensed because of a dehydration regimen. The ribosomes are larger and more robust and can be seen because of their high concentration of protein and rRNA (here, the phosphorus adds additional contrast) (Fig. 2). Even the bilayer of the membrane can be seen, because of the inherent contrast of the phosphorus in the

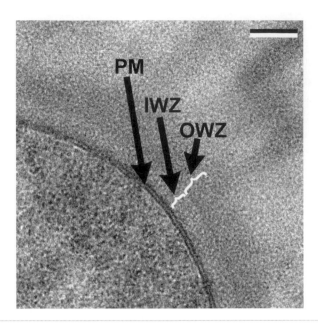

Fig. 2. Frozen hydrated thin section of a *B. subtilis* cell wall, showing the plasma membrane (PM), the inner-wall zone (IWZ) and the outer-wall zone (OWZ). Here, the IWZ corresponds to region 1 and the OWZ to region 2 of the freeze-substituted wall in Fig. 1. Region 3 is not seen. This IWZ and OWZ have different dimensions from the regions in Fig. 1 and are visualized entirely, due to the density of the constituent macromolecules. Bar, 50 nm. Reproduced from Matias & Beveridge (2005) with permission from Blackwell Publishing.

phospholipids. Most important for this chapter, though, is the cell wall. Here, we get a clear idea of the mass distribution within the Gram-positive wall of *B. subtilis* (Fig. 2). Immediately above the bilayered membrane is a low-density space with little contrast. Experiments have shown that this is a periplasmic space (Matias & Beveridge, 2005). Above this is a more densely contrasted region that represents the peptidoglycan–teichoic acid network of the wall. Unlike freeze-substitutions, there is not an outermost fibrous region (cf. Figs 1 and 2).

Correlation of freeze-substitution and frozen hydrated images

How can we reconcile the differences seen in Figs 1 and 2, remembering that the cell in Fig. 1 has been dehydrated and decorated with a heavy-metal stain? As it is dehydrated, we would expect that there would be a certain amount of contraction of the wall regions in freeze-substitution images (Fig. 1), because the structures are no longer hydrated. We would also expect reactive groups to be labelled. On the other hand, frozen hydrated structure would not be condensed and this is why the thickness of each wall region is greater in Fig. 2 than in Fig. 1. This same image shows a periplasmic space, whereas Fig. 1 does not. The periplasm has contracted and condensed in Fig. 1, but it has been preserved in its natural state in Fig. 2. Accordingly, the periplasm in Fig. 1 is more concentrated and (it seems) more reactive, as it stains strongly. In Fig. 2, the periplasm has not condensed and it has low density. The conclusion, then, is that the natural state of the periplasm in these cells is as a relatively low-density matrix of highly reactive biomatter occupying a definite periplasmic space, defined by the membrane and the middle region. Presumably, the periplasm in this space consists of new wall polymers, secreted proteins (and their associated chaperones) and both periplasmic enzymes and oligosaccharides (Matias & Beveridge, 2005).

The region above this periplasmic space is wider in frozen sections than in freeze-substitutions (cf. Figs 1 and 2) and, as this is the hydrated structure, the increased width shows its natural state. Both figures reveal it to be of relatively low contrast; freeze-substitutions suggest that there are few available reactive groups and frozen sections suggest that there is little density (Matias & Beveridge, 2005). Therefore, this region, as the stress-bearing region of the wall, has been stretched taut (thereby reducing its mass) and most reactive groups have been utilized to ensure that the network is cemented firmly together. This forms a strong but elastic fabric of wall polymers, mainly peptidoglycan (Yao *et al.*, 1999; Pink *et al.*, 2000).

The outermost fibrous region, which is highly decorated with stain in Fig. 1, is not seen in Fig. 2. Accordingly, this outermost region has so little mass that it cannot be seen, but is highly reactive. This is in accordance with cell-wall turnover, as this is a region where autolysins are breaking down old peptidoglycan, making it soluble. This would reduce its mass whilst, at the same time, generating many new reactive sites due to hydrolysis (Matias & Beveridge, 2005).

How does this new interpretation of wall structure correlate with metal-ion interaction and mineralization?

It is undeniable that metal ions interact strongly with bacterial surfaces and mineralize them (Beveridge & Murray, 1976; Beveridge & Koval, 1981; Beveridge *et al.*, 1983; Beveridge & Fyfe, 1985; Beveridge, 1989c; Fein *et al.*, 1997; Fortin *et al.*, 1998; Fowle *et al.*, 2000; Daughney *et al.*, 2001; Martinez & Ferris, 2001; Yee & Fein, 2001; Yee *et al.*, 2004a). Cell walls adsorb metal ions and minerals are nucleated in Gram-positive cell walls because of metal-ion interaction with the peptidoglycan and secondary polymers (Fig. 3; Beveridge & Murray, 1980; Beveridge *et al.*, 1982). Our new structural observations on Gram-positive walls now show the quantity of hydrated biomatter that resides in the wall for metal interaction (Fig. 2). They also reveal the positions of the most reactive and likely regions for metal-ion interaction and mineral growth (Fig. 1). The outermost fibrous region would be the most accessible reactive region and, here, there would be little problem of metal-ion access and of mineral-particle growth. As

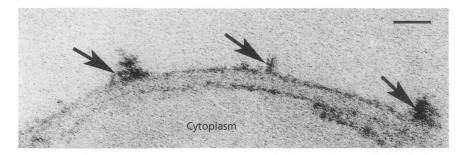

Fig. 3. Conventional thin section of a *B. subtilis* cell that has been subjected to 50 mM FeCl₃ treatment for 15 min at 22 °C. The iron has begun to precipitate from solution onto the cell wall (arrows). Notice that most iron is associated with the wall surface and with the periplasm. No stains other than the iron have been used on this cell. Bar, 50 nm.

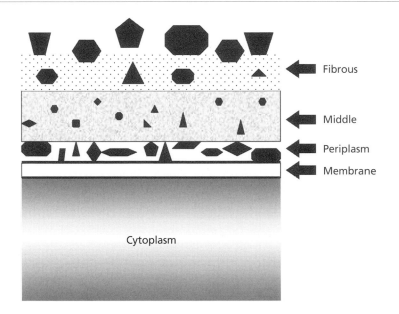

Fig. 4. Diagram to explain the regions of the Gram-positive cell wall where metal ions will interact with reactive sites and develop into fine-grained minerals. The fibrous region refers to region 3, the middle region to region 2 and the periplasm to region 1 of Fig. 1. The membrane is the plasma membrane. The black, crystalline shapes refer to developing minerals and their size represents mineral abundance.

many wall polymers are in the act of being solubilized, these polymers and their precipitates would be sloughed from the cells, but could continue to grow and mature into bona fide mineral phases in the external milieu. The middle region has less reactivity, as it is highly cross-linked. Because turgor pressure stretches this region almost to breaking point, the peptidoglycan would be in a relatively 'open' configuration (Pink *et al.*, 2000) so that most metal ions could penetrate through. The inner region (or periplasmic space) is a highly reactive, loose gel of polymers and proteins that would both interact with and precipitate metal ions. Here, though, as the periplasm resides within a confined space between the plasma membrane and the middle wall region, mineral growth would be restricted to the accessible space.

These correlations and interpretations allow certain predictions to be made as to where environmental metal ions should interact with Gram-positive bacterial surfaces. Fig. 4 shows a model of these predictions, with most metal minerals associated with the outermost region and the periplasmic space. Gram-negative surfaces are structurally more complicated (Beveridge, 1999), but they have also been imaged via freeze-substitution and frozen hydrated sections (Matias *et al.*, 2003) and are, therefore, also able to be correlated. Limited space for this chapter prohibits us from doing so.

POTENTIOMETRIC PROPERTIES OF CELL SURFACES

The surface charge of any microbial cell is controlled by the density, distribution and protonation state of ionizable sites in the cell wall. These chemical sites are predominantly carboxyl, phosphoryl, amino and hydroxyl groups and they deprotonate with increasing pH, imparting a net negative charge on the surface. The deprotonation of, say, a carboxylate can be described by the following equation:

$$B\text{-}COOH \leftrightarrow B\text{-}COO^- + H^+ \qquad \text{(equation 1)}$$

where B-COOH is a protonated carboxyl group on the bacterial surface and B-COO$^-$ is its deprotonated form. The log equilibrium constant (pK_a) for the reaction in equation 1 is then defined as:

$$K_a = [H^+] \times [B\text{-}COO^-] / [B\text{-}COOH] \qquad \text{(equation 2)}$$
$$pK_a = \ log K_a \qquad \text{(equation 3)}$$

In simple terms, the pK_a can be considered as the pH at which a significant number of those groups or ligands become negatively charged. High-resolution acid–base titrations (HRABT), combined with a suitable modelling approach, can be used to determine ligand concentrations and their corresponding pK_a values (e.g. Fein *et al.*, 1997; Cox *et al.*, 1999). This approach works because the adsorption and release of H$^+$ from such ligands causes a shift in the pH away from the expected norm during titration (Fig. 5). This difference, known as the charge excess, can be used to calculate ligand concentrations and their pK_a values. These calculations can be performed by using a number of approaches, such as the surface complexation method with FITEQL (Fein *et al.*, 1997; Borrok *et al.*, 2004a), the linear programming method (LPM; Cox *et al.*, 1999), the fully optimized continuous (FOCUS; Smith & Ferris, 2001; Martinez *et al.*, 2002) method or the Donnan shell model (Plette *et al.*, 1995; Martinez *et al.*, 2002; Yee *et al.*, 2004b).

HRABT analysis of the Gram-negative bacterium *Pseudomonas aeruginosa* PAO1, modelled by using the LPM approach, is shown in Fig. 6. Here, five distinct proton-binding sites (and their corresponding concentrations) are revealed. Based on model compounds, each pK_a can be assigned to a most probable ligand type. Carboxyl groups deprotonate over the lower pH range and exhibit a range of pK_a values that occur predominantly between pK_a 2 and 6 (Perdue, 1985). Similarly, phosphoryl groups exhibit intermediate pK_a values that are commonly between pK_a 5·4 and 8 (Martell *et al.*, 1987). Thiols (e.g. cysteine) exhibit pK_a values around 8, amines generally between 8 and 11 and hydroxyls between 9 and 12 (Perdue, 1985; Martell *et al.*, 1987). Considering this, the pK_a groups in Fig. 6 can be assigned as follows: pK_a 4·7, carboxyl; pK_a 5·9, phosphoryl (or carboxyl); pK_a 6·8, phosphoryl; pK_a 8·1, amino (or thiol); pK_a 9·4, amino/hydroxyl.

The pK_a spectrum of each bacterium is controlled by the chemical composition of the cell wall and, thus, each detected pK_a and its assigned ligand type can be attributed to various components of this structure. Carboxyl groups can be associated with proteins, peptide stems of peptidoglycan and (in Gram-negatives) lipopolysaccharides [LPSs; either because of ketodeoxyoctonate (Kdo) in the oligosaccharide (unless it is acetylated; Vinogradov *et al.*, 2002, 2003a, 2004) or because of O side chains, depending on O serotype]. Phosphoryl groups are associated with teichoic acids in Gram-positive cell walls (Beveridge *et al.*, 1982), whereas, in Gram-negative outer membranes (OMs), they are found in the core oligosaccharide and lipid A fraction of the LPS and in the phospholipids of the membrane's inner face (Beveridge, 1999). Thiol groups, whilst uncommon in OM proteins (OMPs), may be found in periplasmic proteins. Amines are found in abundance within the peptidoglycan (peptide stems), as well as in the proteins of the OM and periplasm.

Importantly, HRABT analysis provides us with both the concentration and pK_a values of these potential metal-binding ligands (Fein *et al.*, 1997). Additionally, some metal-adsorption modelling approaches utilize a pK_a spectrum to underpin their metal-complexation models (e.g. Fein *et al.*, 1997).

PHYSICO-CHEMICAL STUDIES

Cell-surface complexity and its impact on potentiometric properties

The nature of the cell wall is complex and diverse, imparting individual surface characteristics to each bacterial species and strain. Each micro-organism displays a unique arrangement of ligands and thus has unique potentiometric and metal-binding properties (e.g. Beveridge & Fyfe, 1985; Martinez & Ferris, 2001; Borrok *et al.*, 2004a).

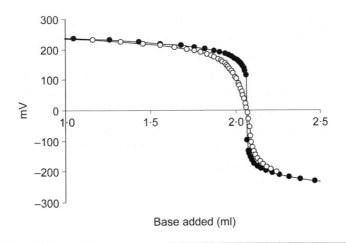

Fig. 5. Raw data from high-resolution acid–base titration (HRABT) analysis. The pH (y axis) is given in mV. ●, Blank titration (no bacteria); ○, titration with bacteria.

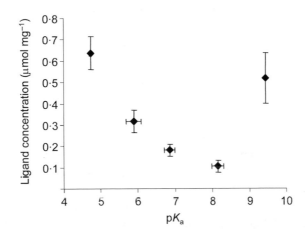

Fig. 6. pK_a spectrum generated from HRABT analysis of *P. aeruginosa* PAO1, modelled by using a linear programming method (LPM). This spectrum was generated from three titrations. Error bars are SD ($\sigma = 1$).

As described earlier, novel cryo-based TEM methods have provided further insights into the distribution of reactive sites within this basic framework. Together with cryoTEM, HRABT adds another surface-analysis tool to our arsenal for deciphering the availability of cell-wall ligands. For example, the cyanobacterium *Calothrix* sp. (strain KC97) surrounds itself in a thick (up to 1 μm) extracellular polysaccharide (PS) sheath, which is above the cell wall. This is ion-permeable and ions are able to interact with functional groups that reside within both the sheath and cell wall. HRABT analysis of intact cells and isolated sheath has shown that only ~15 % of the available ligands are located in the sheath; the remaining 85 % are located in the wall (Phoenix *et al.*, 2002). Thus, the cell-surface reactivity of *Calothrix* sp. can be divided into a dual layer, composed of a highly reactive polar cell wall enclosed within a poorly reactive apolar sheath. The distribution of ionizable groups is likely to be controlled by the organism in a self-serving manner. For example, the motile phase of *Calothrix*, called hormogonia, does not possess a sheath and the electronegative wall is the exposed surface, making the cyanobacterium hydrophilic. This property seems ideally suited to a free-swimming phase, which must be in close contact with its aqueous milieu. However, the benthic, non-motile phase assembles a sheath around cells, making them more hydrophobic and able to stick to inanimate surfaces. The distribution of ionizable groups over the cell surface also has implications for metal complexation. The higher concentration of ionizable groups on the cell wall ensures that this structure remains the main sink for metals (Yee *et al.*, 2004a).

There are several other general types of polymeric structures besides sheaths that reside above cell walls (Whitfield & Valvano, 1993), such as capsules and exopolymeric substances (frequently found in microbial biofilms). All of these structures, including S layers [i.e. paracrystalline arrays of (glyco)protein; Sleytr & Beveridge, 1999] aid in adjusting the availability of surface groups on bacteria, and such polymeric controls on ligand distribution can differ according to bacterial genus, species and (even) strain. HRABT analysis of eight different strains of *Shewanella* sp. demonstrated that those that expressed O side-chain LPS (i.e. smooth strains) commonly exhibited higher total ligand concentrations than those with just core oligosaccharide (i.e. rough strains) (V. R. Phoenix, A. A. Korenevsky, T. J. Beveridge, Y. A. Gorby & F. G. Ferris, unpublished data). Furthermore, comparison of individual pK_a values revealed that *Shewanella* strains possessing O side chains were relatively enriched in ligands with a pK_a of ~5, suggesting that these chains contained available carboxyl groups. Their existence was corroborated by structural nuclear magnetic resonance (NMR) analyses of selected rough and smooth LPS strains (Vinogradov *et al.*, 2002, 2003a, b, 2004). From these data, one may speculate that strains that exhibit O side chains should display higher metal-binding capacity than their rough counterparts.

The complex nature of cell surfaces, however, ensures that we cannot assume that all organisms exhibiting O side chains will contain higher ligand concentrations than rough strains. This may be due to more significant differences in ligand concentrations in other components of the cell wall, such as OMPs, or may simply be due to a low concentration of O side chains or O side chains that are extremely short (and therefore do not have many ligands on them). This is exemplified by comparing two strains of *P. aeruginosa*, one that expresses O side chains (PAO1) and one that does not (rd7513). Importantly, the density of O side chains on PAO1 is typically quite low, with only

~20 % of the LPS containing O side chains (Kropinski *et al.*, 1987). When the total ligand concentrations for each strain were evaluated by using HRABT analysis, both PAO1 and rd7513 revealed ligand concentrations that were not statistically different (V. R. Phoenix & T. J. Beveridge, unpublished data). This reflects the low density of O side chains dispersed over PAO1's surface, which was insufficient to significantly enhance the total ligand concentration.

With this in mind, a combined HRABT plus NMR approach can be further exploited to reveal the density of certain surface polymers. As described above, although some LPS molecules may express O side chains, many will not and it can therefore be difficult to estimate total O side-chain concentration over the cell surface. We have combined NMR and HRABT analyses of isolated LPS to approximate O side-chain coverage on *Shewanella alga* BrY. In this example, NMR analysis of the lipid A and LPS core indicated a 1:5 stoichiometric relationship between carboxyl and phosphoryl groups. Additionally, NMR of the O side chain revealed that this component contained one carboxyl group per repeating structural unit (no other ionizable groups were present). However, HRABT analysis of the same LPS revealed a 1:1 stoichiometry between carboxyl and phosphoryl groups. Thus, considering a 1:5 stoichiometry in the lipid A and core and the single carboxyl per repeat unit in the O side chain (from NMR analysis), each LPS polymer must contain, on average, an O side chain of four repeating structural units (V. R. Phoenix, A. A. Korenevsky, T. J. Beveridge, Y. A. Gorby & F. G. Ferris, unpublished data).

When performed on whole cells, HRABT analyses probe functional groups embedded deep within the cell wall (although exactly how deep is uncertain). This provides additional information about metal-adsorption properties, as metal ions, like protons, will be able to migrate into the wall matrix. However, HRABT analysis may not be suitable for evaluating all cell-interface processes. For example, when considering surface charge to determine electrostatic interactions with mineral surfaces [e.g. during adhesion, an approach that measures the charge properties of the very outermost fractions of the cell surface, such as zeta-potential analysis, may be more suitable (see later section for more details)]. This is because changes in the composition of these outermost cell-wall polymers can have a marked impact on the zeta potential of a bacterium, and yet have a considerably smaller impact on the HRABT-determined properties. This is especially true if the compositions of ionizable groups embedded deeper within the cell wall are similar for the organisms in question. For example, in a recent study of several strains of *Shewanella* sp., the strains displayed a notable diversity in zero point of charge (ZPC) as determined from zeta-potential analysis, whereas the ZPCs from HRABT analysis were similar (V. R. Phoenix, A. A. Korenevsky, T. J. Beveridge, Y. A. Gorby & F. G. Ferris, unpublished data). Thus, as highlighted

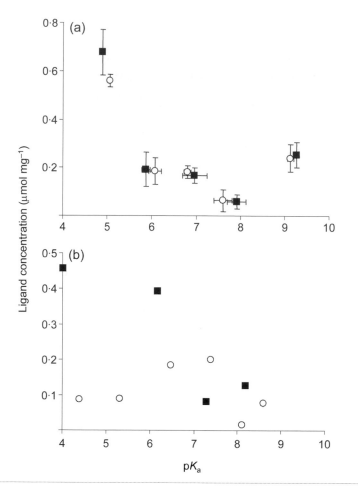

Fig. 7. pK_a spectra from HRABT analysis (modelled by using an LPM approach). (a) pK_a spectra of whole cells of *S. algae* BrY (■) and *S. oneidensis* MR-1 (○) ($n = 3$, $\sigma = 1$). (b) pK_a spectra from LPS isolated from *S. algae* BrY (■) and *S. oneidensis* MR-1 (○).

throughout this chapter, it can be pertinent to understand not only the concentration and types of reactive groups within the cell wall, but also their distribution.

This is further emphasized by comparing HRABT analyses of whole cells with those of polymers isolated from the outer surface of the cell wall. HRABT-determined pK_a spectra for *S. alga* BrY and *Shewanella oneidensis* MR-1 are shown in Fig. 7(a); the two organisms clearly display quite similar pK_a spectra. However, pK_a spectra obtained from LPS isolated from MR-1 and BrY are very different (Fig. 7b). Because LPS is located on the outer face of the OM, it has a significant impact on the microbe's interaction with the surrounding environment. Considering the difference in LPS pK_a spectra, it is unsurprising that these two strains display quite different zeta potentials and mineral-adhesion properties (A. A. Korenevsky & T. J. Beveridge, unpublished data). This would not have been anticipated from the pK_a spectra of whole cells. Note the approximate 1:1 stoichiometry between the carboxyl (pK_a ~4) and phosphoryl (pK_a 6–8) groups in the LPS of BrY used, as described above, to determine the O side-chain density on this bacterium. Furthermore, the dominance of phosphoryl groups (pK_a 6–8) in the LPS of MR-1 (Fig. 7b) has also been corroborated by NMR (Vinogradov *et al.*, 2003a) and further emphasizes the compatibility of these two methods. The overall lower ligand concentrations exhibited by the LPS of MR-1 are also noteworthy. These are probably due to the presence of capsular polysaccharide (CPS) of low ligand concentration associated with the LPS (Korenevsky *et al.*, 2002).

Several studies have noted changes in potentiometric and metal-adsorption properties in Gram-positive and -negative bacteria in response to environmental conditions (Daughney *et al.*, 2001; Borrok *et al.*, 2004a; Haas, 2004). For example, *B. subtilis* walls exchange their major secondary polymer (teichoic acid) for teichuronic acid under phosphate starvation (Beveridge, 1989a, c). However, for the purposes of environmental modelling (to predict the transport and fate of metals in aqueous systems), it is desirable to describe the potentiometric and metal-adsorption properties of a wide range of bacteria by using a few bulk average parameters (Yee & Fein, 2001; Borrok *et al.*, 2004a, b). This is viable, providing the error associated with using universal parameters is smaller than other errors inherent in environmental modelling (e.g. uncertainty in cell-number distribution throughout the system). However, understanding the diversity of potentiometric and metal-adsorption properties displayed by different species is still pertinent because it (i) allows better determination of the average universal parameters and their associated error and (ii) provides high-resolution data for specific environments where more accurate modelling may be required.

Hydrophobicity studies

Many studies using electron microscopy of natural biofilms or planktonic cells from soils, sediments, mine-tailing wastes, hot springs, etc. have revealed that particulate mineral deposits are commonly associated with microbial cells (Fig. 8; Xue *et al.*, 1988; Konhauser *et al.*, 1993, 1994, 2004; Konhauser & Ferris, 1996; Small *et al.*, 1999; Martinez *et al.*, 2003). These minerals may arise not only because of the interaction of metal ions with ionized functional groups on the cell surface, but also as a result of the binding of pre-formed, finely dispersed minerals, such as colloidal silica, metal oxides and clays. Laboratory experiments on the interaction of dissimilatory metal-reducing bacteria, such as *Shewanella putrefaciens*, with nano-sized particulate iron oxides showed strong association between bacterial surfaces and mineral particles (Fig. 9), so strong that it resulted in irreversible adsorption (Caccavo *et al.*, 1997; Glasauer *et al.*, 2001).

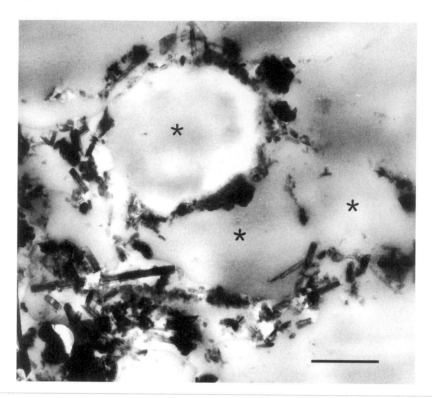

Fig. 8. Thin section of a natural freshwater-stream sediment taken from Table Mountain in Grosse Morne Park, Newfoundland, Canada. No electron-microscopic stains have been used and three cells that are surrounded by minerals have been indicated by asterisks. Bar, 250 nm.

The binding of particulate minerals by cell surfaces is a result of the interplay of electrostatic (or electric double layer), van der Waals and Lewis acid–base interactions. The latter are responsible for such interfacial effects as hydrophobic attraction and hydrophilic repulsion (Van Oss *et al.*, 2001). It has long been recognized that so-called 'hydrophobicity' plays a significant role in the binding of particulate minerals to the bacterial surface, yet it is clearly a most poorly understood aspect of surface physico-chemistry. Hydrophobicity is a relatively non-specific term that has limited utility, as it does not provide any scale of magnitude and is comparative; the term draws no actual line between 'hydrophobic' and 'hydrophilic' surfaces, as one surface is simply more unwettable than the other. The only reliable quantitative criterion for defining the relative terms 'hydrophobicity' and 'hydrophilicity' is the sign and value of free energy of the interfacial interaction between molecules, particles or surfaces immersed in water (Van Oss & Giese, 1995).

Microbial-surface hydrophobicity is determined by the availability and number of apolar functional groups (e.g. $-CH_3$) on cell-wall macromolecules. Although there are

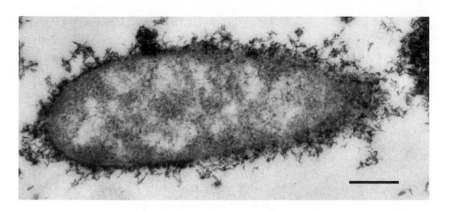

Fig. 9. Conventional thin section of *S. oneidensis* MR-1 that has been reacted with a fine-grained hydrated iron oxide, which has adsorbed to the cell surface. Bar, 200 nm.

no set rules, proteins are considered a major factor for cell-surface hydrophobicity, whilst PSs determine surface hydrophilicity. The influence of these two major components of cell walls on surface properties can be illustrated in the Gram-negative OM, where phospholipids and LPSs assemble into a membrane bilayer and where, extending from the OM face, PSs strongly affect surface physico-chemistry and adhesiveness. Surface PSs (capsule or O side chains of LPS) extending from the cell surface screen hydrophobic OMPs. Smooth strains (expressing O side chains or possibly CPS) are more hydrophilic than their rough counterparts, which possess only core oligosaccharide on their LPS (Williams *et al.*, 1986; Makin & Beveridge, 1996; Williams & Fletcher, 1996; Flemming *et al.*, 1998; DeFlaun *et al.*, 1999; Faille *et al.*, 2002). However, this is not always true; sugar residues of microbial PS are often substituted with apolar methyl and acyl groups (Whitfield & Valvano, 1993), which can contribute to surface hydrophobicity (Makin & Beveridge, 1996).

We have recently conducted studies on a number of different *Shewanella* strains in which a number of different techniques were used to determine cell-surface hydrophobicity, including contact angle measurement (CAM) and hydrophobic interaction chromatography (HIC) (A. A. Korenevsky & T. J. Beveridge, unpublished data). The first method, CAM, yielded quantitative parameters of cell-surface hydrophobicity and assessed the free surface energy by using the Lifshitz–van der Waals/Lewis acid–base approach. Here, measurements showed the overall character of all *Shewanella* surfaces to be hydrophilic and essentially monopolar, with low electron-acceptor but high electron-donor parameters. The second method, HIC, was chosen as the least perturbing cell-surface protocol, as it measures live bacteria in their natural, fully hydrated state (Pembrey *et al.*, 1999). It is based on the microbial interaction with a hydrophobic substrate (octyl Sepharose) and appeared to be much more sensitive to ultrastructural variations of the bacterial surfaces than CAM. The HIC values of *Shewanella* showed a strong relationship with known LPS/PS compositions (A. A. Korenevsky & T. J. Beveridge, unpublished data) and were in good agreement with previous studies where increased cell-surface hydrophobicity was found in strains expressing short O side chains or rough LPS (Hermansson *et al.*, 1982; Williams *et al.*, 1986; Makin & Beveridge, 1996; Williams & Fletcher, 1996; Flemming *et al.*, 1998; Hanna *et al.*, 2003). The higher hydrophobicity of rough strains is considered to be the consequence of increased exposure of hydrophobic OMPs.

In our studies, electrostatic interaction chromatography (ESIC) and microelectrophoresis (zeta potential) were also used and provided information on cell-surface electronegativity (A. A. Korenevsky & T. J. Beveridge, unpublished data). Both types of measurements showed that the presence of smooth LPS or CPS decreased cell-surface electronegativity. Even though encapsulated bacterial strains are often reported to be more electronegative than rough LPS phenotypes, the opposite was found with our *Shewanella* strains. Both methods showed the cells with PS to be significantly less electronegative. This tendency has been reported previously for a number of Gram-negative bacteria (Hermansson *et al.*, 1982; Makin & Beveridge, 1996; Flemming *et al.*, 1998; Razatos *et al.*, 1998; DeFlaun *et al.*, 1999). Our recent LPS/PS structural analyses help to explain these present electronegativity results (Vinogradov *et al.*, 2002, 2003a, b, 2004). The common repeating motif of the *S. alga* BrY O side chain consists of four monosaccharides and contains 3-hydroxybutyric acid and malic acid. Here, the carboxyl group of malic acid is the only available ionizable group in the structure (Vinogradov *et al.*, 2004). A similar situation was seen in the CPS of *S. oneidensis* MR-4; here, the repeating unit contains only one glucuronic acid as an ionizable residue (E. Vinogradov, A. A. Korenevsky & T. J. Beveridge, unpublished data). In contrast, *Shewanella*'s core oligosaccharides were found to be highly phosphorylated (Vinogradov *et al.*, 2002, 2003, 2004b; Moule *et al.*, 2004) and thus very electronegative at circumneutral pH. Accordingly, *Shewanella*'s O side chains and CPS seem to be only weakly charged and are capable of screening the main surface charge located in the core–lipid A region of LPS. Unlike most other shewanellae, *S. oneidensis* MR-1 possesses a surprisingly low surface charge at circumneutral pH, even though it expresses rough LPS. The core of its LPS contains an unusual component, 8-amino-Kdo (Vinogradov *et al.*, 2003), which may account for the lowered surface charge, as the amino group masks the carboxylate. Furthermore, this strain possesses a microcapsule of 20–30 nm that, like the PSs of MR-4, could possess a low charge density (Korenevsky *et al.*, 2002).

It at first seems to be a contradiction that HIC suggests that the more polar (electronegative) strains are more hydrophobic than the less electronegative strains. The answer to this contradiction could reside in our traditional perception of OM structure and the arrangement of surface macromolecules. We too often consider this membrane to be a homogeneous, static assembly of macromolecules that are dispersed randomly over the bilayer's surface. Instead, all molecules are in high motion (usually over extremely short timescales) and it is probable that, at distinct time intervals, a mosaic of molecular patches of definite polarity and charge exists (Sokolov *et al.*, 2001; Korenevsky *et al.*, 2002; Vadillo-Rodríguez *et al.*, 2003, 2005). Given an inanimate surface of consistent hydrophobicity (such as the Sepharose beads used for HIC), the tendency of the OM surface would be to congregate and align compatible macromolecules towards the inanimate surface. Therefore, although the overall *Shewanella* cell surface was found to be hydrophilic by other analyses (e.g. CAM; A. A. Korenevsky & T. J. Beveridge, unpublished data), it still may be capable of hydrophobic interactions through such apolar congregations. This would lead to adhesion to hydrophobic substrata, such as was seen in our HIC experiments.

It is clear that hydrophilicity and hydrophobicity have a strong bearing on the parameters that affect geomicrobiology. The accessibility of metal ions to reactive sites on bacterial surfaces, biomineral development on such surfaces, adsorption of pre-formed fine-grained minerals and adhesion of bacteria to larger mineral phases all depend on the surfaces' polar and apolar properties. In our *Shewanella* study (A. A. Korenevsky & T. J. Beveridge, unpublished data), a strong correlation between such macroscopic surface parameters as surface negativity, relative hydrophobicity and bacterial adhesion to haematite was observed. Rough strains exhibited higher affinity and maximal sorption capacity (by more than an order of magnitude) to haematite when compared with encapsulated strains. It follows, then, that hydrophobic interactions do not make a significant contribution to *Shewanella*'s adhesion to haematite. This is in accord with acid–base titrations, which demonstrated that hydrous ferric oxide interacts directly with carboxyl sites on the surface of *S. putrefaciens* CN32 (Smith & Ferris, 2003; Martinez *et al.*, 2003).

The assessment of the cell-surface hydrophobicity is a challenging task because microbial walls have high chemical and structural complexity and are degraded readily by native enzymes, such as autolysins (Fig. 1). Yet, a good understanding of the bacterial surface, together with a thorough knowledge of the techniques being used in the determination of surface physico-chemistry, can be illuminating and can provide valuable information about the interactions between microbes and the environment. In this chapter, we have integrated together a variety of techniques (from electron microscopy to HRABT to hydrophobicity/hydrophilicity studies) into an amalgam to help explain metal ion–bacterial surface interactions, fine-grained biomineral development and mineral-sorption/adhesion phenomena. Certainly, there is much more to be deciphered, especially as microbes are dynamic systems that can quickly react and respond to changing environmental systems, such as altered redox conditions, nutrient limitation, pH and temperature. Microbial surfaces alter with these environmental fluxes and therefore can present an entirely different set of interfacial parameters. This dynamic cellular behaviour and the range of interdisciplinary fields that are required to study these animate–inanimate systems make geomicrobiology particularly challenging and exciting.

ACKNOWLEDGEMENTS

The research presented in this chapter was made possible through funding provided by a Canadian National Science and Engineering Research Council (NSERC) Discovery grant and a United States Department of Energy Natural and Accelerated Bioremediation Research Program (US-DOE-NABIR) grant to T. J. B. The electron microscopy was performed in the Guelph Regional Integrated Imaging Facility (GRIIF), which is partially funded by an NSERC Major Facility Access grant to T. J. B.

REFERENCES

Beveridge, T. J. (1988). The bacterial surface: general considerations towards design and function. *Can J Microbiol* **34**, 363–372.

Beveridge, T. J. (1989a). Role of cellular design in bacterial metal accumulation and mineralization. *Annu Rev Microbiol* **43**, 147–171.

Beveridge, T. J. (1989b). The structure of bacteria. In *Bacteria in Nature: a Treatise on the Interaction of Bacteria and their Habitats*, vol. 3, pp. 1–65. Edited by E. R. Leadbetter & J. S. Poindexter. New York: Plenum.

Beveridge, T. J. (1989c). Metal ions and bacteria. In *Metal Ions and Bacteria*, pp. 1–29. Edited by T. J. Beveridge & R. J. Doyle. New York: Wiley.

Beveridge, T. J. (1999). Structures of Gram-negative cell walls and their derived membrane vesicles. *J Bacteriol* **181**, 4725–4733.

Beveridge, T. J. (2000). Ultrastructure of Gram-positive cell walls. In *Gram-Positive Pathogens*, pp. 3–10. Edited by V. A. Fischetti, R. P. Novick, J. J. Ferretti, D. A. Portnoy & J. I. Rood. Washington, DC: American Society for Microbiology.

Beveridge, T. J. & Fyfe, W. S. (1985). Metal fixation by bacterial cell walls. *Can J Earth Sci* **22**, 1893–1898.

Beveridge, T. J. & Graham, L. L. (1991). Surface layers of bacteria. *Microbiol Rev* **55**, 684–705.

Beveridge, T. J. & Koval, S. F. (1981). Binding of metals to cell envelopes of *Escherichia coli* K-12. *Appl Environ Microbiol* **42**, 325–335.

Beveridge, T. J. & Murray, R. G. E. (1976). Uptake and retention of metals by cell walls of *Bacillus subtilis*. *J Bacteriol* **127**, 1502–1518.

Beveridge, T. J. & Murray, R. G. E. (1980). Sites of metal deposition in the cell wall of *Bacillus subtilis*. *J Bacteriol* **141**, 876–887.

Beveridge, T. J., Forsberg, C. W. & Doyle, R. J. (1982). Major sites of metal binding in *Bacillus licheniformis* walls. *J Bacteriol* **150**, 1438–1448.

Beveridge, T. J., Meloche, J. D., Fyfe, W. S. & Murray, R. G. E. (1983). Diagenesis of metals chemically complexed to bacteria: laboratory formation of metal phosphates, sulfides, and organic condensates in artificial sediments. *Appl Environ Microbiol* **45**, 1094–1108.

Beveridge, T. J., Hughes, M. N., Lee, H., Leung, K. T., Poole, R. K., Savvaidis, I., Silver, S. & Trevors, J. T. (1997). Metal–microbe interactions: contemporary approaches. *Adv Microb Physiol* **38**, 177–243.

Beveridge, T. J., Moyles, D. & Harris, B. (2005). Electron microscopy. In *Methods for General and Molecular Microbiology*. Edited by C. A. Reddy, T. J. Beveridge, J. A. Breznak, L. Snyder, T. M. Schmidt & G. A. Marzluf. Washington, DC: American Society for Microbiology (in press).

Borrok, D. M., Fein, J. B. & Kulpa, C. F., Jr (2004a). Cd and proton adsorption onto bacterial consortia grown from industrial wastes and contaminated geologic settings. *Environ Sci Technol* **38**, 5656–5664.

Borrok, D., Fein, J. B. & Kulpa, C. F. (2004b). Proton and Cd adsorption onto natural bacterial consortia: testing universal adsorption behavior. *Geochim Cosmochim Acta* **68**, 3231–3238.

Caccavo, F., Jr, Schamberger, P. C., Keiding, K. & Nielsen, P. H. (1997). Role of hydrophobicity in adhesion of the dissimilatory Fe(III)-reducing bacterium *Shewanella alga* to amorphous Fe(III) oxide. *Appl Environ Microbiol* **63**, 3837–3843.

Cox, J. S., Smith, D. S., Warren, L. A. & Ferris, F. G. (1999). Characterizing heterogeneous bacterial surface functional groups using discrete affinity spectra for proton binding. *Environ Sci Technol* **33**, 4514–4521.

Daughney, C. J., Fowle, D. A. & Fortin, D. (2001). The effect of growth phase on proton and metal adsorption by *Bacillus subtilis*. *Geochim Cosmochim Acta* **65**, 1025–1035.

DeFlaun, M. F., Oppenheimer, S. R., Streger, S., Condee, C. W. & Fletcher, M. (1999). Alterations in adhesion, transport, and membrane characteristics in adhesion-deficient pseudomonad. *Appl Environ Microbiol* **65**, 759–765.

Dubochet, J., McDowall, A. W., Menge, B., Schmid, E. N. & Lickfeld, K. G. (1983). Electron microscopy of frozen-hydrated bacteria. *J Bacteriol* **155**, 381–390.

Dubochet, J., Adrian, M., Chang, J. J., Homo, J. C., Lepault, J., McDowall, A. W. & Schultz, P. (1988). Cryo-electron microscopy of vitrified specimens. *Q Rev Biophys* **21**, 129–228.

Faille, C., Jullien, C., Fontaine, F., Bellon-Fontaine, M.-N., Slomianny, C. & Benezech, T. (2002). Adhesion of *Bacillus* spores and *Escherichia coli* cells to inert surfaces: role of surface hydrophobicity. *Can J Microbiol* **48**, 728–738.

Fein, J. B., Daughney, C. J., Yee, N. & Davis, T. A. (1997). A chemical equilibrium model for metal adsorption onto bacterial surfaces. *Geochim Cosmochim Acta* **61**, 3319–3328.

Ferris, F. G. & Beveridge, T. J. (1986). Site specificity of metallic ion binding in *Escherichia coli* K-12 lipopolysaccharide. *Can J Microbiol* **32**, 52–55.

Ferris, F. G., Fyfe, W. S. & Beveridge, T. J. (1988). Metallic ion binding by *Bacillus subtilis*: implications for the fossilization of microorganisms. *Geology* **16**, 149–152.

Flemming, C. A., Palmer, R. J., Jr, Arrage, A. A., van der Mei, H. C. & White, D. C. (1998). Cell surface physicochemistry alters biofilm development of *Pseudomonas aeruginosa* lipopolysaccharide mutants. *Biofouling* **13**, 213–231.

Fortin, D., Ferris, F. G. & Beveridge, T. J. (1998). Surface-mediated mineral development by bacteria. *Rev Mineral Geochem* **35**, 161–180.

Fowle, D. A., Fein, J. B. & Martin, A. M. (2000). Experimental study of uranyl adsorption onto *Bacillus subtilis*. *Environ Sci Technol* **34**, 3737–3741.

Glasauer, S., Langley, S. & Beveridge, T. J. (2001). Sorption of Fe (hydr)oxides to the surface of *Shewanella putrefaciens*: cell-bound fine-grained minerals are not always formed de novo. *Appl Environ Microbiol* **67**, 5544–5550.

Graham, L. L. & Beveridge, T. J. (1990). Effect of chemical fixatives on accurate preservation of *Escherichia coli* and *Bacillus subtilis* structure in cells prepared by freeze-substitution. *J Bacteriol* **172**, 2150–2159.

Graham, L. L. & Beveridge, T. J. (1994). Structural differentiation of the *Bacillus subtilis* 168 cell wall. *J Bacteriol* **176**, 1413–1421.

Haas, R. J. (2004). Effects of cultivation conditions on acid–base titration properties of *Shewanella putrefaciens*. *Chem Geol* **209**, 67–81.

Hanna, A., Berg, M., Stout, V. & Razatos, A. (2003). Role of capsular colanic acid in adhesion of uropathogenic *Escherichia coli*. *Appl Environ Microbiol* **69**, 4474–4481.

Hermansson, M., Kjelleberg, S., Korhonen, T. K. & Stenström, T.-A. (1982). Hydrophobic and electrostatic characterization of surface structures of bacteria and its relationship to adhesion to an air-water interface. *Arch Microbiol* **131**, 308–312.

Hobot, J. A., Carlemalm, E., Villiger, W. & Kellenberger, E. (1984). Periplasmic gel: new concept resulting from the reinvestigation of bacterial cell envelope ultrastructure by new methods. *J Bacteriol* **160**, 143–152.

Konhauser, K. O. & Ferris, F. G. (1996). Diversity of iron and silica precipitation by microbial mats in hydrothermal waters, Iceland: implications for Precambrian iron formations. *Geology* **24**, 323–326.

Konhauser, K. O., Fyfe, W. S., Ferris, F. G. & Beveridge, T. J. (1993). Metal sorption and mineral precipitation by bacteria in two Amazonian river systems: Rio Solimões and Rio Negro, Brazil. *Geology* **21**, 1103–1106.

Konhauser, K. O., Schultze-Lam, S., Ferris, F. G., Fyfe, W. S., Longstaffe, F. J. & Beveridge, T. J. (1994). Mineral precipitation by epilithic biofilms in the Speed River, Ontario, Canada. *Appl Environ Microbiol* **60**, 549–553.

Konhauser, K. O., Jones, B., Phoenix, V. R., Ferris, G. & Renaut, R. W. (2004). The microbial role in hot spring silicification. *Ambio* **33**, 552–558.

Korenevsky, A. A., Vinogradov, E., Gorby, Y. & Beveridge, T. J. (2002). Characterization of the lipopolysaccharides and capsules of *Shewanella* spp. *Appl Environ Microbiol* **68**, 4653–4657.

Koval, S. F. & Beveridge, T. J. (1999). Microscopy, electron. In *Encyclopedia of Microbiology*, vol. 3, pp. 276–287. Edited by J. S. Lederberg. San Diego, CA: Academic Press.

Kropinski, A. M. B., Lewis, V. & Berry, D. (1987). Effect of growth temperature on the lipids, outer membrane proteins, and lipopolysaccharides of *Pseudomonas aeruginosa* PAO. *J Bacteriol* **169**, 1960–1966.

Lee, J.-U. & Beveridge, T. J. (2001). Interaction between iron and *Pseudomonas aeruginosa* biofilms attached to Sepharose surfaces. *Chem Geol* **180**, 67–80.

Makin, S. A. & Beveridge, T. J. (1996). The influence of A-band and B-band lipopolysaccharide on the surface characteristics and adhesion of *Pseudomonas aeruginosa* to surfaces. *Microbiology* **142**, 299–307.

Martell, A. E., Smith, R. M. & Motekaitis, R. J. (1987). NIST critically selected stability constants of metal complexes database, version 4.0. College Station, TX: Texas A&M University.

Martinez, R. E. & Ferris, F. G. (2001). Chemical equilibrium modeling techniques for the analysis of high-resolution bacterial metal sorption data. *J Colloid Interface Sci* **243**, 73–80.

Martinez, R. E., Smith, D. S., Kulczycki, E. & Ferris, F. G. (2002). Determination of intrinsic bacterial surface acidity constants using a Donnan shell model and a continuous pK_a distribution method. *J Colloid Interface Sci* **253**, 130–139.

Martinez, R. E., Smith, D. S., Pedersen, K. & Ferris, F. G. (2003). Surface chemical heterogeneity of bacteriogenic iron oxides from a subterranean environment. *Environ Sci Technol* **37**, 5671–5677.

Matias, V. R. F. & Beveridge, T. J. (2005). Cryo-electron microscopy reveals native polymeric cell wall structure in *Bacillus subtilis* 168 and the existence of a periplasmic space. *Mol Microbiol* **56**, 240–251.

Matias, V. R. F., Al-Amoudi, A., Dubochet, J. & Beveridge, T. J. (2003). Cryo-transmission electron microscopy of frozen-hydrated sections of *Escherichia coli* and *Pseudomonas aeruginosa*. *J Bacteriol* **185**, 6112–6118.

Moule, A. L., Galbraith, L., Cox, A. D. & Wilkinson, S. G. (2004). Characterization of a tetrasaccharide released on mild acid hydrolysis of LPS from two rough strains of *Shewanella* species representing different DNA homology groups. *Carbohydr Res* **339**, 1185–1188.

Paul, T. R., Graham, L. L. & Beveridge, T. J. (1993). Freeze-substitution and conventional electron microscopy of medically-important bacteria. *Rev Med Microbiol* **4**, 65–72.

Pembrey, R. S., Marshall, K. C. & Schneider, R. P. (1999). Cell surface analysis techniques: what do cell preparation protocols do to cell surface properties? *Appl Environ Microbiol* **65**, 2877–2894.

Perdue, E. M. (1985). The acidic functional groups of humic substances. In *Humic Substances in Soil, Sediment, and Water: Geochemistry, Isolation, and Characterization*, pp. 493–526. Edited by G. R. Aiken, D. M. McKnight, R. L. Wershaw & P. MacCarthy. New York: Wiley.

Phoenix, V. R., Martinez, R. E., Konhauser, K. O. & Ferris, F. G. (2002). Characterization and implications of the cell surface reactivity of *Calothrix* sp. strain KC97. *Appl Environ Microbiol* **68**, 4827–4834.

Pink, D., Moeller, J., Quinn, B., Jericho, M. & Beveridge, T. (2000). On the architecture of the Gram-negative bacterial murein sacculus. *J Bacteriol* **182**, 5925–5930.

Plette, A. C. C., van Riemsdijk, W. H., Benedetti, M. F. & van der Wal, A. (1995). pH dependent charging behaviour of isolated cell walls of a Gram-positive soil bacterium. *J Colloid Interface Sci* **173**, 354–363.

Razatos, A., Ong, Y.-L., Sharma, M. M. & Georgiou, G. (1998). Molecular determinants of bacterial adhesion monitored by atomic force microscopy. *Proc Natl Acad Sci U S A* **95**, 11059–11064.

Sleytr, U. B. & Beveridge, T. J. (1999). Bacterial S-layers. *Trends Microbiol* **7**, 253–260.

Small, T. D., Warren, L. A., Roden, E. E. & Ferris, F. G. (1999). Sorption of strontium by bacteria, Fe(III) oxide, and bacteria–Fe(III) oxide composites. *Environ Sci Technol* **33**, 4465–4470.

Smith, D. S. & Ferris, F. G. (2001). Proton binding by hydrous ferric oxide and aluminum oxide surfaces interpreted using fully optimized continuous pK_a spectra. *Environ Sci Technol* **35**, 4637–4642.

Smith, D. S. & Ferris, F. G. (2003). Specific surface chemical interactions between hydrous ferric oxide and iron-reducing bacteria determined using pK_a spectra. *J Colloid Interface Sci* **266**, 60–67.

Sokolov, I., Smith, D. S., Henderson, G. S., Gorby, Y. A. & Ferris, F. G. (2001). Cell surface electrochemical heterogeneity of the Fe(III)-reducing bacteria *Shewanella putrefaciens*. *Environ Sci Technol* **35**, 341–347.

Umeda, A., Ueki, Y. & Amako, K. (1987). Structure of the *Staphylococcus aureus* cell wall determined by the freeze-substitution method. *J Bacteriol* **169**, 2482–2487.

Vadillo-Rodríguez, V., Busscher, H. J., Norde, W., de Vries, J. & van der Mei, H. C. (2003). On relations between microscopic and macroscopic physicochemical properties of bacterial cell surfaces: an AFM study on *Streptococcus mitis* strains. *Langmuir* **19**, 2372–2377.

Vadillo-Rodríguez, V., Busscher, H. J., van der Mei, H. C., de Vries, J. & Norde, W. (2005). Role of lactobacillus cell surface hydrophobicity as probed by AFM in adhesion to surfaces at low and high ionic strength. *Colloids Surf B Biointerfaces* **41**, 33–41.

Van Oss, C. J. & Giese, R. F. (1995). The hydrophilicity and hydrophobicity of clay minerals. *Clays Clay Mineral* **43**, 474–477.

Van Oss, C. J., Giese, R. F. & Docoslis, A. (2001). Water, treated as the continuous liquid in and around cells. *Cell Mol Biol* **47**, 721–733.

Vinogradov, E., Korenevsky, A. & Beveridge, T. J. (2002). The structure of the carbohydrate backbone of the LPS from *Shewanella putrefaciens* CN32. *Carbohydr Res* **337**, 1285–1289.

Vinogradov, E., Korenevsky, A. & Beveridge, T. J. (2003a). The structure of the rough-type lipopolysaccharide from *Shewanella oneidensis* MR-1, containing 8-amino-8-deoxy-Kdo and an open-chain form of 2-acetamido-2-deoxy-D-galactose. *Carbohydr Res* **338**, 1991–1997.

Vinogradov, E., Korenevsky, A. & Beveridge, T. J. (2003b). The structure of the O-specific polysaccharide chain of the *Shewanella algae* BrY lipopolysaccharide. *Carbohydr Res* **338**, 385–388.

Vinogradov, E., Korenevsky, A. A. & Beveridge, T. J. (2004). The structure of the core region of the lipopolysaccharide from *Shewanella algae* BrY, containing 8-amino-3,8-dideoxy-D-*manno*-oct-2-ulosonic acid. *Carbohydr Res* **339**, 737–740.

Whitfield, C. & Valvano, M. A. (1993). Biosynthesis and expression of cell-surface polysaccharides in Gram-negative bacteria. *Adv Microb Physiol* **35**, 135–246.

Williams, V. & Fletcher, M. (1996). *Pseudomonas fluorescens* adhesion and transport through porous media are affected by lipopolysaccharide composition. *Appl Environ Microbiol* **62**, 100–104.

Williams, P., Lambert, P. A., Haigh, C. G. & Brown, M. R. W. (1986). The influence of the O and K antigens of *Klebsiella aerogenes* on surface hydrophobicity and susceptibility to phagocytosis and antimicrobial agents. *J Med Microbiol* **21**, 125–132.

Xue, H.-B., Stumm, W. & Sigg, L. (1988). The binding of heavy metals to algal surfaces. *Water Res* **22**, 917–926.

Yao, X., Jericho, M., Pink, D. & Beveridge, T. (1999). Thickness and elasticity of Gram-negative murein sacculi measured by atomic force microscopy. *J Bacteriol* **181**, 6865–6875.

Yee, N. & Fein, J. (2001). Cd adsorption onto bacterial surfaces: a universal adsorption edge? *Geochim Cosmochim Acta* **65**, 2037–2042.

Yee, N., Benning, L. G., Phoenix, V. R. & Ferris, F. G. (2004a). Characterization of metal–cyanobacteria sorption reactions: a combined macroscopic and infrared spectroscopic investigation. *Environ Sci Technol* **38**, 775–782.

Yee, N., Fowle, D. A. & Ferris, F. G. (2004b). A Donnan potential model for metal sorption onto *Bacillus subtilis*. *Geochim Cosmochim Acta* **68**, 3657–3664.

Horizontal gene transfer of metal homeostasis genes and its role in microbial communities of the deep terrestrial subsurface

Jonna Coombs and Tamar Barkay

Department of Biochemistry and Microbiology, Cook College, Rutgers University, 76 Lipman Dr., New Brunswick, NJ 08901, USA

INTRODUCTION

Both basic and applied science issues drive our interests in the microbiology of the deep terrestrial subsurface. As an environment that is disconnected from the Earth's surface, the deep subsurface is less subject to variations in temperature and light and, in unsaturated zones, to intense gradients across interfaces created at the microscale level. These characteristics dictate an average growth rate that is very slow, up to thousands of years per cell division (Kieft & Brockman, 2001), and an ecosystem where change occurs over very long time scales (Fredrickson & Onstott, 2001). Thus, the subsurface is one of the most extreme environments on Earth, and identifying what limits life in the subsurface has value as a model for life on other planets (Chapelle *et al.*, 2002; Nealson & Cox, 2002). The inadvertent release of contaminants from industrial processing plants and storage tanks, as well as the possibility of permanently depositing nuclear wastes deep below the Earth's surface (Pedersen, 2001), raise questions about how microbial activities might exacerbate or mitigate contamination problems in the subsurface.

The terrestrial subsurface is the habitat for diverse microbial communities that, together with the oceanic subsurface, may be the habitat for the largest proportion of Earth's biomass (Whitman *et al.*, 1998). As subsurfaces are characterized by a range of physical and chemical properties, from fully aerated sedimentary shallow aquifers to deep igneous rocks devoid of oxygen and elevated in temperatures, their microbial communities are equally varied (Fredrickson & Fletcher, 2001). Microbial life in the subsurface is greatly constrained by temperature, pressure, limited space and availability of water and scarce resources of electron donors, acceptors and micro-

SGM symposium 65: Micro-organisms and Earth systems – advances in geomicrobiology.
Editors G. M. Gadd, K. T. Semple & H. M. Lappin-Scott. Cambridge University Press. ISBN 0 521 86222 1 ©SGM 2005

nutrients, challenging microbial life to its limit (Colwell, 2001). Nevertheless, studies over the last 30 years have revealed metabolically and phylogenetically diverse microbial communities in the subsurface (Amy *et al.*, 1992; Balkwill *et al.*, 1997).

Microbial biomass and diversity in the subsurface

Microbial biomass has been estimated by direct and viable counts and by the quantification of total phospholipid fatty acids (PLFA) (Kieft *et al.*, 1997; Ringelberg *et al.*, 1997). Biomass estimates show variability that corresponds to the heterogeneity of the geological strata that were sampled. Direct counts range from 10^7 cells (g soil)$^{-1}$ in the sediments of the Atlantic coastal plain in North America, Rainier Mesa in Nevada and Witwatersrand Basin in South Africa to 10^4 cells g^{-1} in deep sediments of the western USA, and groundwater samples rarely contain more the 10^4 cells ml^{-1} (summarized by Fredrickson & Onstott, 2001). Thus, the microbial biomass of both the soil and the aqueous subsurface is orders of magnitude lower than those of the corresponding surface environments.

Representatives belonging to the major prokaryotic lineages have been detected in the subsurface, as revealed by culturing methods (Balkwill *et al.*, 1997) and by molecular signatures of both PLFA and 16S rRNA clone libraries (Chandler *et al.*, 1998; Feris *et al.*, 2004). Interesting unique observations emerge, however, when community structure and diversity are examined within the context of the unique spatial properties of the subsurface. For example, in the vadose zone, the area located between the top soil and the water table, water availability is a limiting resource not so much due to desiccation (water potentials of > -0.1 MPa, sufficient for hydration, are common), but mostly because water is trapped in small spaces, creating discontinuous environments limiting transport of microbes, nutrients and toxicants (Kieft & Brockman, 2001). This spatial discontinuity of microbial niches determines microbial distribution and diversity patterns and limits microbial interactions to microniches. For example, Takai *et al.* (2003b) demonstrated a varied distribution of methanogens in the transition from low-sulfate and organic- and methane-rich shale to high-sulfate and methane- and organic-poor sandstone, thus relating community structure to geochemical gradients and lithologies. Zhou *et al.* (2002) reported a much higher diversity in surface relative to subsurface soils, but both had a high degree of species evenness rather than species dominance, suggesting non-competitive diversity patterns. The authors proposed that, in soils typified by discontinuity, microbial growth is limited by lack of diffusion of essential substrates rather than by competition (Zhou *et al.*, 2002, 2004).

Microbial metabolism in the subsurface

Diverse modes of microbial metabolism exist in the subsurface. Heterotrophic metabolism is supported in aquifers recharged by surface water containing soluble organic carbon, where the consumption of limited oxygen leads to anaerobic conditions and the dominance of anaerobic respiratory pathways. Organic carbon accreted during the slow process of sediment formation is another energy source in the subsurface (Krumholz *et al.*, 1997; Colwell, 2001), and microbial degradation of natural petroleum reserves is an example of heterotrophic anaerobic metabolism with far-reaching consequences for oil quality and quantity (Aitken *et al.*, 2004; Head *et al.*, 2003).

Dissimilatory iron reduction is a dominant respiratory pathway in anoxic aquifers (Lovley *et al.*, 2004). Iron reducers representing common (Petrie *et al.*, 2003) and novel (Coates *et al.*, 2001) members of the *Geobacteraceae* are often detected in such environments. Furthermore, the novel iron reducer *Rhodoferax ferrireducens*, the first non-phototrophic species of its genus, was isolated from subsurface sediment (Finneran *et al.*, 2003). Iron reducers are important targets for bioremediation efforts in the subsurface because they can use other electron acceptors, among them uranium (Lovley *et al.*, 2004) and vanadium (Ortiz-Bernad *et al.*, 2004), inducing the formation of insoluble precipitates. These activities, stimulated *in situ* by the injection of readily oxidizable substrates such as acetate (Anderson *et al.*, 2003) or ethanol (North *et al.*, 2004) into contaminated aquifers, result in the precipitation of metal and radionuclide contaminants. Invariably, such treatments stimulate growth of *Geobacteraceae* in the treated subsurface communities (Anderson *et al.*, 2003; North *et al.*, 2004).

Sulfate reduction is another common respiratory pathway in the subsurface (Wong *et al.*, 2004), driven by sulfate in groundwater and possibly by the activity of pyrite-oxidizing bacteria (Ulrich *et al.*, 1998). Sulfide that accumulates during sulfate reduction may complex with metals and radionuclides (Neal *et al.*, 2004) and retard their mobility in the subsurface, thus making stimulation of sulfate-reducing bacteria another strategy for bioremediation in the subsurface.

Chemolithoautotrophic metabolism is a major mode of microbial metabolism in the subsurface, supporting vast communities of methanogens (Chapelle *et al.*, 2002) and possibly acetogens (Pedersen, 2001) in environments with very low levels of organic substrates, such as deep aquifers or igneous rocks. Pedersen (2001) has proposed the hydrogen-driven biosphere hypothesis to explain microbial life in the latter. According to this hypothesis, hydrogen and carbon dioxide drive methanogens and acetogens, which then support the activities of acetoclastic methanogens and acetate-utilizing iron and sulfate reducers, resulting in the formation of organic polymers that, upon their degradation, are converted to hydrogen and carbon dioxide. A continuous source of hydrogen in the subsurface is required to support this hypothesis. The issue of whether hydrogen can (Stevens & McKinley, 1995; Freund *et al.*, 2002) or cannot (Anderson *et al.*, 1998) be produced in the subsurface by the reaction of water with minerals is presently undecided. The prevalence and diversity of chemolithoautotrophic metabolism among subsurface microbes is also highlighted by the isolation of a thermophilic hydrogen- and sulfur-utilizing chemolithoautotroph from a thermal aquifer, *Sulfurihydrogenibium subterraneum*, belonging to the order *Aquificales* (Takai *et al.*, 2003a). Hydrogen-driven chemoautotrophy by aerobes may be an important, as-yet unexplored, process in vadose zones where little organic matter is available for heterotrophic processes. This lack of information may reflect the fact that the focus of subsurface microbiology research has been on anaerobic processes in groundwater aquifers because of the higher biomass and metabolic rates relative to those in the vadose zone.

In this chapter, we address the issue of the interactions of subsurface micro-organisms with heavy metals and how they are affected by the exchange of genetic material. Thus, we touch on the issues of metal homeostasis and genetic diversity in subsurface microbiology, two topics barely investigated to date.

THE INTERACTIONS OF SUBSURFACE BACTERIA WITH HEAVY METALS

The contamination of the deep subsurface with mixtures of radionuclides, metals and organic solvents that may leach into groundwater aquifers is one of the most detrimental legacies of the Cold War. Immobilization of these contaminants may be the only feasible approach to solving this problem, and micro-organisms that convert inorganic contaminants to less soluble precipitates play a prominent role in *in situ* immobilization strategies for the subsurface (NABIR, 2001). As these strategies depend on the activity of microbes, the presence of a mixture of toxicants may result in the inhibition of reactions essential for immobilization. It is for that reason that strains of *Deinococcus* spp. with high levels of resistance to ionizing and gamma radiation have been engineered with the ability to degrade organic contaminants and withstand metal toxicity (Brim *et al.*, 2000, 2003), and that the toxicity of metals and actinides to bacteria with a potential in bioremediation has been evaluated (Reed *et al.*, 1999; Ruggiero *et al.*, 2005).

A study to determine the level of metal resistance among subsurface aerobic heterotrophic bacteria was initiated, reasoning that these microbes play a role in facilitating microbial metabolism in the subsurface (Benyehuda *et al.*, 2003). The microbes tested were from the subsurface microbial culture collection (SMCC) that is maintained at Florida State University, USA, and included 261 isolates from the Savannah River Site (SRS) in South Carolina, USA (borehole P24) and 89 strains from the Hanford site in Washington state (borehole YB-02). The SRS isolates belonged to the α-, β- and γ-proteobacteria, as well as to the high-G+C Gram-positive group. The Hanford strains contained representatives of these taxonomic groups, as well as the low-G+C Gram-positive group. Resistance to Pb(II), Hg(II) and Cr(VI) was determined by disk-inhibition assays on solid growth media and by comparison with the response of well-characterized reference resistant and sensitive strains (Benyehuda *et al.*, 2003). Results were analysed for the relationship of metal resistance to the properties of the tested microbial communities and the environments from which they were isolated. The major findings were:

(i) Resistance to Pb(II) and Cr(VI) was common among subsurface strains from SRS and Hanford sediments, while fewer, mostly Gram-positive strains, were resistant to Hg(II).

(ii) With the exception of a high level of metal tolerance among *Arthrobacter* spp., there was no relationship between the phylogeny of the microbes and their metal-resistance patterns. This is not surprising, as metal resistance is often specified by mobile elements such as plasmids and transposons (Silver & Phung, 1996; Kholodii *et al.*, 2002). Some subsurface *Arthrobacter* isolates proved to be exceptionally resistant to Cr(VI) and Hg(II). Other researchers have also reported high levels of metal tolerance among soil *Arthrobacter* spp. (Roane, 1999; Megharaj *et al.*, 2003) and the high abundance of this genus in soils impacted by mixed-waste contamination (Fredrickson *et al.*, 2004). Thus, further investigation of *Arthrobacter*–metal interactions is highly warranted.

(iii) Resistances to Hg(II) and Pb(II) were more common in the SRS collection than in the Hanford collection (ANOVA; $P<0.05$) and multiple metal resistance was also higher for the SRS, with 33 % of all strains resistant to more than one metal in this group compared with 23 % for the Hanford group (Fig. 1). Thus, toxic metals influenced the evolution of resistance more effectively in the SRS community than in the Hanford community. Varying geological and geochemical factors may explain this difference. For example, metal toxicity is mitigated by low redox potential and high clay and organic matter (Collins & Stotzky, 1989; Giller *et al.*, 1998), and the clay content of Hanford sediment is higher than in the more sandy SRS sediment. These results illustrate that, in complex environments, microbe–metal interactions are greatly impacted by environmental factors, most likely by controlling bioavailability and thus metal toxicity. For more details of this study, see Benyehuda *et al.* (2003).

This study, as well as others addressing the issue of survival and activities of microbes in mixed-waste-contaminated subsurfaces, has focused on aerobic bacteria. However, current *in situ* immobilization efforts target the activities of metal- and radionuclide-reducing anaerobic bacteria (Anderson *et al.*, 2003; Istok *et al.*, 2004). The few who

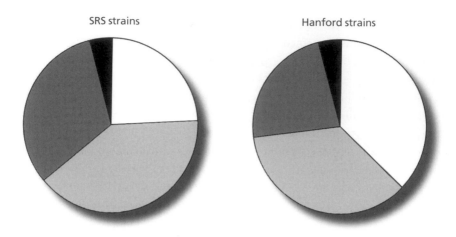

Fig. 1. Multi-resistance to Pb(II), Hg(II) and Cr(VI) among bacteria from the subsurface. The proportion of strains resistant to none (white segments), one (pale grey), two (dark grey) or all three (black) of the test metals among strains from the SRS and the Hanford sites are shown. Redrawn with permission from Benyehuda *et al.* (2003).

have examined the response of anaerobes to metal toxicity reported conflicting results. *Desulfovibrio desulfuricans* G20, a model organism for the immobilization of metals as sulfides, was susceptible to micromolar concentrations of Cu(II), Zn(II) and Pb(II) when a medium designed to minimize metal complexation was used (Sani *et al.*, 2003). Mixed cultures of sulfate reducers were inhibited by Cr(VI) (Smith & Gadd, 2000) and Cu(II) and Zn(II) (Utgikar *et al.*, 2003). Likewise, *Shewanella* spp., studied for their role in immobilizing metals and radionuclides by reducing them to insoluble forms, were affected by U(VI) (Wade & DiChristina, 2000) and Cr(VI) (Viamajala *et al.*, 2004). These observations clearly suggest that susceptibility to metals and thus the acquisition of metal resistance by microbes in the subsurface is critical to the success of bioremediation in environments contaminated by mixed wastes.

THE EVOLUTION OF METAL HOMEOSTASIS GENES BY HORIZONTAL GENE TRANSFER (HGT) IN SUBSURFACE MICROBIAL COMMUNITIES

Gene transfer among microbes in the subsurface environment

Genes encoding resistance to heavy metals are often transferred among micro-organisms in much the same way that antibiotic-resistance genes travel through microbial populations, by horizontal gene transfer (HGT) mechanisms. Like antibiotics, which are frequently produced by soil organisms, heavy metals are compounds that subsurface organisms are likely to encounter as part of their environment.

Bioavailable heavy metals are produced naturally by the geochemical weathering of ores and mobilized with the movement of groundwater. Anthropogenic contamination, however, may increase the concentrations of toxic metals in a given ecological niche manyfold. The presence of resistance genes on mobile genetic elements within the subsurface community is therefore a distinct advantage. The occurrence of HGT in topsoils, natural waters and in association with the internal and external surfaces of plants and animals is well recognized. However, HGT has barely been examined in the deep subsurface. Because population densities (Normander *et al.*, 1998; Licht *et al.*, 1999) and active metabolism (Smets *et al.*, 1993) stimulate HGT, while most deep subsurface environments are notorious for low population densities and metabolic rates (Balkwill, 1989; Kieft & Brockman, 2001), the deep subsurface may be the least conducive environment for genetic exchange. To examine HGT and its role in the evolution of metal resistance in the subsurface, we have looked for evidence of the horizontal inheritance of genes encoding P_{IB}-type ATPases in bacteria from subsurface sediments of the SRS (Coombs & Barkay, 2004).

P_{IB}-type ATPases and their roles in metal homeostasis and HGT

P_{IB}-type ATPases are membrane-associated ion pumps that are responsible for maintaining metal homeostasis by mediating the transport of heavy metals using the energy generated by the hydrolysis of ATP. Those that are specific for monovalent cations [Cu(I) and/or Ag(I)] are found in the three domains of life, while those specific to divalent cations [Zn(II), Cd(II) and/or Pb(II)] are only found among the prokaryotes. These metal pumps can function in either the import of essential ions or the export of ions that have reached harmful levels in the cell cytoplasm, depending on the orientation of the protein in the membrane (Rosen, 2002).

Several P_{IB}-type ATPases have been shown to be associated with mobile genetic elements, from Gram-positive organisms such as *Lactococcus lactis* (O'Sullivan *et al.*, 2001), *Staphylococcus aureus* (Nucifora *et al.*, 1989) and *Arthrobacter* spp. (K. Jerke and C. Nakatsu, personal communication) and from Gram-negative organisms such as *Ralstonia metallidurans* (Borremans *et al.*, 2001) and *Stenotrophomonas maltophilia* (Alonso *et al.*, 2000). However, the occurrence of HGT and its effects on the evolution of this locus within a specific microbial community had not previously been examined. We have targeted the genes encoding P_{IB}-type ATPases for a study on the role of HGT in the evolution of metal homeostasis among subsurface bacteria because of the importance of metal ion homeostasis for survival in the harsh environment of the metal-contaminated subsurface. Isolates from the SMCC were selected for study because of the large number of Pb(II)-resistant bacteria in the SRS community (Benyehuda *et al.*, 2003).

Evolution of P$_{IB}$-type ATPases by HGT in a subsurface microbial community

We have used a retrospective approach, based on the recognition of genomic indicators for evolution by HGT. We reasoned that, because HGT is estimated to occur at rates of 31 kb per million years (Lawrence & Ochman, 1997) and subsurface micro-organisms reproduce very slowly, prospective approaches to HGT detection, such as microcosm incubations, may be of little relevance to subsurface communities. However, because genomic data may be interpreted in different ways, calling into question the validity of all molecular signatures of HGT (Eisen, 2000), we employed multiple methods in combination while determining whether or not a gene encoding a P$_{IB}$-type ATPase was horizontally transferred. These methods included (i) examining the congruence of the P$_{IB}$-type ATPase phylogeny with that of the 16S rRNA gene, (ii) looking for unusual sequence composition (G+C content) of the P$_{IB}$-type ATPase when compared with that of the host genome and (iii) looking for shared indels (insertion/deletion events) among P$_{IB}$-type ATPase genes from different organisms.

To obtain P$_{IB}$-type ATPase genes (*zntA/cadA/pbrA*-like genes) from the SRS aerobic heterotrophs, a nested PCR approach was developed. Novel primer sets for PCR amplification were designed by aligning conserved domains in *zntA/cadA/pbrA*-like genes that were available in databases. The first PCR targeted the phosphatase domain and the ATP-binding domain and the second reaction used conserved sequences in the transmembrane metal-binding domain and the ATP-binding domain. Using nine PCR primer pairs, amplification products of *zntA/cadA/pbrA*-like genes from 48 of 105 Pb(II)- resistant subsurface strains were obtained and sequenced. These sequences and the DNA sequences of 16S rRNA coding genes of the corresponding hosts were then used to determine whether HGT has contributed to the evolution of *zntA/cadA/pbrA*-like genes among the subsurface bacteria. For more details about this approach, see Coombs & Barkay (2004). Phylogenetic incongruence using both parsimony (heuristic) and distance (neighbour-joining) methods indicated that, in four of the isolates, *zntA/cadA/pbrA*-like genes evolved by HGT (Fig. 2), and three of these were supported by unusual sequence composition, i.e. G+C content and the presence of indels. All transfers were among the *β*- and/or *γ*-proteobacteria, which were the predominant groups among the Pb(II)-resistant subsurface bacteria from which sequence data were obtained. Two of these transfers were to *Comamonas* spp., and comparison of clustering patterns between the *zntA/cadA/pbrA* and the 16S rRNA trees suggested that, in one case (in strain BO669), transfer could have occurred from another *Comamonas* sp., and in the other (in strain BO173), the origin could not be clearly identified (Fig. 2). A third HGT event, the acquisition of a *Pseudomonas*-like P$_{IB}$-type ATPase by *Ralstonia* sp. B0665, was not supported by additional sequence features, but the phylogenetic evidence as indicated by the bootstrap support value was very strong. Finally, a P$_{IB}$-type

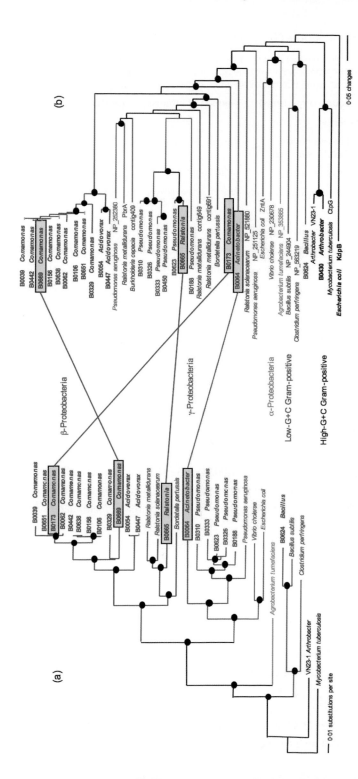

Fig. 2. Phylogenetic evidence for HGT among metal-resistant subsurface strains from the SRS site. Neighbour-joining trees of the deduced amino acid sequence of the *zntA/cadA/pbrA*-like gene (a) and the nucleotide sequence of the 16S rRNA gene (b) are shown. Incongruencies between the positions of sequences from the same organism are indicated by slanted lines connecting boxed branches in the two trees. Names of subsurface strains are in bold. Filled circles indicate bootstrap-supported nodes. Reproduced with permission from Coombs & Barkay (2004).

Table 1. Evidence to support HGT in the subsurface

Supporting evidence is provided either by phylogenetic incongruence of a target gene when compared to the 16S rRNA gene phylogeny of the host organism or by the presence of mobile genetic elements carrying functional genes (Plasmid). Two-letter abbreviations are used for US states.

Site of isolation	Source organism(s)	Gene(s) of interest	Reference
Phylogenetic incongruence			
Savannah River, SC	Comamonas spp., Pseudomonas spp.	P_{IB}-type ATPases	Coombs & Barkay (2004)
Savannah River, SC	γ-Proteobacteria	tRNA(Leu)(UAA)	Vepritskiy et al. (2002)
South Glens Falls, NY	Gram-negative bacteria	Naphthalene dioxygenase	Herrick et al. (1997)
Plasmid			
Metal-plating lagoon	Pseudomonas sp.	czc, ncc	Smets et al. (2003)
Savannah River, SC	Heterotrophic bacteria	Metal-resistance genes	Fredrickson et al. (1988)
South Glens Falls, NY	Gram-negative bacteria	Naphthalene dioxygenase	Ghiorse et al. (1995)
Savannah River, SC	Sphingomonas sp. F199	Catabolic genes	Romine et al. (1999)
Various	Sphingomonas spp.	Dibenzo-p-dioxin, dibenzofuran and naphthalene sulfonates degradation	Basta et al. (2004)
Savannah River, SC	Sphingomonas sp. F199	Catechol 2,3-dioxygenase	Stillwell et al. (1995)
Savannah River, SC	Gram-negative	Cryptic	Brockman et al. (1989)
Savannah River, SC	Sphingomonas spp.	2,3-Dihydroxybiphenyl 1,2-dioxygenase and catechol 2,3-dioxygenase	Kim et al. (1996)

ATPase from *Acinetobacter* sp. strain B0064 grouped phylogenetically within a β-proteobacterial clade. The G+C content supported the phylogenetic evidence, indicating that this could have been a recent gene acquisition by the *Acinetobacter* strain.

These results indicate that HGT has occurred, albeit at low frequencies, during the evolution of metal homeostasis genes among subsurface bacteria. Other observations also support these conclusions (Table 1). Plasmid-borne genes for metal resistance and the degradation of hydrocarbons have been obtained from subsurface isolates, and phylogenetic analyses have suggested transfer of hydrocarbon-degradation genes in a shallow aquifer contaminated with coal tar (Herrick *et al.*, 1997). The demonstration of conjugation in microcosms simulating low-nutrient subsurface soils (Smets *et al.*, 2003) suggests that HGT can affect the evolution and genetic diversity of subsurface soil communities.

Here, we used the primary DNA sequences of *zntA/cadA/pbrA*-like genes to deduce the evolutionary pathway of an environmentally important function, metal homeostasis, among subsurface bacteria. While it is tempting to conclude from this study that HGT occurs in the subsurface, such a conclusion is impossible without clear evidence that the studied strains evolved in the subsurface. Without such evidence, the alternative possibility that transfer occurred prior to deposition in the subsurface cannot be ruled out. Evidence for evolution *in situ* currently exists for a collection of *Arthrobacter* spp. from the Yakima Barricade, where a coherence exists between the 16S rRNA- and *recA*-based phylogenies and the geological strata from which the strains originated, suggesting a long-term evolution, possibly as long as 8 million years, in the subsurface (van Waasbergen *et al.*, 2000). Examining microbial communities from this and other similar environments may reveal HGT and other processes that affect genetic diversity as they have occurred in the subsurface.

Evidence for HGT of P_{IB}-type ATPases in complete prokaryotic genomes

In order to evaluate the observed frequency of HGT among subsurface microbes, we analysed genes encoding P_{IB}-type ATPases of 188 bacterial and 22 archaeal genomes. As a clear phylogenetic distinction between P_{IB}-type ATPases specifying mono- and divalent pumps does not currently exist, our analysis encompassed all *zntA* and *copA*-like sequences. Only P_{IB}-type ATPases loci that exactly matched the amino acid sequence of the phosphatase and the transmembrane metal-binding domains of enzymes with documented activity were included in the collection. The resulting collections of 311 P_{IB}-type ATPases and the 16S rRNA genes of the corresponding genomes were subjected to phylogenetic analysis and cases of incongruence between the two phylogenies were identified. When incongruence was detected, supportive evidence was sought by examining sequence composition as described above. In addition, sequences proximal, i.e. within 5 kbp, were examined for the presence of regulatory genes that might have been co-transferred with the P_{IB}-type ATPase genes. When found, these were subjected to the incongruence test as above to determine their phylogenetic affiliation (Coombs & Barkay, 2005).

Twelve instances of phylogenetic incongruence were detected, six of which were transfers across subclasses within the *Proteobacteria* (Table 2). This is not surprising, because our collection of P_{IB}-type ATPases was dominated by proteobacteria. However, the remaining transfer events were across a longer phylogenetic distance. In two cases, *zntA* loci from low-G+C Gram-positive bacteria were found in the genomes of the γ-proteobacteria *Stenotrophomonas maltophilia* and *Legionella pneumophila* and a transfer of an ε-proteobacterial *copA* to *Ureaplasma parvum*, a low-G+C Gram-positive, was also noted. These findings highlight an extensive involvement of Gram-positive bacteria in gene exchange. Evidence that *zntA* and *copA* were transferred to *Deinococcus radiodurans* from α-proteobacteria emerged from the

Table 2. HGT of the *zntA* and *copA* genes based on evidence from complete microbial genomes

When the genome contained more than one *zntA* or *copA* gene, the locus number, as designated in the annotated genome, is included (e.g. *D. radiodurans* ZntA locus 1).

Genome	Lineage of likely donor*	HGT supported by:		
		Unusual G+C content†	Proximal regulatory gene‡	Shared indel
ZntA				
Deinococcus radiodurans 1	α-Proteobacteria	No	No	No
Stenotrophomonas maltophilia 1	Low-G+C Gram-positive	Yes	Yes (low-G+C)	No
Legionella pneumophila 2	Low-G+C Gram-positive	No	No	No
Pseudomonas aeruginosa 3	β-Proteobacteria	No	No	No
Azotobacter vinelandii 1	β-Proteobacteria	No	No	No
'*Pyrococcus abyssi*'	Cyanobacteria	No	Yes (cyanobacteria)	No
CopA				
Salmonella typhimurium 3	α-Proteobacteria	No	Yes (α-proteobacteria)	No
Deinococcus radiodurans	α-Proteobacteria	Yes	No	No
Ureaplasma parvum	ε-Proteobacteria	No	No	No
Azotobacter vinelandii 3	α-Proteobacteria	No	No	Yes
'*Microbulbifer flagellatus*' 1	β-Proteobacteria	No	No	No
Pseudomonas aeruginosa 4	β-Proteobacteria	No	No	No
Pyrobaculum aerophilum	Euryarchaea	Yes	No	Yes

*As revealed by the incongruent phylogenetic clustering of the gene relative to that of the 16S rRNA gene (see text).

†Unusual G+C content is defined here as differing by more than one standard deviation from the mean for the genome.

‡Proximal is defined here as a gene that is within 5 kb of a gene encoding a ZntA or a CopA. The phylogenetic affiliation of the regulatory gene is indicated in parentheses.

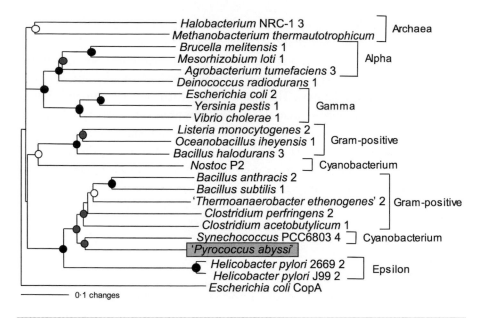

Fig. 3. Phylogenetic evidence for the transfer of a *zntA* gene from a cyanobacterium to the euryarchaeon '*Pyrococcus abyssi*' (boxed). Circles at each node indicate the level of bootstrap support obtained when analysed by both parsimony and distance methods: black, > 80 %; grey, > 50 %; white, supported at > 50 % by one method only.

incongruence analysis and separate locations of these two loci in the genomes of both *D. radiodurans* and α-proteobacteria suggested that two independent transfer events were involved. A single transfer event of *copA* between closely related euryarchaeota was detected in the genome of *Pyrobaculum aerophilum* (Table 2) and, most excitingly, a *zntA* of cyanobacterial origin was present in the genome of the archaeon '*Pyrococcus abyssi*' (Fig. 3). Supportive evidence confirming HGT was available for six of the transfers that were revealed by incongruent phylogenies (Table 2) and in three cases these included the presence of a gene with homology to regulatory elements that are known to control expression of *zntA/cadA/pbrA* loci. Phylogenetic analysis of these regulatory genes showed that they likely shared an origin with the *zntA* or *copA* genes they accompanied. This latter criterion also confirmed the cross-domain transfer between archaea and cyanobacteria.

Thus, it seems that, as we have found with the subsurface strains, the evolution of genes encoding P_{IB}-type ATPases in sequenced genomes has been subjected to HGT but that their inheritance has mostly proceeded vertically. This is in contrast to the well-documented dominance of HGT in the evolution of other traits that enhance fitness to toxicants, such as resistance to mercury (Kholodii *et al.*, 2002) and antibiotics. As P_{IB}-type ATPases mediate metal homeostasis, they may be considered more essential to core metabolism than phenotypes that are exclusively involved in detoxification, thus enhancing stable genomic inheritance. The frequency of transfer detected among the microbial genomes, 12 of 311, was slightly lower than that among the subsurface bacteria, 4 of 48 (see above). This difference was most likely due to differences in the composition of the datasets. The genome study encompassed a broader phylogenetic range, and therefore we were able to detect transfers across large phylogenetic distances, whereas the subsurface study detected transfer among more closely related organisms. However, it is possible that the frequency of HGT among the subsurface strains was underestimated. The more closely related microbes are phylogenetically, the more likely they are to exchange genetic material (Lawrence & Ochman, 1997), but the less likely are the transfer events to leave a detectable molecular footprint in the new host genome (Eisen, 2000).

HGT gene microarray

DNA and expression microarrays are a powerful tool in biological research, and applications to the study of microbial community structure (Small *et al.*, 2001) and function (Taroncher-Oldenburg *et al.*, 2003; Rhee *et al.*, 2004) have been documented. Zhou and his collaborators have identified three types of microarrays in microbial ecology (Zhou & Thompson, 2002). Phylogenetic oligonucleotide arrays (POA) consist of probes homologous to 16S rRNA genes and are used to study community composition and its response to environmental change. Functional gene arrays (FGA) are designed to evaluate the metabolic potential of a community by probing for genes that specify major biogeochemical reactions, including those essential for biodegradation and bioremediation. Community genome arrays (CGA) target genes of pure isolates from a specific environment. Our discovery of molecular signatures indicative of HGT in the genomes of subsurface bacteria (Coombs & Barkay, 2004) prompted us to develop a fourth type of microarray, possibly a variation on the CGA, the horizontal gene transfer array (HGT array). This array was designed to answer the question: 'what are the genetic elements that transfer metal-resistance genes among subsurface bacteria?'

The HGT array includes 158 oligonucleotide (70-mer) probes specific for genes that encode replication/incompatibility (*inc/rep*) loci in 86 broad-host-range (BHR) plasmids belonging to 13 distinct plasmid groups and 100 probes for metal-resistance genes. The linkage of metal resistance on specific plasmids is suggested by positive signals obtained following hybridization of Cy3- or Cy5-labelled plasmid DNA extracted from subsurface isolates with the array. The emerging patterns classify plasmids according to their incompatibility groupings and linkage with metal-resistance genes.

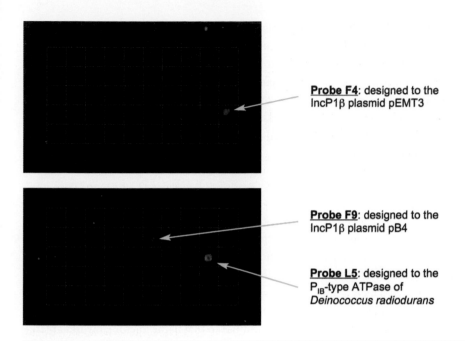

Probe F4: designed to the IncP1β plasmid pEMT3

Probe F9: designed to the IncP1β plasmid pB4

Probe L5: designed to the P$_{IB}$-type ATPase of *Deinococcus radiodurans*

Fig. 4. HGT array hybridized with Cy3-labelled plasmid from the subsurface strain *Comamonas* sp. B0173. Positive hybridization signals appear as green spots.

Analysis of the four SRS subsurface isolates that were shown to have inherited Pb(II) resistance by HGT (Fig. 2) indicated that at least two of them carried multiple small plasmids. Plasmid DNA extracts of these strains and of additional metal-resistant isolates from contaminated subsurface sediments in Oak Ridge, TN, USA, were hybridized with the array following optimization and testing with 26 exact-match reference plasmids (J. Coombs and T. Barkay, in preparation). Of these, a plasmid extract from *Comamonas* sp. B0173, a strain that inherited its *zntA*/*cadA*/*pbrA* gene by transfer from an unknown donor (Coombs & Barkay, 2004), hybridized to two probes homologous to *rep*/*inc* loci of plasmids belonging to IncP1β and to a P$_{IB}$-type ATPase probe (Fig. 4). This finding indicates that inheritance of Pb(II) resistance in strain B0173 likely occurred by conjugal transfer of an IncP1β BHR Pb(II)-resistance plasmid. Array studies with a plasmid extract from the other SRS isolate are currently in progress.

The three strains from the contaminated sediments in Oak Ridge, TN, are a part of a large collection of Gram-positive bacteria where metal-resistance patterns correlated well with plasmid carriage (Patty Sobecky, personal communication). Strain *Bacillus* sp. U26 hybridized to *rep*/*inc* probes from a group of characterized *Bacillus* loci and hybridized weakly to probes for arsenic-resistance genes. Interestingly, plasmid DNA from another *Bacillus* strain, strain V6, hybridized to *rep*/*inc* probes homologous to those of plasmids that had been described previously in both Gram-positive and -negative bacteria. In addition, the plasmid preparation hybridized to arsenic-resistance probes. These results suggest either that two plasmids exist in strain V6 or that a single arsenic-resistance plasmid has two different origins of replication, and imply that strain V6 carries plasmids of a broader diversity than has previously been described in other Gram-positive bacteria. Although our dataset is currently very small, it emphasizes that interesting, previously uncharacterized metal-resistance plasmids exist in subsurface soil bacteria.

CONCLUSIONS

The studies reported here focused on two issues that are critical to the activities of microbial communities in the subsurface, metal homeostasis and HGT. Critical, because both of these are considerations that affect strategies for controlling the transport of metals and radionuclides in contaminated subsurface environments. Results showed a high frequency of resistance to divalent cations and a modest, yet significant, inheritance of a gene encoding metal homeostasis by HGT. This frequency of HGT of metal homeostasis genes was similar to that in the microbial world at large as suggested by the analysis of sequenced microbial genomes. While these results suggest that HGT may have contributed to the survival of microbes in the harsh subsurface environment, they could not determine whether transfer occurred *in situ*. This question could only be addressed by examining signatures of HGT in the genomes of subsurface microbes from communities whose evolution in the subsurface can be documented unequivocally, enabling the histories of microbial speciation to be related to geological processes. This opportunity may exist in certain ecological niches within the subsurface, which are permanently isolated from the surface and where microbial migration is restricted by the discontinuity of niches that support life.

ACKNOWLEDGEMENTS

The research described here was funded by the Natural and Accelerated Bioremediation Research (NABIR) program, Biological and Environmental Research (BER), US Department of Energy (grant no. DE-FG02-99ER62864).

REFERENCES

Aitken, C. M., Jones, D. M. & Larter, S. R. (2004). Anaerobic hydrocarbon biodegradation in deep subsurface oil reservoirs. *Nature* **431**, 291–294.

Alonso, A., Sanchez, P. & Martínez, J. L. (2000). *Stenotrophomonas maltophilia* D457R contains a cluster of genes from gram-positive bacteria involved in antibiotic and heavy metal resistance. *Antimicrob Agents Chemother* **44**, 1778–1782.

Amy, P. S., Haldeman, D. L., Ringelberg, D., Hall, D. H. & Russell, C. (1992). Comparison of identification systems for classification of bacteria isolated from water and endo-lithic habitats within the deep subsurface. *Appl Environ Microbiol* **58**, 3367–3373.

Anderson, R. T., Chapelle, F. H. & Lovley, D. R. (1998). Evidence against hydrogen-based microbial ecosystems in basalt aquifers. *Science* **281**, 976–977.

Anderson, R. T., Vrionis, H. A., Ortiz-Bernad, I. & 10 other authors (2003). Stimulating the *in situ* activity of *Geobacter* species to remove uranium from the groundwater of a uranium-contaminated aquifer. *Appl Environ Microbiol* **69**, 5884–5891.

Balkwill, D. L. (1989). Numbers, diversity, and morphological characteristics of aerobic, chemoheterotrophic bacteria in deep subsurface sediments from a site in South Carolina. *Geomicrobiol J* **7**, 33–52.

Balkwill, D. L., Reeves, R. H., Drake, G. R., Reeves, J. Y., Crocker, F. H., King, M. B. & Boone, D. R. (1997). Phylogenetic characterization of bacteria in the subsurface microbial culture collection. *FEMS Microbiol Rev* **20**, 201–216.

Basta, T., Keck, A., Klein, J. & Stolz, A. (2004). Detection and characterization of conjugative degradative plasmids in xenobiotic-degrading *Sphingomonas* strains. *J Bacteriol* **186**, 3862–3872.

Benyehuda, G., Coombs, J., Ward, P. M., Balkwill, D. & Barkay, T. (2003). Metal resistance among aerobic chemoheterotrophic bacteria from the deep terrestrial subsurface. *Can J Microbiol* **49**, 151–156.

Borremans, B., Hobman, J. L., Provoost, A., Brown, N. L. & van der Lelie, D. (2001). Cloning and functional analysis of the *pbr* lead resistance determinant of *Ralstonia metallidurans* CH34. *J Bacteriol* **183**, 5651–5658.

Brim, H., McFarlan, S. C., Fredrickson, J. K., Minton, K. W., Zhai, M., Wackett, L. P. & Daly, M. J. (2000). Engineering *Deinococcus radiodurans* for metal remediation in radioactive mixed waste environments. *Nat Biotechnol* **18**, 85–90.

Brim, H., Venkateswaran, A., Kostandarithes, H. M., Fredrickson, J. K. & Daly, M. J. (2003). Engineering *Deinococcus geothermalis* for bioremediation of high-temperature radioactive waste environments. *Appl Environ Microbiol* **69**, 4575–4582.

Brockman, F. J., Denovan, B. A., Hicks, R. J. & Fredrickson, J. K. (1989). Isolation and characterization of quinoline-degrading bacteria from subsurface environments. *Appl Environ Microbiol* **55**, 1029–1032.

Chandler, D. P., Brockman, F. J., Bailey, T. J. & Fredrickson, J. K. (1998). Phylogenetic diversity of Archaea and Bacteria in a deep subsurface paleosol. *Microb Ecol* **36**, 37–50.

Chapelle, F. H., O'Neill, K., Bradley, P. M., Methe, B. A., Ciufo, S. A., Knobel, L. L. & Lovley, D. R. (2002). A hydrogen-based subsurface microbial community dominated by methanogens. *Nature* **415**, 312–315.

Coates, J. D., Bhupathiraju, V. K., Achenbach, L. A., McLnerney, M. J. & Lovley, D. R. (2001). *Geobacter hydrogenophilus*, *Geobacter chapellei* and *Geobacter grbiciae*, three new, strictly anaerobic, dissimilatory Fe(III)-reducers. *Int J Syst Evol Microbiol* **51**, 581–588.

Collins, Y. E. & Stotzky, G. (1989). Factors affecting the toxicity of heavy metals to microbes. In *Metal Ions and Bacteria*, pp. 31–90. Edited by T. J. Beveridge & R. J. Doyle. New York: Wiley.

Colwell, F. S. (2001). Constrains on the distribution of microorganisms in subsurface environments. In *Subsurface Microbiology and Biogeochemistry*, pp. 71–95. Edited by J. K. Fredrickson & M. Fletcher. New York: Wiley-Liss.

Coombs, J. M. & Barkay, T. (2004). Molecular evidence for the evolution of metal homeostasis genes by lateral gene transfer in bacteria from the deep terrestrial subsurface. *Appl Environ Microbiol* **70**, 1698–1707.

Coombs, J. M. & Barkay, T. (2005). New findings on the evolution of metal homeostasis genes: evidence from comparative genome analysis of bacteria and archaea. *Appl Environ Microbiol* (in press).

Eisen, J. A. (2000). Horizontal gene transfer among microbial genomes: new insights from complete genome analysis. *Curr Opin Genet Dev* **10**, 606–611.

Feris, K. P., Hristova, K., Gebreyesus, B., Mackay, D. & Scow, K. M. (2004). A shallow BTEX and MTBE contaminated aquifer supports a diverse microbial community. *Microb Ecol* **48**, 589–600.

Finneran, K. T., Johnsen, C. V. & Lovley, D. R. (2003). *Rhodoferax ferrireducens* sp. nov., a psychrotolerant, facultatively anaerobic bacterium that oxidizes acetate with the reduction of Fe(III). *Int J Syst Evol Microbiol* **53**, 669–673.

Fredrickson, J. K. & Fletcher, M. (2001). *Subsurface Microbiology and Biogeochemistry*. New York: Wiley-Liss.

Fredrickson, J. K. & Onstott, T. C. (2001). Biogeochemical and geological significance of subsurface microbiology. In *Subsurface Microbiology and Biogeochemistry*, pp. 3–37. Edited by J. K. Fredrickson & M. Fletcher. New York. Wiley-Liss.

Fredrickson, J. K., Hicks, R. J., Li, S. W. & Brockman, F. J. (1988). Plasmid incidence in bacteria from deep subsurface sediments. *Appl Environ Microbiol* **54**, 2916–2923.

Fredrickson, J. K., Zachara, J. M., Balkwill, D. L., Kennedy, D., Li, S. M., Kostandarithes, H. M., Daly, M. J., Romine, M. F. & Brockman, F. J. (2004). Geomicrobiology of high-level nuclear waste-contaminated vadose sediments at the Hanford site, Washington state. *Appl Environ Microbiol* **70**, 4230–4241.

Freund, F., Dickinson, J. T. & Cash, M. (2002). Hydrogen in rocks: an energy source for deep microbial communities. *Astrobiology* **2**, 83–92.

Ghiorse, W. C., Herrick, J. B., Sandoli, R. L. & Madsen, E. L. (1995). Natural selection of PAH-degrading bacterial guilds at coal-tar disposal sites. *Environ Health Perspect* **103** (Suppl. 5), 107–111.

Giller, K. E., Witter, E. & McGrath, S. P. (1998). Toxicity of heavy metals to microorganisms and microbial processes in agricultural soils: a review. *Soil Biol Biochem* **30**, 1389–1414.

Head, I. M., Jones, D. M. & Larter, S. R. (2003). Biological activity in the deep subsurface and the origin of heavy oil. *Nature* **426**, 344–352.

Herrick, J. B., Stuart-Keil, K. G., Ghiorse, W. C. & Madsen, E. L. (1997). Natural horizontal transfer of a naphthalene dioxygenase gene between bacteria native to a coal tar-contaminated field site. *Appl Environ Microbiol* **63**, 2330–2337.

Istok, J. D., Senko, J. M., Krumholz, L. R., Watson, D., Bogle, M. A., Peacock, A., Chang, Y. J. & White, D. C. (2004). In situ bioreduction of technetium and uranium in a nitrate-contaminated aquifer. *Environ Sci Technol* **38**, 468–475.

Kholodii, G., Gorlenko, Z., Mindlin, S., Hobman, J. & Nikiforov, V. (2002). Tn*5041*-like transposons: molecular diversity, evolutionary relationships and distribution of distinct variants in environmental bacteria. *Microbiology* **148**, 3569–3582.

Kieft, T. L. & Brockman, F. J. (2001). Vadose zone microbiology. In *Subsurface Microbiology and Biogeochemistry*, pp. 141–169. Edited by J. K. Fredrickson & M. Fletcher. New York: Wiley-Liss.

Kieft, T. L., Wilch, E., O'Connor, K., Ringelberg, D. B. & White, D. C. (1997). Survival and phospholipid fatty acid profiles of surface and subsurface bacteria in natural sediment microcosms. *Appl Environ Microbiol* **63**, 1531–1542.

Kim, E., Aversano, P. J., Romine, M. F., Schneider, R. P. & Zylstra, G. J. (1996). Homology between genes for aromatic hydrocarbon degradation in surface and deep-subsurface *Sphingomonas* strains. *Appl Environ Microbiol* **62**, 1467–1470.

Krumholz, L. R., McKinley, J. P., Ulrich, G. A. & Suflita, J. M. (1997). Confined subsurface microbial communities in Cretaceous rock. *Nature* **386**, 64–66.

Lawrence, J. G. & Ochman, H. (1997). Amelioration of bacterial genomes: rates of change and exchange. *J Mol Evol* **44**, 383–397.

Licht, T. R., Christensen, B. B., Krogfelt, K. A. & Molin, S. (1999). Plasmid transfer in the animal intestine and other dynamic bacterial populations: the role of community structure and environment. *Microbiology* **145**, 2615–2622.

Lovley, D. R., Holmes, D. E. & Nevin, K. P. (2004). Dissimilatory Fe(III) and Mn(IV) reduction. *Adv Microb Physiol* **49**, 219–286.

Megharaj, M., Avudainayagam, S. & Naidu, R. (2003). Toxicity of hexavalent chromium and its reduction by bacteria isolated from soil contaminated with tannery waste. *Curr Microbiol* **47**, 51–54.

NABIR (2001). *The NABIR strategic plan*. Natural and Accelerated Bioremediation Research Program, Berkeley Lab, US Department of Energy.

Neal, A. L., Amonette, J. E., Peyton, B. M. & Geesey, G. G. (2004). Uranium complexes formed at hematite surfaces colonized by sulfate-reducing bacteria. *Environ Sci Technol* **38**, 3019–3027.

Nealson, K. H. & Cox, B. L. (2002). Microbial metal-ion reduction and Mars: extraterrestrial expectations? *Curr Opin Microbiol* **5**, 296–300.

Normander, B., Christensen, B. B., Molin, S. & Kroer, N. (1998). Effect of bacterial distribution and activity on conjugal gene transfer on the phylloplane of the bush bean (*Phaseolus vulgaris*). *Appl Environ Microbiol* **64**, 1902–1909.

North, N. N., Dollhopf, S. L., Petrie, L., Istok, J. D., Balkwill, D. L. & Kostka, J. E. (2004). Change in bacterial community structure during in situ biostimulation of subsurface sediment cocontaminated with uranium and nitrate. *Appl Environ Microbiol* **70**, 4911–4920.

Nucifora, G., Chu, L., Misra, T. K. & Silver, S. (1989). Cadmium resistance from *Staphylococcus aureus* plasmid pI258 *cadA* gene results from a cadmium-efflux ATPase. *Proc Natl Acad Sci U S A* **86**, 3544–3548.

Ortiz-Bernad, I., Anderson, R. T., Vrionis, H. A. & Lovley, D. R. (2004). Vanadium respiration by *Geobacter metallireducens*: novel strategy for *in situ* removal of vanadium from groundwater. *Appl Environ Microbiol* **70**, 3091–3095.

O'Sullivan, D., Ross, R. P., Twomey, D. P., Fitzgerald, G. F., Hill, C. & Coffey, A. (2001). Naturally occurring lactococcal plasmid pAH90 links bacteriophage resistance and mobility functions to a food-grade selectable marker. *Appl Environ Microbiol* **67**, 929–937.

Pedersen, K. (2001). Diversity and activity of microorganisms in deep igneous rock aquifers of the fennoscandian shield. In *Subsurface Microbiology and Biogeochemistry*, pp. 97–139. Edited by J. K. Fredrickson & M. Fletcher. New York: Wiley-Liss.

Petrie, L., North, N. N., Dollhopf, S. L., Balkwill, D. L. & Kostka, J. E. (2003). Enumeration and characterization of iron(III)-reducing microbial communities from acidic subsurface sediments contaminated with uranium(VI). *Appl Environ Microbiol* **69**, 7467–7479.

Reed, D. T., Vojta, Y., Quinn, J. W. & Richmann, M. K. (1999). Radiotoxicity of plutonium in NTA-degrading *Chelatobacter heintzii* cell suspensions. *Biodegradation* **10**, 251–260.

Rhee, S. K., Liu, X., Wu, L., Chong, S. C., Wan, X. & Zhou, J. (2004). Detection of genes involved in biodegradation and biotransformation in microbial communities by using 50-mer oligonucleotide microarrays. *Appl Environ Microbiol* **70**, 4303–4317.

Ringelberg, D. B., Sutton, S. & White, D. C. (1997). Biomass, bioactivity and biodiversity: microbial ecology of the deep subsurface: analysis of ester-linked phospholipid fatty acids. *FEMS Microbiol Rev* **20**, 371–377.

Roane, T. M. (1999). Lead resistance in two bacterial isolates from heavy metal-contaminated soils. *Microb Ecol* **37**, 218–224.

Romine, M. F., Stillwell, L. C., Wong, K. K., Thurston, S. J., Sisk, E. C., Sensen, C., Gaasterland, T., Fredrickson, J. K. & Saffer, J. D. (1999). Complete sequence of a 184-kilobase catabolic plasmid from *Sphingomonas aromaticivorans* F199. *J Bacteriol* **181**, 1585–1602.

Rosen, B. P. (2002). Transport and detoxification systems for transition metals, heavy metals and metalloids in eukaryotic and prokaryotic microbes. *Comp Biochem Physiol A Mol Integr Physiol* **133**, 689–693.

Ruggiero, C. E., Boukhalfa, H., Forsythe, J. H., Lack, J. G., Hersman, L. E. & Neu, M. P. (2005). Actinide and metal toxicity to prospective bioremediation bacteria. *Environ Microbiol* **7**, 88–97.

Sani, R. K., Peyton, B. M. & Jandhyala, M. (2003). Toxicity of lead in aqueous medium to *Desulfovibrio desulfuricans* G20. *Environ Toxicol Chem* **22**, 252–260.

Silver, S. & Phung, L. T. (1996). Bacterial heavy metal resistance: new surprises. *Annu Rev Microbiol* **50**, 753–789.

Small, J., Call, D. R., Brockman, F. J., Straub, T. M. & Chandler, D. P. (2001). Direct detection of 16S rRNA in soil extracts by using oligonucleotide microarrays. *Appl Environ Microbiol* **67**, 4708–4716.

Smets, B. F., Rittmann, B. E. & Stahl, D. A. (1993). The specific growth rate of *Pseudomonas putida* PAW1 influences the conjugal transfer rate of the TOL plasmid. *Appl Environ Microbiol* **59**, 3430–3437.

Smets, B. F., Morrow, J. B. & Arango Pinedo, C. (2003). Plasmid introduction in metal-stressed, subsurface-derived microcosms: plasmid fate and community response. *Appl Environ Microbiol* **69**, 4087–4097.

Smith, W. L. & Gadd, G. M. (2000). Reduction and precipitation of chromate by mixed culture sulphate-reducing bacterial biofilms. *J Appl Microbiol* **88**, 983–991.

Stevens, T. O. & McKinley, J. P. (1995). Lithoautotrophic microbial ecosystems in deep basalt aquifers. *Science* **270**, 450–455.

Stillwell, L. C., Thurston, S. J., Schneider, R. P., Romine, M. F., Fredrickson, J. K. & Saffer, J. D. (1995). Physical mapping and characterization of a catabolic plasmid from the deep-subsurface bacterium *Sphingomonas* sp. strain F199. *J Bacteriol* **177**, 4537–4539.

Takai, K., Kobayashi, H., Nealson, K. H. & Horikoshi, K. (2003a). *Sulfurihydrogenibium subterraneum* gen. nov., sp. nov., from a subsurface hot aquifer. *Int J Syst Evol Microbiol* **53**, 823–827.

Takai, K., Mormile, M. R., McKinley, J. P., Brockman, F. J., Holben, W. E., Kovacik, W. P., Jr & Fredrickson, J. K. (2003b). Shifts in archaeal communities associated with lithological and geochemical variations in subsurface Cretaceous rock. *Environ Microbiol* **5**, 309–320.

Taroncher-Oldenburg, G., Griner, E. M., Francis, C. A. & Ward, B. B. (2003). Oligonucleotide microarray for the study of functional gene diversity in the nitrogen cycle in the environment. *Appl Environ Microbiol* **69**, 1159–1171.

Ulrich, G. A., Martino, D., Burger, K., Routh, J., Grossman, E. L., Ammerman, J. W. & Suflita, J. M. (1998). Sulfur cycling in the terrestrial subsurface: commensal interactions, spatial scales, and microbial heterogeneity. *Microb Ecol* **36**, 141–151.

Utgikar, V. P., Tabak, H. H., Haines, J. R. & Govind, R. (2003). Quantification of toxic and inhibitory impact of copper and zinc on mixed cultures of sulfate-reducing bacteria. *Biotechnol Bioeng* **82**, 306–312.

van Waasbergen, L. G., Balkwill, D. L., Crocker, F. H., Bjornstad, B. N. & Miller, R. V. (2000). Genetic diversity among *Arthrobacter* species collected across a heterogeneous series of terrestrial deep-subsurface sediments as determined on the basis of 16S rRNA and *recA* gene sequences. *Appl Environ Microbiol* **66**, 3454–3463.

Vepritskiy, A. A., Vitol, I. A. & Nierzwicki-Bauer, S. A. (2002). Novel group I intron in the tRNA^{Leu}(UAA) gene of a gamma-proteobacterium isolated from a deep subsurface environment. *J Bacteriol* **184**, 1481–1487.

Viamajala, S., Peyton, B. M., Sani, R. K., Apel, W. A. & Petersen, J. N. (2004). Toxic effects of chromium(VI) on anaerobic and aerobic growth of *Shewanella oneidensis* MR-1. *Biotechnol Prog* **20**, 87–95.

Wade, R. & DiChristina, T. J. (2000). Isolation of U(VI) reduction-deficient mutants of *Shewanella putrefaciens*. *FEMS Microbiol Lett* **184**, 143–148.

Whitman, W. B., Coleman, D. C. & Wiebe, W. J. (1998). Prokaryotes: the unseen majority. *Proc Natl Acad Sci U S A* **95**, 6578–6583.

Wong, D., Suflita, J. M., McKinley, J. P. & Krumholz, L. R. (2004). Impact of clay minerals on sulfate-reducing activity in aquifers. *Microb Ecol* **47**, 80–86.

Zhou, J. & Thompson, D. K. (2002). Challenges in applying microarrays to environmental studies. *Curr Opin Biotechnol* **13**, 204–207.

Zhou, J., Xia, B., Treves, D. S., Wu, L. Y., Marsh, T. L., O'Neill, R. V., Palumbo, A. V. & Tiedje, J. M. (2002). Spatial and resource factors influencing high microbial diversity in soil. *Appl Environ Microbiol* **68**, 326–334.

Zhou, J., Xia, B., Huang, H., Palumbo, A. V. & Tiedje, J. M. (2004). Microbial diversity and heterogeneity in sandy subsurface soils. *Appl Environ Microbiol* **70**, 1723–1734.

Biosilicification: the role of cyanobacteria in silica sinter deposition

Liane G. Benning,[1] Vernon R. Phoenix[2] and
Bruce W. Mountain[3]

[1]Earth and Biosphere Institute, School of Earth and Environment, University of Leeds, UK

[2]Molecular and Cellular Biology, University of Guelph, Canada

[3]Institute of Geological and Nuclear Sciences, Wairakei Research Centre, Taupo, New Zealand

INTRODUCTION

The contribution of micro-organisms to amorphous silica precipitation in modern geothermal hot-spring environments has been the topic of intense study in the last three to four decades. Here, we present a review on the field and laboratory studies that have specifically addressed bacterial silicification, with a special focus on cyanobacterial silicification. Studies related to the biogenic silicification processes in diatoms, radiolarians and sponges are not discussed, despite the fact that, in the modern oceans (which are undersaturated with respect to silica), the diagenetic 'ripening' of such biogenic silica controls the global silica cycle (Dixit et al., 2001). It is well-known that the amorphous silica in these organisms (particularly in size, shape and orientation) is controlled primarily by the templating functions of glycoproteins and polypeptides (e.g. silaffin and silicatein). For information on these issues, we refer the reader to the extensive reviews by Simpson & Volcani (1981), Kröger et al. (1997, 2000), Baeuerlein (2000), Perry & Keeling-Tucker (2000), Hildebrand & Wetherbee (2003) and Perry (2003). In addition, in terrestrial environments, a large pool of amorphous silica is cycled through higher plants (grasses and trees) that are believed to use silicification as a protection mechanism against pathogens and insects. Information on these processes can be found in the papers by Chen & Lewin (1969), Sangster & Hodson (1986) and Perry & Fraser (1991).

In this review, we will focus solely on microbial silicification processes, which have been studied extensively in active geothermal hot-spring environments. These are characterized by geothermal waters supersaturated with respect to amorphous silica

SGM symposium 65: Micro-organisms and Earth systems – advances in geomicrobiology.
Editors G. M. Gadd, K. T. Semple & H. M. Lappin-Scott. Cambridge University Press. ISBN 0 521 86222 1 ©SGM 2005

derived from water–rock interaction at depth. The link between microbes and the surface manifestations of sinter formation (both carbonate- and silica-based) was first documented in Yellowstone National Park at the end of the 19th century (Weed, 1889). Since that time, active geothermal systems have been studied widely, due to their importance as geothermal energy sources and as a proxy to understanding the formation of epithermal ore deposits, which constitute the deep-seated hydrothermal features beneath active systems. In the last quarter of the 20th century, a multitude of studies have been carried out to quantify the formation of silica and carbonate terraces in active systems, with a view towards understanding whether micro-organisms play an active or passive role in their formation (Walter *et al.*, 1972; Ferris *et al.*, 1986; Schultze-Lam *et al.*, 1995; Cady & Farmer, 1996; Konhauser & Ferris, 1996; Jones *et al.*, 1998, 2001; Konhauser *et al.*, 2001; Mountain *et al.*, 2003). Most of these studies have focused on the relationships between microbes and the resulting morphology and structure of modern siliceous sinters. They provide insights into the driving forces for sinter formation in contemporary deposits and are thus relevant to processes in Archaean and early Proterozoic settings, where microbes may have become encased and thus preserved as microfossils (Konhauser, 2000; Cady, 2001; Toporski *et al.*, 2002).

There have also been numerous experimental laboratory microbial silicification studies. In single-step batch experiments, it has been shown clearly that the affinity of aqueous silica to bind to a microbial surface is low, regardless of whether the micro-organisms are equilibrated with solutions supersaturated or undersaturated with respect to amorphous silica (Fein *et al.*, 2002; Phoenix *et al.*, 2003; Yee *et al.*, 2003). Such single-step experiments do not reliably mimic the processes leading to the significant silica accumulation observed in hot springs. Other experimental studies have used high concentrations of either organosilicon solvents, such as tetraethylorthosilicate, or inorganic silica concentrations and/or a variety of temperatures and pressures to induce silicification in the presence of micro-organisms and demonstrated that a complex interplay exists between the precipitation of silica and the formed textures and structures (e.g. Oehler & Schopf, 1971; Leo & Barghoorn, 1976; Walters *et al.*, 1977; Francis *et al.*, 1978; Ferris *et al.*, 1988; Westall *et al.*, 1995; Konhauser *et al.*, 2001; Toporski *et al.*, 2002; Mountain *et al.*, 2003). These studies offered important insights into the diagenetic-related fossilization processes and sinter textural development, but they cannot provide mechanistic data pertaining to molecular-level interactions between micro-organisms and silica accumulating in environments such as hot springs or the ancient oceans.

Many studies of microbial silicification in active hot springs have shown that silicification rates are rapid, but that the silicification process is controlled by purely abiotic driving forces [i.e. boiling, cooling, evaporation, waves and splash; see Mountain *et al.*

(2003) and references therein]. Microscopic analysis of silicified micro-organisms from active hot springs shows that the microbial surface may act as a nucleation site for silica precipitation (Schultze-Lam *et al.*, 1995; Konhauser & Ferris, 1996; Jones *et al.*, 2000; Phoenix *et al.*, 2000; Mountain *et al.*, 2003). Recent studies that exposed cyanobacteria repeatedly to freshly prepared, supersaturated, polymerizing silica solution (a pseudo-flow-through setting) have shown that extensive biomineralization, similar to that observed in hot springs, can be induced (Phoenix *et al.*, 2000; Benning & Mountain, 2004; Benning *et al.*, 2004a, b), with similar structures and textures to those observed in the field (Benning & Mountain, 2004; Fig. 1). Based on detailed microscopic and, more recently, spectroscopic measurements of samples from such laboratory experiments, it is now believed that the accumulation of amorphous silica on the surface of cyano-bacteria is controlled solely by silica nanoparticle aggregation, but that the contri-bution of the microbial sheaths or cell walls in this aggregation process is considered significant (see below; Phoenix *et al.*, 2000; Benning *et al.* 2004a, b). Lastly, recent studies by van der Meer *et al.* (2002) and Pancost *et al.* (2005) have shown that specific biomarker lipids can be preserved in natural modern silica sinters. Such biomarker studies can provide insight into the complex community structure of thermophilic and hyperthermophilic micro-organisms (including both archaea and bacteria) that are present during silica sinter formation. The knowledge of what biomolecules remain preserved in the rock record may provide a means to extrapolate back in time and thus to better understand processes in ancient rocks.

In the following pages, we describe the current understanding of the abiotic and biotic processes occurring in geothermal environments through a review of (i) the chemistry of silica and the thermodynamic and kinetic aspects of precipitation, (ii) the role of specific components of the microbial cell surface and (iii) the pathways of silica-colloid interaction and aggregation on cell surfaces. Only such a synergistic approach can provide a quantitative model for the reactions that drive microbial silicification and that lead ultimately to sinter formation and fossil microbial preservation.

THE CHEMISTRY OF SILICA

Soluble silica or monomeric orthosilicic acid (H_4SiO_4) is composed of a silicon atom coordinated tetrahedrally to four hydroxyl groups. Amorphous silica is defined as a non-stoichiometric, inorganic polymer made up of a mixture of SiO_2 and H_2O units in various ratios. Monomeric silica remains stable in solution at 25 °C, as long as its concentration is below the equilibrium concentration for amorphous hydrated silica [at 25 °C, approx. 100–125 parts per million (p.p.m.); Iler, 1979]. In most natural waters, the concentration of dissolved silica is low (between 1 and 100 µM; Treguer *et al.*, 1995) and, specifically in marine settings, the silica concentrations are regulated by the growth of diatoms and radiolarians. In contrast, in the surface expression of active geothermal

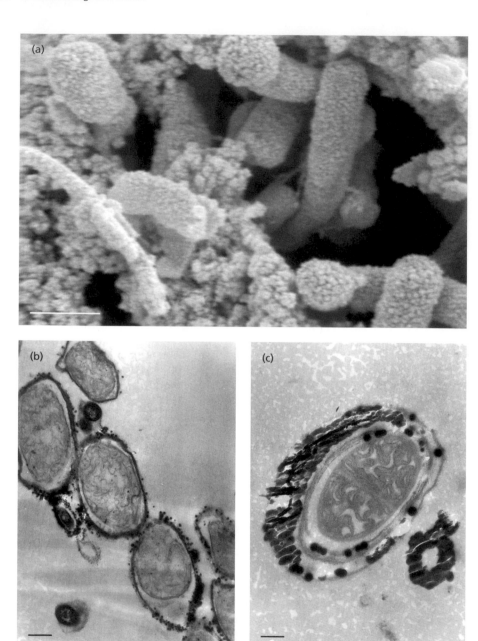

Fig. 1. Silicified microbes from the New Zealand geothermal hot springs. (a) High-resolution field emission gun scanning electron micrograph showing silica nanoparticles attached to microbial cells from the Rotokawa Geothermal Pool; bar, 500 nm. (b) Transmission electron micrograph of silicified micro-organisms from the Wairakei Geothermal Field; bar, 1 μm. (c) Transmission electron micrograph of fully silicified micro-organism from the Wairakei Geothermal Field; bar, 500 nm. Note the small (30–200 nm) silica particles that form aggregates on the surface of the bacterial sheath.

systems, where temperatures are higher (approx. 30–100 °C), the dissolved silica concentration in effluent solutions often exceeds the equilibrium solubility of amorphous silica. Total silica in hot-spring effluents can be as high as 1000 p.p.m. and this represents a level many times higher than saturation, even at 100 °C. Subsurface, geothermal fluids may be undersaturated with respect to amorphous silica but, upon reaching the surface, drastic changes in temperature and other physico-chemical parameters will induce the autocatalytic polycondensation/polymerization of silica monomers, because these changes will induce amorphous silica saturation to be surpassed. Field experimental determination of precipitation rates showed that the ratio of monomeric to polymeric silica in the effluent solution plays an important role in controlling silica-precipitation rates (Carroll *et al.*, 1998).

In a purely inorganic system, the polycondensation process follows a series of steps that progress from the polymerization of initial monomers to form dimers, trimers etc. and, finally, to the formation of highly soluble, critical nuclei of approximately 3 nm in size, which correspond to approximately 800–900 silicon atoms and have an approximate molecular mass of around 50 kDa (Iler, 1980; Perry, 2003; Icopini *et al.*, 2005). This initial step occurs via the condensation of two silicic acid molecules and the expulsion of water:

$$H_4SiO_4 + H_4SiO_4 \rightarrow (HO)_3Si\text{–}O\text{–}Si(OH)_3 + H_2O \qquad \text{(equation 1)}$$

Once the silicic acid molecules condense and Si–O–Si siloxane bonds form, cyclic ring structures will grow and other monomers, dimers etc. will react preferentially with these nuclei via Ostwald ripening. Dove & Rimstidt (1994) showed that the surface free energy of such a particle, σ [erg cm^{-2} (1 erg = 10^{-7} J)], can be linked with the bulk precipitate $\Delta G_{f(\text{bulk solid})}$ and the particle surface area (A, cm^2) to give the free energy of the particle, $\Delta G_{f(\text{particle})}$. This in turn can be expressed as a function of the particle radius (r, cm; assuming spherical morphology) and the molar volume (V_m, cm^3 mol^{-1}) via:

$$\Delta G_{f(\text{particle})} = [-4\pi r^3 \Delta G^0 / 2V_m] + [4 \times 10^{10} \pi r^2 \sigma] \qquad \text{(equation 2)}$$

and, from equation 2, an expression for the solubility of a single particle can be derived (Dove & Rimstidt, 1994). Alexander (1975) calculated the surface free energy for amorphous silica in equilibrium with a solution to be approximately 45 erg cm^{-2} (4·5 × 10^{-6} J cm^{-2}). This number increases dramatically with particle size and ordering of the silica phase, reaching a value of 120 erg cm^{-2} (12 × 10^{-6} J cm^{-2}) for quartz (Rimstidt & Cole, 1983), thus confirming that smaller and less ordered particles will dissolve as larger particles grow. Once formed, the critical nuclei will grow to form either large nanoparticles (from several hundred nanometres up to a micrometre) or will aggregate to form three-dimensional complex structures (Iler, 1979, 1980; Perry, 2003).

Based on the data of Gunnarsson & Arnórsson (2000), the equilibrium amorphous silica solubility at temperatures from 20 to 95 °C lies between 100 and approximately 330 p.p.m. Conventionally, the equation representing the equilibrium between silica and water is written as:

$$SiO_2 \text{ (s)} + 2H_2O \rightarrow H_4SiO_4 \text{ (aq)} \qquad \text{(equation 3)}$$

with the equilibrium constant K expressed as the activities (a) of the species:

$$K = a(H_4SiO_4)/a(SiO_2) \cdot a^2(H_2O) \qquad \text{(equation 4)}$$

This reaction is valid for all thermodynamic calculations, but fails to take into account kinetic effects, as well as the variations in the hydration states of silica. For example, the ratio between SiO_2 and H_2O in the aqueous species, as well as in the solid, often differs from the ideal 1 : 2, due to hydrogen-bonded waters of hydration in the stoichiometry. In addition, in equations 3 and 4, aqueous deprotonated and polynuclear species (e.g. $H_2SiO_4^{2-}$, $H_3SiO_4^-$ and $H_6Si_2O_7^{2-}$) are not taken into account although, in some cases, such species may contribute to up to 40 % of the dissolved silica (Aplin, 1987). Equation 3 is particularly important in geothermal systems where the geothermal solutions are supersaturated with respect to amorphous silica, and polymerization and precipitation are induced due to changes in physical and hydrodynamic conditions.

From a thermodynamic point of view, the precipitation of amorphous silica is driven by cooling, evaporation, boiling, solution mixing and changes in pH. These factors all strongly affect the saturation level of amorphous silica. For a general precipitation rate, $Rate_{ppt}$, an equation of the type:

$$Rate_{ppt} = -d[nH_4SiO_4]/dt = -A \times k_{ppt}[a(SiO_2) \cdot a^2(H_2O)] \qquad \text{(equation 5)}$$

can be written, where n is the no. moles H_4SiO_4, A is the interfacial area (in m^2) and k_{ppt} is the pH-dependent precipitation-rate constant (Iler, 1979; Rimstidt & Barnes, 1980; Carroll et al., 1998). In solutions that are close to saturation, a nucleation barrier that needs to be surpassed for the first nuclei to form inhibits the precipitation process. Nielsen (1959) modelled the growth of such nuclei and showed that the flux of monomers towards such nuclei is related to the collision rate, the Boltzman constant, temperature and the free energy of formation of a critical nucleus. Because quartz has a higher surface free energy than amorphous silica and its nucleation is inhibited, it follows that the nucleation and growth of amorphous silica in geothermal systems are more favoured.

In most geothermal systems, the amorphous silica that precipitates is composed of opal-A, a phase that displays varying degrees of ordering of the SiO_4 rings, as well as varying amounts of structural SiO_2 units and degrees of hydration. Opal-A (nominally $SiO_2 . nH_2O$) is a poorly ordered, highly hydrated phase that displays only one weak, broad Bragg diffraction band. Other silica phases observed in geothermal silica-dominated systems are considered good indicators of a diagenetic ageing/altering process. During this ageing, opal-A is transformed into opal-CT, opal-C, moganite, cristobalite, chalcedony and ultimately quartz (Herdianita *et al.*, 2000). The main factors influencing this transformation to more stable counterparts are time, re-equilibration with high-temperature or high-pH solutions, dehydroxylation/drying cycles or diagenetic recrystallization. Opal-A can contain between 1 and 13 % water in its structure; this water is present either as network water or as liquid water in interstices bound to either internal silanols or defect sites of surface silanols (Langer & Flörke, 1974; Knauth & Epstein, 1982). During this diagenetic transformation/ripening process, this water is expelled gradually and this is accompanied by a gradual change in *d*-spacing for the main Bragg peak from 4·12 to 4·04 Å (0·412–0·404 nm). This process has been used to derive an indicator of structural ripening, as well as a measure of depth of burial and age. During the ageing and transformation process, water content drops and particle density increases to $2·3 \text{ g cm}^{-3}$ (from as low as $1·5 \text{ g cm}^{-3}$). At the same time, porosity (initially between 35 and 60 %) can be reduced by more than half to a value below 30 % [see Herdianita *et al.* (2000) and references therein].

In an effluent solution, the saturation state and thus the precipitation rate of amorphous silica are dependent on a variety of parameters that include thermal gradients, time, changes in pH, concentration of inorganic cations (i.e. Al and trace elements), organics and ionic strength. Furthermore, this rate depends on the presence of nucleation sites/surfaces, as well as hydrodynamic parameters such as evaporation, waves, splash etc. As a result, the precipitated, amorphous silica phases will be highly variable from site to site and the resulting morphology and textures will depend strongly on these precipitation regimes (with resulting morphologies of the precipitated silica varying from nanometre-sized spheroidal particles to flat sheets to bulk silica).

The first precipitated opal-A is usually made up of nanometre-sized spheroids that are later filled in by silica cement to form bulk silica structures. Its formation is a dynamic process and even 'fresh' sinter features can appear homogeneous; thus, they are sometimes difficult to distinguish from aged sinters. This has been a major stumbling block when purely morphological and structurally preserved biosignatures observed in modern sinters have been used to relate and extrapolate to processes in ancient rocks, where subsequent diagenetic or metamorphic processes have homogenized and altered the structures, mineral ordering and composition. Specifically, in ancient rocks, the preservation of biogenic material is hampered by the fact that, in most cases, the only preserved features are the mould or casings surrounding the microbes, whilst the cell walls or sheaths have been lost. However, recent geothermal sinters have revealed that some specific biomarkers (specifically, bacterial and archaeal lipids) can be preserved. This may be the approach to elucidate the preservation of biota in ancient rocks lacking unequivocal morphological indicators (van der Meer *et al.*, 2002; Toporski *et al.*, 2002; Pancost *et al.*, 2005).

CYANOBACTERIAL SURFACE PROPERTIES AND FUNCTION

The structure and composition of cyanobacterial cell walls display a number of characteristics that are atypical of Gram-negative bacteria. They exhibit a thick, highly cross-linked peptidoglycan layer (similar to that of Gram-positive organisms) that makes the cell wall notably stronger (Drews & Weckesser, 1982; Hoiczyk & Hansel, 2000). Additionally, biomolecules commonly found in the cyanobacterial outer membrane, such as atypical fatty acids and carotenoids, are uncommon in other Gram-negative bacteria (Schrader *et al.*, 1981; Resch & Gibson, 1983; Jurgens & Mantele, 1991).

As with all bacteria, each component of the cyanobacterial cell envelope plays a specific role. The inner component, the cytoplasmic membrane, behaves as a highly selective barrier, allowing vital nutrients into the cell and excreting waste material out of the cell. The outer component of the Gram-negative cell wall is the outer membrane (an asymmetrical bilayer composed of lipopolysaccharide and phospholipid). This bilayer also acts as a selective barrier, facilitating the transport of low-molecular-mass molecules via proteins known as porins. Housed between the two membranes are the peptidoglycan, which provides rigidity, strength and shape, and the periplasm, which contains functionally important enzymes. Significantly, these components of the cell envelope are highly functional, complex and sensitive. One may then ask how these metabolically vital, and in some cases delicate, layers would respond to encrustation in silica precipitates.

For some Gram-negative bacteria, the outer membrane is the outermost component of the organism and it acts as the interface between the cell and the external environment. In other cases, the organism surrounds itself in an extracellular polysaccharide capsule or sheath. The structure of this extracellular layer can vary considerably, ranging from

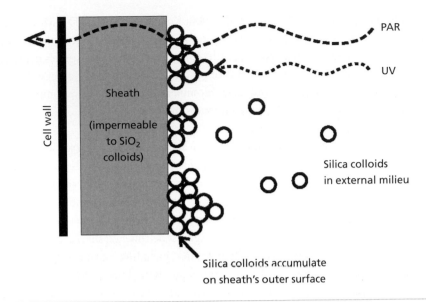

Fig. 2. Summary schematic illustrating extracellular silicification of an ensheathed cyanobacterium. Silica colloids accumulate on the outer surface of the sheath, due to the impermeability of the sheath to 'large' particles. In addition, it is shown how silica inhibits damaging UV light from reaching the cell, whilst the photosynthetically active radiation (PAR) can pass through the silica layer with less attenuation.

diffuse to dense and fibrous. Dense, fibrous polysaccharide layers known as sheaths are particularly common in cyanobacteria (e.g. Rippka *et al.*, 1979). Moreover, the cyano-bacterial sheath is known to be devoid of metabolically vital components and it is thus likely that the organism can withstand a higher degree of damage to this layer than the rest of the cell envelope. The exact role of the cyanobacterial sheath is not well under-stood but, as it encloses the more delicate components of the cell envelope, one of its primary objectives may be to help prevent damage to these components. It has been shown that a coating of extracellular polysaccharide can protect bacteria against dehydration (Dudman, 1977; Scott *et al.*, 1996; Hoiczyk, 1998; Tsuneda *et al.*, 2003) and predation (Dudman, 1977), or it can aid in adhesion to a solid substrate (Dudman, 1977; Scott *et al.*, 1996). More specifically, some cyanobacterial sheaths can contain the UV light-absorbing pigment scytonemin, aiding cyanobacterial resistance to solar radiation (Garcia-Pichel & Castenholz, 1991). Of particular relevance to this chapter is the ability of the cyanobacterial sheath to protect the cell from detrimental bio-mineralization (Phoenix *et al.*, 2000; Benning *et al.*, 2004a) and to aid in the aggregation of silica nanoparticles (Benning *et al.*, 2004b).

Sheathed cyanobacteria are found in abundance in hot-spring systems, where it has been shown that silica accumulation is restricted to the outer surface of the sheath on living cyanobacteria (Fig. 1; Phoenix *et al.*, 2000; Konhauser *et al.*, 2001; Mountain *et al.*, 2003). This is likely to occur because the polysaccharide meshwork of the sheath enables it to act as a filter against colloidal silica (Fig. 2). Permeability studies demonstrated that the sheath of *Calothrix* sp. was impermeable to particles of at least 11 nm in diameter, thus preventing the colloids from biomineralizing the sensitive components of the cell wall [Phoenix *et al.* (2000) and references therein].

Interestingly, in the Archaean oceans, which were enriched in silica (Siever, 1992) and inhabited by cyanobacteria, silica biomineralization was likely to have occurred (Cloud, 1965). This is particularly true for the shallow waters, in which intermittently exposed environments were inhabited by stromatolitic communities and evaporation may have controlled silica precipitation. It is possible that the sheath developed/evolved in the early oceans as a response and protection against detrimental silica accumulation on the cell wall. This is supported by several studies, including transmission electron microscope- and synchrotron-based Fourier-transform IR analysis, which have demonstrated that, in response to increased silica exposure, the sheath of cyanobacteria thickens (Phoenix *et al.*, 2000; Benning *et al.*, 2004a, b). Again, this indicates that the sheath can act as a protective layer against silicification. Naturally, when cyanobacteria are exposed continuously to supersaturated solutions of silica, silicification eventually becomes too extensive and this may be detrimental to the cyanobacteria. Phoenix *et al.* (2000) have shown that even quite thick silica crusts (approx. 5 μm thick for a 10 μm diameter cell) did not appear to be detrimental to the cells. However, whether there is a maximum amount of extracellular silicification that cyanobacteria can withstand has yet to be determined.

One mechanism to overcome extensive silicification may be the release of transient motile phases (hormogonia) (Herdman & Rippka, 1988) from the ends of heavily encrusted filaments and this may provide a pathway for survival. Benning *et al.* (2004a, b) have followed bacterially mediated silica accumulation both *in situ* and *in vivo* via the changes in the IR signature induced by the increase in silica concentration on the organic framework of single bacterial cells. This approach allowed the quantification of the actual and not the apparent bacterial silica-accumulation process, and they showed that the role of the sheath is twofold. Initially, the cells react to exposure to a silica-rich solution by producing more sheath polysaccharide as protection. As this thicker sheath acts as a good ('sticky') substrate for further inorganically precipitated silica-colloid aggregation, silica precipitation is enhanced, with detrimental effects to the cell.

This ability of cyanobacteria to survive and grow continually, despite extensive extracellular silicification in modern hot springs and presumably also in the ancient oceans, may have provided additional advantages to the microbes. This is because amorphous silica biomineralization has been demonstrated to act as an effective screen against UV radiation (Phoenix *et al.*, 2001). This study has shown that damaging wavelengths of UV-B and particularly UV-C are absorbed strongly by amorphous silica, whilst photosynthetically active radiation (400–700 nm) will pass through the silica with significantly less adsorption (Fig. 2). This process enables cyanobacteria to photosynthesize in environments subjected to elevated UV, a protective mechanism particularly relevant to the Archaean (Phoenix *et al.*, 2001), where highly detrimental levels of UV irradiated the Earth's surface (Kasting, 1987). Furthermore, Heijnen *et al.* (1992) have shown that habitation of micro-niches in bentonite clays can protect other bacterial forms from predation by grazing protozoa and, thus, silica encrustation may similarly protect cyanobacteria by making them inaccessible or inedible to protozoa. Biomineralization also plays a key role in the formation of siliceous stromatolitic communities, both modern and ancient, by increasing their structural integrity and thus longevity (Konhauser *et al.*, 2001). It has also been speculated that, because the amorphous silica matrix is highly hydrated, it may afford the organisms an additional protection layer against dehydration. Interestingly, these potential advantages are similar to the functions of the sheath, and it thus seems that the sheath and enshrouding silica biominerals may work collectively to protect the organisms within.

CYANOBACTERIAL BIOMINERALIZATION PATHWAYS AND COLLOID AGGREGATION

Benning *et al.* (2004a, b) have followed the processes leading to cyanobacterial silicification by using *in situ* IR microspectroscopy and imaging and have quantified the complex interplay between the cyanobacterial cell components, specifically the sheath, and the polymerizing silica solution. The progression of nucleation, growth and aggregation of nanometre-sized silica particles and their effect on cyanobacterial feedback have been described as a three-stage progression. In the first stage, in response to the presence of polymerizing silica, the cyanobacteria will increase the formation of new polysaccharide polymers, i.e. they will thicken their sheath. Concomitantly, silica will form branched polymers that, upon collapse, will bind to the carbon backbone of the hydrated polysaccharide sheath via hydrogen bridges. This step can be expressed as:

$$[>ROH] + [>ROH] \rightarrow 2[>ROH] \qquad \text{(equation 6a)}$$

and

$$[>ROH] + [\equiv Si-OH] \rightarrow [\equiv Si-OR<] + H_2O \qquad \text{(equation 6b)}$$

where >ROH represents the surface-hydroxylated sugar polymer in the sheath and $\equiv$Si–OH is the monomeric silica attached to a surface. Equation 6(b) implies a possible site-specific silica accumulation, with the silica monomers bound via hydrogen bridges to another OH-containing radical. The sheath polysaccharides are the obvious candidates for this step. Benning *et al.* (2004b) used a kinetic approach to show that this reaction proceeds via a diffusion-limited mechanism (see below), in which polymerizing silica units in the supersaturated aqueous environment begin to coalesce and aggregate on the 'fresh' microbial-sheath surface. This process is enhanced once a silane group is attached to the bacterial sheath via hydrogen bonds and, thus, a further increase in Si load may lead to the formation of a thin, fully hydrated silica network. Subsequently, other silane bonds may form independently of the sheath and this process will become uncoupled from the formation of the silica–carbohydrate hydrogen bonds. This can be expressed as:

$$[\equiv Si-OR<] + [\equiv Si-OH] \rightarrow [\equiv Si-O-Si\equiv] + [ROH<] \qquad \text{(equation 7)}$$

At this stage, the formation of additional polysaccharides will no longer compete with the polymerization of silica, a fact supported by the change in IR spectra, which become dominated by the more ionic Si–O–Si bonds. The last stage is the formation of inorganic silane bonds. This process has been shown to be governed by a reaction-limited process that leads to the growth of purely inorganic Si–O–Si bonds via the formation of an oxo bridge (Si–O–Si), whilst one water molecule is expelled. This is similar to the purely inorganic process described in equation 1. When surface attachment is considered, this step can be described via:

$$[\equiv Si-OH] + [OH-Si\equiv] \rightarrow [\equiv Si-O-Si\equiv] + H_2O \qquad \text{(equation 8)}$$

In this way, a silica network made of corner-sharing $[SiO_4]$ tetrahedra is obtained when all Si–O groups have reacted and the critical silica nuclei have formed. Their further growth and aggregation will follow and no other connection to the polysaccharide sheath is needed.

In general, for colloid or polymer growth and aggregation, two restrictive regimes have been defined: diffusion-limited aggregation (DLA) and reaction-limited aggregation (RLA) (Everett, 1988; Hunter, 1996; Jamtveit & Meakin, 1999). In the diffusion-limited case, the limiting step is the movement of two polymer units toward each other prior to encounter and formation of a cluster (or aggregate). In such reactions, monomers or oligomers collide and combine instantaneously, producing a relatively porous aggregate. For the formation of critical nuclei of silica, the DLA process has been confirmed experimentally (Lin *et al.*, 1990; Martin *et al.*, 1990; Pontoni *et al.*, 2002). For polysaccharide polymers, however, such data are unavailable, although Rees (1977) showed that glucose polymers – specifically amylose – grow by a DLA process. On the other hand, in RLA, the concentrations of the encountered reactant pairs are maintained at equilibrium and thus condensation occurs more slowly. In addition, a significant repulsive barrier exists, such that the 'sticking probability' upon oligomer–oligomer interaction is small (Everett, 1988; Gedde, 1995). For silica, the RLA process results in a more compact aggregate structure during slow condensation (Martin, 1987; Lin *et al.*, 1990).

Based on theoretical calculations for nucleation, crystallization, growth and aggregation of mineral phases and organic polymers, Hulbert (1969) and Gedde (1995) have derived hypothetical constants for the mechanistic constant *n*, which represents a parameter that is related to specific mechanisms and geometric shape of the final mineral particles or polymer. The two types of mechanisms (DLA and RLA) and several different shapes (needles, plates, spheres, fibres, sheaths etc.) have been investigated and values for *n* have been deduced. In general, *n* increases with increasing 'dimensionality' of the resulting particle/polymer and, in heterogeneous systems, a change in mechanism often occurs. Hulbert (1969) and Gedde (1995) have concluded that a particle/polymer of low geometry (one- or two-dimensional; e.g. fibre or sheath), forming via a DLA process, will have values of *n* varying between 0·5 and 2, with the highest values representing two-dimensional growth. On the other hand, if a spherical (three-dimensional) entity grows or aggregates via a DLA mechanism, the value of *n* will vary between 1·5 and 2·5, with the higher numbers indicating a switch to an RLA mechanism. Lastly, if the same three-dimensional spherical entity grows or aggregates via a purely RLA mechanism, values of 3–4 are expected for *n*.

It is well-known that the polysaccharide components of the sheath of *Calothrix* sp. (composed primarily of neutral sugars; Weckesser *et al.*, 1988) is usually found in the form of amylose. Amylose is a linear polymer of glucose units joined by repeating covalent C–O bonds, and it normally forms complex aggregates of linear geometry (Rees, 1977). In the cyanobacteria silica system, a low geometric ordering of newly formed polymers was corroborated by Benning *et al.* (2004b), who have derived *n* values for the first step in polysaccharide growth of 0·8–1·1, thus confirming a one-dimensional DLA growth for the carbohydrate polymers. For the second step, which is dominated by the attachment of the formed silica nuclei to the cyanobacterial sheath, Benning *et al.* (2004b) derived a value for *n* of 1·8–2·2, indicating a DLA mechanism for three-dimensional growth. Finally, for the last step, an *n* value of 3·5–3·8 was derived, clearly indicating three-dimensional growth via an RLA mechanism. It needs to be noted, however, that polysaccharide and silica polymers can both form structures of mixed or changing geometry during growth or aggregation and that the data derived by Benning *et al.* (2004b), which were based on a kinetic approach, may not fully describe all steps in this complex process. This is particularly true because, in most polymers and colloid systems, the nucleation and aggregation behaviour in solution is affected strongly by pH, ionic strength, temperature, organic concentration and type. However, for silica nucleation and growth, the aggregation steps quantified in the laboratory are expected to be similar to processes in modern geothermal hot springs and thus can be used as analogues to model processes in natural environments.

CONCLUSIONS

Cyanobacteria are a major group of phototrophic prokaryotes that play an important role in the textural development of silica sinters in modern geothermal environments. Processes that are analogous to those observed in modern hot springs may also have been active during the fossilization of microbes and the formation of siliceous stromatolites in the Archaean. These, in turn, can provide a proxy for the biogeochemical conditions of the early biosphere. A large number of field observations and experimental laboratory studies, as well as a few molecular-dynamic simulations, have led to the conclusion that, in active geothermal hot springs, cyanobacteria play no active role in the initial silica polymerization. It was shown that covalent bonds between silica and bacterial cell-wall or sheath components are not favoured and that the nucleation of silica from supersaturated aqueous solutions is driven by purely inorganic polycondensation reactions, which are strongly pH-, ionic strength-, temperature- and saturation state-dependent. Nevertheless, many field and experimental microscopic observations showed clear evidence of a link between silica sinter structures/textures and micro-organisms via the deposition of silica nanospheres onto the microbial surfaces. Ultimately, this process promotes the incorporation of micro-organisms into the sinter structure and leads to the preservation of microbial colonies as fossils. However, despite the large variety of studies carried out so far, the question remains as to the exact role of the microbial surface in the processes that lead to silica precipitation. We argue here that the formation of silica sinters is a multi-step process that is governed primarily by inorganically driven polycondensation of silica monomers and the formation of silica nanoparticles. This is followed by the microbially enhanced aggregation of the silica nanospheres into larger assemblages. In the first step, experimental and theoretical evidence indicates that polymerization of silica monomers leads to the formation of branching clusters that eventually collapse to form a spherical particle. Although the parameters and mechanisms controlling this collapse are unclear, some evidence indicates that it may be the dehydroxylation of silanol or silane clusters. When cyanobacteria and their complex surface structures are present, it appears that the polycondensation rates and, thus, silica nanoparticle nucleation rates, are not enhanced. In addition, and more importantly, it is believed that the precipitation of silica does not affect cyanobacterial metabolism or duplication rates. However, cyanobacteria do react by increasing the amount of extracellular polysaccharide that they produce. Once silicification is advanced and thus unavoidable, the thicker polysaccharide sheath will enhance the aggregation of the inorganically nucleated silica nanoparticles into larger silica assemblages. This process occurs while they are alive, but continued silicification leads to cell death, lysis and finally fossilization.

The surface features seen in active geothermal systems are often regarded as ideal model systems for study, because they can shed light onto processes occurring in the shallow-subsurface portions of the geothermal systems that are linked to deep-seated epithermal ore deposits. In the streams, pools and sinters forming in modern geothermal systems, a vast array of mesophilic and thermophilic organisms thrive at high temperatures and varied pH, as well as high toxic-metal concentrations that are usually detrimental to microbial growth. The knowledge gained from studying such communities and their interaction with the minerals precipitating from the super-saturated solutions can give valuable insights into processes of biomineralization. In addition, our understanding of processes related to the evolution of early life forms in the Archaean and early Proterozoic has been extrapolated from observations (both structural and chemical) of bacterial–mineral interactions in modern hot-spring environments or from morphological observations of Precambrian silicified micro-fossils and stromatolites.

From field and laboratory observations, reaction pathways for the formation of silica sinters in modern or ancient hot springs have been determined. The abiotic versus biotic components of the silica biomineralization reaction, and thus the role of micro-organisms in this process, were defined. These studies have shown that the silicification process follows a series of interlinked but unavoidable steps, starting with the microbes reacting to highly supersaturated silica solutions by increasing their production of exopolymeric material. Simultaneously with this process, but driven inorganically, silica polycondensation is proceeding, with monomers condensing to dimers, trimers etc., leading to the formation of critical silica nanospheres. The newly formed, 'sticky', exopolymeric sugars do not affect this polymerization, but will subsequently enhance the aggregation of the inorganically formed nanospheres on the microbial surface. Finally, this will invariably lead to the full silicification of the organic microbial frameworks and the formation of microfossils that can thus be preserved in modern silica sinter environments and that provide a modern analogue to processes in the ancient past.

ACKNOWLEDGEMENTS

Support for L. G. B. from the Leverhulme Trust (ref. #F/00122/F) and the Natural Environment Research Council (GR9/04623) is gratefully acknowledged. V. R. P. acknowledges the kind support of Professor Terry Beveridge. B. W. M. acknowledges funding from the Foundation for Research in Science and Technology (contract C05X0201) and from the NSOF Extremophiles Programme.

REFERENCES

Alexander, G. B. (1975). The effect of particle size on the solubility of amorphous silica in water. *J Phys Chem* **61**, 1563–1564.

Aplin, K. R. (1987). The diffusion of dissolved silica in dilute aqueous solution. *Geochim Cosmochim Acta* **51**, 2147–2151.

Baeuerlein, E. (editor) (2000). *Biomineralization: from Biology to Biotechnology and Medical Application*. Weinheim, Germany: Wiley-VCH.

Benning, L. G. & Mountain, B. W. (2004). The silicification of microorganisms: a comparison between *in situ* experiments in the field and in the laboratory. In *Proceedings of the 11th International Symposium on Water–Rock Interactions*, pp. 3–10. Edited by R. Wanty, R. Seal & A. A. Balkema. London: Taylor & Francis.

Benning, L. G., Phoenix, V. R., Yee, N. & Tobin, M. J. (2004a). Molecular characterization of cyanobacterial silicification using synchrotron infrared micro-spectroscopy. *Geochim Cosmochim Acta* **68**, 729–741.

Benning, L. G., Phoenix, V. R., Yee, N. & Konhauser, K. O. (2004b). The dynamics of cyanobacterial silicification: an infrared micro-spectroscopic investigation. *Geochim Cosmochim Acta* **68**, 743–757.

Cady, S. L. (2001). Paleobiology of the Archean. *Adv Appl Microbiol* **50**, 3–35.

Cady, S. L. & Farmer, J. D. (1996). Fossilization processes in siliceous thermal springs: trends in preservation along thermal gradients. *Ciba Found Symp* **202**, 150–173.

Carroll, S., Mroczek, E., Alai, M. & Ebert, M. (1998). Amorphous silica precipitation (60 to 120 °C): comparison of laboratory and field rates. *Geochim Cosmochim Acta* **62**, 1379–1396.

Chen, C. H. & Lewin, J. (1969). Silicon as a nutrient element for *Equisetum arvense*. *Can J Bot* **47**, 125–131.

Cloud, P. E. (1965). Significance of the Gunflint (Precambrian) microflora. *Science* **148**, 27–35.

Dixit, S., Van Capellen, P. & van Bennekom, A. J. (2001). Processes controlling solubility of biogenic silica and pore water build-up of silicic acid in marine sediments. *Mar Chem* **73**, 333–352.

Dove, P. M. & Rimstidt, J. D. (1994). Silica–water interactions. In *Silica: Physical Behaviour, Geochemistry, and Materials Applications* (Reviews in Mineralogy 29), pp. 259–308. Edited by P. J. Heaney, C. T. Prewitt & G. V. Gibbs. Washington, DC: Mineralogical Society of America.

Drews, G. & Weckesser, J. (1982). Function, structure and composition of cell walls and external layers. In *The Biology of Cyanobacteria*, pp. 333–357. Edited by N. G. Carr & B. A. Whitton. Oxford: Blackwell Scientific Publications.

Dudman, W. F. (1977). The role of surface polysaccharides in natural environments. In *Surface Carbohydrates of the Prokaryotic Cell*, pp. 357–414. Edited by I. Sutherland. London: Academic Press.

Everett, D. H. (1988). *Basic Principles of Colloid Science*. London: Royal Society of Chemistry.

Fein, J. B., Scott, S. & Rivera, N. (2002). The effect of Fe on Si adsorption by *Bacillus subtilis* cell walls: insights into non-metabolic bacterial precipitation of silicate minerals. *Chem Geol* **182**, 265–273.

Ferris, F. G., Beveridge, T. J. & Fyfe, W. S. (1986). Iron–silica crystallite nucleation by bacteria in a geothermal sediment. *Nature* **320**, 609–611.

Ferris, F. G., Fyfe, W. S. & Beveridge, T. J. (1988). Metallic ion binding by *Bacillus subtilis*: implications for the fossilization of microorganisms. *Geology* **16**, 149–152.

Francis, S., Margulis, L. & Barghoorn, E. S. (1978). On the experimental silicification of microorganisms. II. On the time of appearance of eukaryotic organisms in the fossil record. *Precambrian Res* **6**, 65–100.

Garcia-Pichel, F. & Castenholz, R. W. (1991). Characterization and biological implications of scytonemin, a cyanobacterial sheath pigment. *J Phycol* **27**, 395–409.

Gedde, U. W. (1995). *Polymer Physics*. London: Chapman & Hall.

Gunnarsson, I. & Arnórsson, S. (2000). Amorphous silica solubility and the thermodynamic properties of $H_4SiO_4^\circ$ in the range of 0° to 350 °C at P_{sat}. *Geochim Cosmochim Acta* **64**, 2295–2307.

Heijnen, C. E., Hok-A-Hin, C. H. & van Veen, J. A. (1992). Improvements to the use of bentonite clay as a protective agent, increasing survival levels of bacteria introduced into soil. *Soil Biol Biochem* **24**, 533–538.

Herdianita, N. R., Browne, P. R. L., Rodgers, K. A. & Campbell, K. A. (2000). Mineralogical and morphological changes accompanying ageing of siliceous sinter. *Mineral Deposita* **35**, 48–62.

Herdman, M. & Rippka, R. (1988). Cellular differentiation: hormogonia and baeocytes. *Methods Enzymol* **167**, 232–242.

Hildebrand, M. & Wetherbee, R. (2003). Components and control of silicification in diatoms. *Prog Mol Subcell Biol* **33**, 11–57.

Hoiczyk, E. (1998). Structural and biochemical analysis of the sheath of *Phormidium uncinatum*. *J Bacteriol* **180**, 3923–3932.

Hoiczyk, E. & Hansel, A. (2000). Cyanobacterial cell walls: news from an unusual prokaryotic envelope. *J Bacteriol* **182**, 1191–1199.

Hulbert, S. F. (1969). Models for solid-state reactions in powder compacts: a review. *J Br Ceramic Soc* **6**, 11–20.

Hunter, R. J. (1996). *Introduction to Modern Colloid Science*. New York: Oxford University Press.

Icopini, G. A., Brantley, S. L. & Heaney, P. J. (2005). Kinetics of silica oligomerization and nanocolloid formation as a function of pH and ionic strength at 25 °C. *Geochim Cosmochim Acta* **69**, 293–303.

Iler, R. K. (1979). *The Colloid Chemistry of Silica and Silicates*. Ithaca, NY: Cornell University Press.

Iler, R. K. (1980). Isolation and characterization of particle nuclei during the polymerization of silicic acid to colloidal silica. *J Colloid Interface Sci* **75**, 138–148.

Jamtveit, B. & Meakin, P. (editors) (1999). *Growth, Dissolution and Pattern Formation in Geosystems*. Dordrecht: Kluwer.

Jones, B., Renaut, R. W. & Rosen, M. R. (1998). Microbial biofacies in hot-spring sinters: a model based on Ohaaki Pool, North Island, New Zealand. *J Sediment Res* **68**, 413–434.

Jones, B., Renaut, R. W. & Rosen, M. R. (2000). Stromatolites forming in acidic hot-spring waters, North Island, New Zealand. *Palaios* **15**, 450–475.

Jones, B., Renaut, R. W. & Rosen, M. R. (2001). Taphonomy of silicified filamentous microbes in modern geothermal sinters – implications for identification. *Palaios* **16**, 580–592.

Jurgens, U. J. & Mantele, W. (1991). Orientation of carotenoids in the outer membrane of *Synechocystis* PCC 6714 (cyanobacteria). *Biochim Biophys Acta* **1067**, 208–212.

Kasting, J. F. (1987). Theoretical constraints on oxygen and carbon dioxide concentrations in the Precambrian atmosphere. *Precambrian Res* **34**, 205–229.

Knauth, L. P. & Epstein, S. (1982). The nature of water in hydrous silica. *Am Mineral* **67**, 510–520.

Konhauser, K. O. (2000). Hydrothermal bacterial biomineralization: potential modern-day analogues for banded iron-formations. In *Marine Authigenesis: From Global to Microbial* (Society for Sedimentary Geology Special Publication no. 66), pp. 133–145. Edited by C. R. Glenn, J. Lucas and L. Prévôt. Tulsa, OK: Society for Sedimentary Geology.

Konhauser, K. O. & Ferris, F. G. (1996). Diversity of iron and silica precipitation by microbial mats in hydrothermal waters, Iceland: implications for Precambrian iron formations. *Geology* **24**, 323–326.

Konhauser, K. O., Phoenix, V. R., Bottrell, S. H., Adams, D. G. & Head, I. M. (2001). Microbial–silica interactions in Icelandic hot spring sinter: possible analogues for some Precambrian siliceous stromatolites. *Sedimentology* **48**, 415–433.

Kröger, N., Lehmann, G., Rachel, R. & Sumper, M. (1997). Characterization of a 200-kDa diatom protein that is specifically associated with a silica-based substructure of the cell wall. *Eur J Biochem* **250**, 99–105.

Kröger, N., Deutzmann, R., Bergsdorf, C. & Sumper, M. (2000). Species-specific polyamines from diatoms control silica morphology. *Proc Natl Acad Sci U S A* **97**, 14133–14138.

Langer, K. & Flörke, O. W. (1974). Near infrared absorption spectra (4000–9000 cm^{-1}) of opals and the role of "water" in these $SiO_2 . nH_2O$ minerals. *Fortschr Miner* **52**, 17–51.

Leo, R. F. & Barghoorn, E. S. (1976). Silicification of wood. *Bot Mus Leafl Harv Univ* **25**, 1–47.

Lin, M. Y., Lindsay, H. M., Weitz, D. A., Ball, R. C., Klein, R. & Meakin, P. (1990). Universal reaction-limited colloid aggregation. *Phys Rev A* **41**, 2005–2020.

Martin, J. E. (1987). Slow aggregation of colloidal silica. *Phys Rev A* **36**, 3415–3426.

Martin, J. E., Wilcoxon, J. P., Schaefer, D. & Odinek, J. (1990). Fast aggregation of colloidal silica. *Phys Rev A* **41**, 4379–4391.

Mountain, B. W., Benning, L. G. & Boerema, J. A. (2003). Experimental studies on New Zealand hot spring sinters: rates of growth and textural development. *Can J Earth Sci* **40**, 1643–1667.

Nielsen, A. E. (1959). The kinetics of crystal growth in barium sulfate precipitation. II. Temperature dependence and mechanism. *Acta Chem Scand* **13**, 784–802.

Oehler, J. H. & Schopf, J. W. (1971). Artificial microfossils: experimental studies of permineralization of blue-green algae in silica. *Science* **174**, 1229–1231.

Pancost, R. D., Pressley, S., Coleman, J. M., Benning, L. G. & Mountain, B. W. (2005). Lipid biomolecules in silica sinters: indicators of microbial biodiversity. *Environ Microbiol* **7**, 66–77.

Perry, C. C. (2003). Silicification: the processes by which organisms capture and mineralize silica. *Rev Mineral Geochem* **54**, 291–327.

Perry, C. C. & Fraser, M. A. (1991). Silica deposition and ultrastructure in the cell wall of *Equisetum arvense*: the importance of cell wall structures and flow control in biosilicification. *Philos Trans R Soc Lond B Biol Sci* **334**, 149–157.

Perry, C. C. & Keeling-Tucker, T. (2000). Biosilicification: the role of the organic matrix in structural control. *J Biol Inorg Chem* **5**, 537–550.

Phoenix, V. R., Adams, D. G. & Konhauser, K. O. (2000). Cyanobacterial viability during hydrothermal biomineralization. *Chem Geol* **169**, 329–338.

Phoenix, V. R., Konhauser, K. O., Adams, D. G. & Bottrell, S. H. (2001). Role of bio-mineralization as an ultraviolet shield: implications for Archean life. *Geology* **29**, 823–826.

Phoenix, V. R., Konhauser, K. O. & Ferris, F. G. (2003). Experimental study of iron and silica immobilization by bacteria in mixed Fe–Si systems: implications for microbial silicification in hot springs. *Can J Earth Sci* **40**, 1669–1678.

Pontoni, D., Narayanan, T. & Rennie, A. R. (2002). Time-resolved SAXS study of nucleation and growth of silica colloids. *Langmuir* **18**, 56–59.

Rees, D. A. (1977). *Polysaccharide Shapes*. London: Chapman & Hall.

Resch, C. M. & Gibson, J. (1983). Isolation of the carotenoid-containing cell wall of three unicellular cyanobacteria. *J Bacteriol* **155**, 345–350.

Rimstidt, J. D. & Barnes, H. L. (1980). The kinetics of silica–water reactions. *Geochim Cosmochim Acta* **44**, 1683–1699.

Rimstidt, J. D. & Cole, D. R. (1983). Geothermal mineralization. I. The mechanism of formation of the Beowawe, Nevada, siliceous sinter deposit. *Am J Sci* **283**, 861–875.

Rippka, R., Deruelles, J., Waterbury, J. B., Herdman, M. & Stanier, R. Y. (1979). Generic assignments, strain histories and properties of pure cultures of cyanobacteria. *J Gen Microbiol* **111**, 1–61.

Sangster, A. G. & Hodson, M. J. (1986). Silica in higher plants. *Ciba Found Symp* **121**, 90–111.

Schrader, M., Drews, G. & Weckesser, J. (1981). Chemical analysis on cell wall constituents of the thermophilic cyanobacterium *Synechococcus* PCC6716. *FEMS Microbiol Lett* **11**, 37–40.

Schultze-Lam, S., Ferris, F. G., Konhauser, K. O. & Wiese, R. G. (1995). In-situ silicification of an Icelandic hot spring microbial mat: implications for microfossil formation. *Can J Earth Sci* **32**, 2021–2026.

Scott, C., Fletcher, R. L. & Bremer, G. B. (1996). Observations of the mechanisms of attachment of some marine fouling blue-green algae. *Biofouling* **10**, 161–173.

Siever, R. (1992). The silica cycle in the Precambrian. *Geochim Cosmochim Acta* **56**, 3265–3272.

Simpson, T. L. & Volcani, B. E. (editors) (1981). *Silicon and Siliceous Structures in Biological Systems*. New York: Springer.

Toporski, J. K. W., Steele, A., Westall, F., Thomas-Keprta, K. L. & McKay, D. S. (2002). The simulated silicification of bacteria – new clues to the modes and timing of bacterial preservation and implications for the search for extraterrestrial microfossils. *Astrobiology* **2**, 1–26.

Treguer, P., Nelson, D. M., van Bennekom, A. J., DeMaster, D. J., Leynaert, A. & Queguiner, B. (1995). The silica balance in the world ocean – a reestimate. *Science* **268**, 375–379.

Tsuneda, S., Aikawa, H., Hayashi, H., Yuasa, A. & Hirata, A. (2003). Extracellular polymeric substances responsible for bacterial adhesion onto solid surface. *FEMS Microbiol Lett* **223**, 287–292.

van der Meer, M. T., Schouten, S., Hanada, S., Hopmans, E. C., Damsté, J. S. & Ward, D. M. (2002). Alkane-1,2-diol-based glycosides and fatty glycosides and wax esters in *Roseiflexus castenholzii* and hot spring microbial mats. *Arch Microbiol* **178**, 229–237.

Walter, M. R., Bauld, J. & Brock, T. D. (1972). Siliceous algal and bacterial stromatolites in hot spring and geyser effluents of Yellowstone National Park. *Science* **178**, 402–405.

Walters, C. C., Margulis, L. & Barghoorn, E. S. (1977). On the experimental silicification of microorganisms. I. Microbial growth on organosilicon compounds. *Precambrian Res* **5**, 241–248.

Weckesser, J., Hofmann, K., Jürgens, U. J., Whitton, B. A. & Raffelsberger, B. (1988). Isolation and chemical analysis of the sheaths of the filamentous cyanobacteria *Calothrix parietina* and *C. scopulorum*. *J Gen Microbiol* **134**, 629–634.

Weed, W. H. (1889). Formation of travertine and siliceous sinter by the vegetation of hot springs. In *United States Geological Survey Ninth Annual Report* (1887–1888), pp. 613–676. Washington, DC: United States Geological Survey.

Westall, F., Boni, L. & Guerzoni, E. (1995). The experimental silicification of microorganisms. *Palaeontology* **38**, 495–528.

Yee, N., Phoenix, V. R., Konhauser, K. O., Benning, L. G. & Ferris, F. G. (2003). The effect of cyanobacteria on silica precipitation at neutral pH: implications for bacterial silicification in geothermal hot springs. *Chem Geol* **199**, 83–90.

Metabolic diversity in the microbial world: relevance to exobiology

Kenneth H. Nealson and Radu Popa

Department of Earth Sciences, University of Southern California, Los Angeles, CA 90089-0740, USA

INTRODUCTION

Metabolic diversity is used here as a physiological or ecological concept referring to the metabolic repertoire available to any group of organisms: in this case, microbes. At least for now, metabolic diversity is conceptually distinct from genetic diversity, although one imagines that, as both concepts are understood in greater depth, the relationships between them will become clear. The metabolic repertoire encompasses, for the most part, the entire range of redox-related energy sources that are available on our planet, from photochemistry to organic and inorganic redox chemistry. Earthly microbes have 'learned' to harvest the energy of nearly every useful and abundant redox couple, revealing a nutritional versatility that to some extent could be used to describe what the planet has to offer energetically. To turn this around, one can imagine that, if energy sources were defined for Earth, one might well predict what kinds of metabolism should have evolved to exploit them and, in fact, for the most part, this would lead to the correct answer. Metabolic diversity is further accentuated by various symbioses, syntrophisms and community interactions (intracellular, intercellular and inter-population), leading to the establishment of communities with seemingly new and unexpected abilities. The functional diversity of the prokaryotic world is thus expressed in terms of its redox chemistry and, with regard to geobiology, this redox chemistry/metabolic connection defines a wide variety of relevant reactions, many of which involve phase changes of the interacting molecules (i.e. between solid, liquid and gas phase). In some sense, it is the world of 'accidental biominerals'; a world where minerals are formed, altered or dissolved as a result of the forever-starved prokaryotic world. In contrast, many eukaryotic microbes also exert significant effects on the geological landscape, and these work by completely different metabolic modes – forming minerals by directed, energy-consuming reactions, minerals that have altered and continue to alter the face of our planet. All in all, the way that life has adapted to the energy regime of our planet allows the exobiological prediction that the same should be true of any other abode where life resides. Understanding energy types and flows may well provide a critical pathway for life detection, on or off our own planet.

MICROBIAL DIVERSITY

Genetic diversity

Genetic diversity is estimated by any of a number of methods, often related to DNA sequence similarities and differences. For the sake of this discussion, the microbial world has been divided into two groups, the prokaryotes and eukaryotes (Fig. 1), with the former group consisting of the *Bacteria* and the *Archaea* (Woese, 2004; Woese *et al.*, 1984, 1990) and the latter including a wide array of single-celled, microscopic eukaryotes, collectively called the protists or protoctista (Whittaker & Margulis, 1978). This grouping is based primarily on sequence comparisons of 16S rRNA genes and includes sequences of both cultivated and as-yet uncultivated microbes. Both the prokaryotic and eukaryotic microbes are underrepresented with regard to cultivated members, so that any discussion of metabolic diversity of these groups is of necessity strongly biased towards a very small subset of the whole. For example, it is estimated that only 0·1 to 1 % of the prokaryotes in most environments have been cultivated (Whitman *et al.*, 1998). In fact, there are many phylum-level bacterial taxonomic groups that are represented by DNA sequences of 16S rRNA genes alone: i.e. groups for which no species have yet been cultivated (Pace *et al.*, 1986). While it is obviously inappropriate to make ecological extrapolations based on 1 % or less of the population, in many cases it is the best that has been possible.

Metabolic diversity

Metabolic diversity, on the other hand, can be assessed by measurement of the reactants and products and quantified in the laboratory and the field. Just what organisms are responsible for a given activity is often discovered after the process has been demonstrated in natural samples. Thus, the connection between geochemists and microbiologists has been a strong one for many years and, in some cases, the geochemists have observed the processes for which the causative microbes were later discovered and characterized. Some examples of this are the processes of anaerobic methane oxidation (Boetius *et al.*, 2000; Orphan *et al.*, 2001b), anaerobic ammonia oxidation (Jetten *et al.*, 1998) and microbial iron and manganese reduction (Lovley & Phillips, 1988; Myers & Nealson, 1988), all of which were known in the geochemical literature prior to the microbiology being demonstrated. The metabolic diversity of the prokaryotes, how-

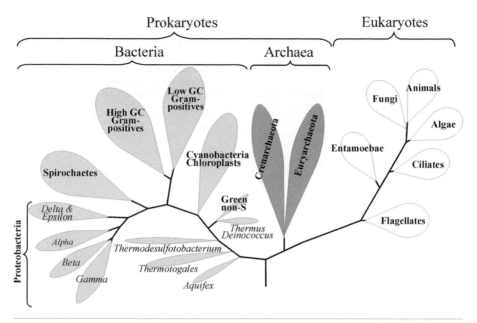

Fig. 1. Microbial diversity according to molecular phylogeny. This diagram shows some of the range of microbes as estimated by sequence similarity of 16S rRNA gene sequences. This genetically diverse assemblage is dominated by bacteria and archaea, but includes a large number of very important eukaryotic microbes such as carbonate- and silica-depositing algae, bacteriovores (ciliates and flagellates) and photosynthetic algae.

ever, extends beyond individual abilities, as some of the truly unique things done are accomplished not by single microbes, but by collaborative efforts between microbes that specialize in the intercellular metabolite transfer called syntrophism, thus providing even more functional diversity than might be expected from single cells. In fact, many of the metabolic abilities of eukaryotes (photosynthesis, respiration, lithotrophy and nitrogen fixation) are carried out by symbionts acquired from the prokaryotic world (Margulis, 1981).

One way to view metabolic diversity is shown in Table 1, which illustrates the range of redox reactions in which microbes participate. As can be seen, the prokaryotes dominate this scene: only a few organics and no inorganics can be utilized by eukaryotes, and, with the exception of nitrate and possibly fumarate, which can be used by a few anaerobic fungi (Fenchel & Finlay; 1995; Tielens *et al.*, 2002; Zumft, 1997), only one respiratory electron acceptor (oxygen) can be utilized by the eukaryotes. Purists might argue that even oxygen respiration is an acquired trait – one that was 'invented' by prokaryotes and arose in the eukaryotes via symbiosis (Margulis, 1981), making respiration a hallmark trait of the prokaryotes and symbiosis, perhaps, one of the most important parts of eukaryotic evolution. The point is, however, that, with

Table 1. Energy sources (fuels) and oxidants used by life

Here we see some of the range of electron donors and electron acceptors used by all of life. The metabolic diversity of the prokaryotes (which can use all those listed) is dramatically expanded in comparison with the rather constrained eukaryotes (which can use those entries in bold only). Asterisks denote those molecules that form insoluble compounds (minerals) when oxidized or reduced, connecting this redox-driven biochemistry with the geological world. DMSO, Dimethyl sulfoxide; TMAO, trimethylamine *N*-oxide.

Fuels		Oxidants	
Sunlight			Fumarate
	Glucose	**Organics**	DMSO
Organics	**Ethanol**		TMAO
	Formaldehyde	Carbon dioxide*	
	Methanol	Sulfur*	
Hydrogen		Sulfate*	
Ammonia		Arsenate*	
Hydrogen sulfide*		Selenite*	
Sulfur*		Iron*	
Iron*		Manganese*	
Manganese*		Nitrate	
Carbon monoxide*		**Oxygen**	
Arsenite*			

regard to redox chemistry, the prokaryotic microbes are the experts, while the eukaryotes have shunned redox diversity in favour of a high energy yield – energy used to power their own impressive diversity of cellular structure and behaviour. The most pertinent features with regard to this are: (i) the use of inorganic energy sources (lithotrophy) is the realm of the prokaryotes; (ii) the process of anaerobic respiration (i.e. using electron acceptors other than oxygen) is, with few exceptions, the realm of the prokaryotes; (iii) the oxidation and reduction of these inorganic compounds forms a strong link with planetary geology, as many of the reactions either form or dissolve solid-phase minerals and mineraloids during the process (Table 1).

Having extolled prokaryotic diversity, a few words of respect and admiration need to be inserted with regard to the anaerobic eukaryotes. Both protists and fungi are often abundant in many different permanently anaerobic niches (i.e. rumens, termite guts, anaerobic sediments), where they feed on prokaryotes and their breakdown products (Fenchel & Finlay, 1995). These eukaryotes are almost without exception devoid of mitochondria and exist either by the use of hydrogenosomes or via nitrate reduction. Hydrogenosomes, which are common in both anaerobic protists and fungi, are organelles believed to have arisen from mitochondria (Theissen *et al.*, 2003) – they have replaced the enzymes involved in oxidative phosphorylation with those involved in

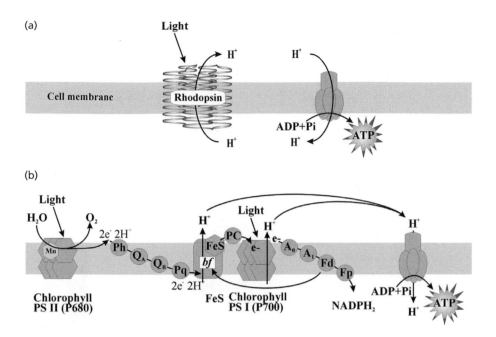

Fig. 2. Mechanisms for harvesting light energy. (a) Rhodopsin-mediated proton translocation as seen in archaea (*Halobacterium halobium*). It should be noted that the proton gradient that is established by this process can be used to power many cellular functions such as transport or flagellar motion or, as shown, to power the synthesis of ATP via the membrane-bound ATPase. (b) Oxygenic photosynthesis in chloroplasts. A_0 and A_1, Acceptors of electrons from P700; *bf*, cytochrome *bf* complex; Fd, ferredoxin; Fp, flavoprotein; PC, plastocyanin; Ph, phaeophytin; Pq, plastoquinone; PS, photosystem; Q_A and Q_B, plastoquinone-binding proteins.

fermentation, thus creating an organelle that makes ATP by substrate-level phosphory-lation and produces dihydrogen gas as a by-product (Martin & Muller, 1998; Muller, 1993).

A final point with regard to diversity relates to the use of light as an energy source. By a factor of about 50 000, visible light dominates our planet in terms of yearly energy flow (Nealson & Conrad, 1999), and it is used in a number of ways. The simplest is via a process called rhodopsin-mediated proton pumping, in which the molecule rhodopsin is used to absorb light and the energy is used directly to pump a proton from the inside to the outside of the cell, charging the cell membrane via an osmotic and electrical gradient (Fig. 2a). This gradient can then be used by the cell either directly or to syn-thesize ATP. This type of metabolism, long known in archaea (Stoeckenius & Bogo-molni, 1982), has more recently been found to be widely distributed in the bacterial world (Beja *et al.*, 2000; de la Torre *et al.*, 2003; Venter *et al.*, 2004). Photosynthesis, the more familiar use of visible light by life, was apparently 'invented' by the bacteria, almost certainly as an anaerobic process, using sulfur compounds or iron as electron

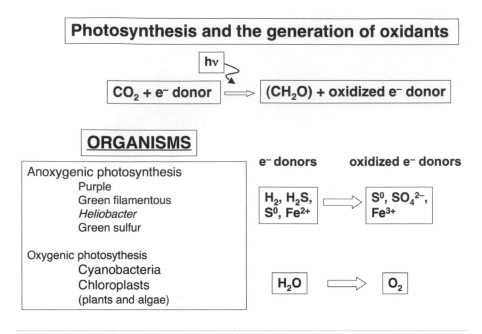

Fig. 3. Generation of oxidants during photosynthesis. The general formula for photosynthesis is shown at the top of the diagram, with the critical difference being the molecules used as electron donors (and the resultant oxidized forms produced).

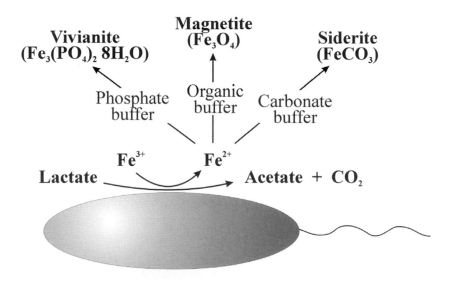

Fig. 4. Accidental biominerals. The redox activity of some bacteria (such as iron reducers) results in the formation of various minerals (vivianite, magnetite, siderite), depending primarily on the chemistry of the environment.

donor (Blankenship, 2002; Xiong *et al.*, 2000). Such anoxygenic photosynthetic bacteria produce oxidized electron donors as a by-product of photosynthesis (Fig. 3) and were the precursors of the now widespread oxygenic photosynthesis (Blankenship, 2002). The more complex oxygenic photosynthesis is thought to have arisen in the cyanobacteria and eventually appeared symbiotically in algae and plants (Margulis, 1981), one of the most important events in the evolution of complex life.

With the above discussion of metabolic diversity, it is easy to miss an important point: namely that there is also an impressive unity to the metabolism of life. Virtually all present-day metabolism on Earth operates in a similar way, with environmental redox equivalents being harvested and used for the reduction of cellular electron or hydrogen carriers which are similar throughout all of life (Nealson, 1997; Nealson & Conrad, 1999). These reduced carriers are then used directly for biosynthesis or for the generation of a membrane potential (combination of a pH and electrical potential) that can be used for a variety of functions (including ATP formation, transport and motility). Thus, despite apparent rampant diversity in terms of resource allocation, the central energy-processing systems of all life, including those of structurally complex large eukaryotes, operate in a similar way. With regard to this issue, there are really only two metabolic means of extracting energy from the environment: chemiosmosis and fermentation. Almost without exception, the interactions between the living world and the mineral world are of the chemiosmotic type (i.e. redox chemistry). While it is difficult to specify exactly when the ability to respire arose in time, one imagines that it is indeed one of the earliest 'inventions' of life (Nealson & Rye, 2004) and, once invented, became a component of virtually all subsequent successful experiments in metabolic evolution.

MINERAL FORMATION: THE GEOBIOLOGY CONNECTION

Prokaryotic mineral formation

As discussed above, the array of electron donors and acceptors that the prokaryotic world uses is, to some extent, the glue that knits microbial metabolism to geobiology. As shown in Table 1, many of the redox-active compounds become a solid (mineral or mineraloid) form – thus, oxidation or reduction of these components often results in a change of state of the component (i.e. between insoluble and soluble), leading to formation or dissolution of minerals. In this sense, the metabolism of the prokaryotes, designed for the purpose of harvesting energy from the environment, is inadvertently linked to the formation and/or dissolution of many minerals. The inadvertent nature of this process is shown in Fig. 4, a summary of results in which a culture of *Shewanella putrefaciens* CN-32, during the process of iron reduction, was shown to produce any of several mineral products depending on the environment in which the bacteria are grown

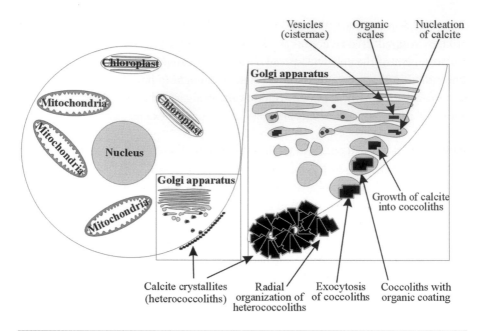

Vesicles (cisternae)
Organic scales
Nucleation of calcite

Chloroplast

Mitochondria

Chloroplast

Golgi apparatus

Nucleus

Mitochondria

Mitochondria

Golgi apparatus

Growth of calcite into coccoliths

Calcite crystallites (heterococcoliths)
Radial organization of heterococcoliths
Exocytosis of coccoliths
Coccoliths with organic coating

Fig. 5. Purposeful biominerals. Biomineralization of coccolithophores (calcareous algae) making crystals of calcite. This diagram is intended to demonstrate the complex and detailed interactions that occur in general in the formation of mineral products by eukaryotes (after Young & Henriksen, 2003).

(Roden & Zachara, 1996). That is, while the bacteria metabolically impact the local mineralogy, the end products produced are controlled by the environment rather than by the bacteria themselves. Thus, to some extent, one can view the formation of minerals by some (perhaps most) of the prokaryotes as metabolic 'accidents', dependent on the metabolic chemistry, but not directed by it. As far as is known, there is no direct advantage to the bacterium of forming a given mineral in comparison to any other: this is determined in fact by the chemistry of the environment in which the bacteria are living. This leaves us in the interesting position of viewing prokaryotic minerals as a metabolic by-product – something done 'accidentally', and something, perhaps, that occurs as an indicator of microbial metabolism, but without the morphological clues usually associated with biominerals or fossils.

This being said, while the formation of a specific mineral may have no advantage for prokaryotes, it may well be that mineral formation in general serves a useful thermodynamic/kinetic purpose. From the point of view of the bacterium, removing one of the soluble products as an insoluble mineral could have a dynamic effect on the chemical abilities of the bacterium. This is a strategy often employed by bacteria living in symbiosis with other bacteria, in which the end product of one is the substrate for another, and this cascade makes chemical reactions proceed at speeds far beyond those predicted if the reaction products were disposed of by diffusion alone. Perhaps the most pertinent example is that of anaerobic methane oxidation, in which the process is really driven by sulfate-reducing bacteria that consume the hydrogen produced via methane oxidation, pushing the reaction forward (Boetius *et al.*, 2000; Orphan *et al.*, 2001a, b).

Eukaryotic mineral formation

The eukaryotes, in contrast, have perfected the art of mineral fabrication, making a wide array of biominerals with many different uses (Dove *et al.*, 2003; Lowenstam, 1981; Lowenstam & Weiner, 1989). Biomineralization is extremely common among the multicellular eukaryotes, many of which need structural elements to grow and function in three dimensions, providing advantages for predation (teeth, bones etc.) and protection (shells, frustules, cell coverings) (Lowenstam, 1981; Lowenstam & Weiner, 1989). These biominerals are often ornate and recognizable structures that provide us with a visible fossil record, allowing the calibration of the molecular record (discussed below), something that is extremely difficult prior to the 'invention' of these biominerals. While it is less widely distributed among the eukaryotic microbes, organisms like coccolithophores and foraminifera (organisms that make carbonate minerals) and diatoms and radiolarians (organisms that make silica minerals) have dramatic impacts on the carbon and silicon cycles on the planet (Lowenstam & Weiner, 1989; Skinner & Jahren, 2004).

It is clear from emerging mechanistic studies that the situation is dramatically different with regard to eukaryotic mineral formation (Dove *et al.*, 2003). Eukaryotic biominerals are genetically directed structures, formed using protein templates that are used to catalyse and direct the synthesis of the specific minerals. The resulting structures are reproducible enough that they can often be used to identify the organism that produced them, often to the level of genus or even species (Fig. 5). This is in marked contrast to prokaryotic mineral formation and dissolution, which is primarily a function of environment rather than genetics (Fig. 3). As opposed to the prokaryotes, which form minerals while gaining metabolic energy, the formation of eukaryotic biominerals is 'costly' in the sense of requiring specific templates, energy for synthesis and often auxiliary systems for transport and assembly. Finally, redox chemistry is not a fundamental part of eukaryotic mineral formation, as redox changes do not occur and no energy is gained. Given that the eukaryotes are unable to view minerals as either viable electron donors or acceptors, their ability to form or dissolve redox-active minerals is extremely limited.

A final note with regard to the minerals produced by prokaryotes and eukaryotes: because protein templates and other structural aids are not involved in prokaryotic mineral formation, these minerals are often 'pure', sulfides, carbonates etc., while those of eukaryotes are 'hybrids', composed of biological material interspersed with

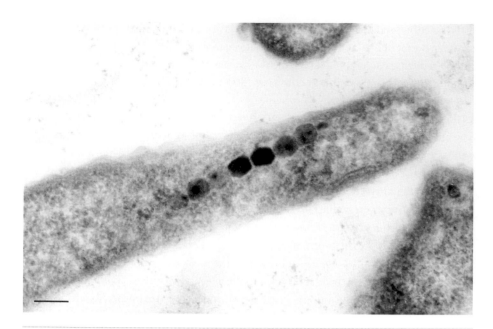

Fig. 6. Chains of magnetosomes in *Magnetospirillum magnetotacticum* AMB1 as seen in a high-magnification TEM section (image provided courtesy of Dr Virginia Souza-Egipsy). Bar 100 nm.

crystalline minerals (Dove *et al.*, 2003). These adaptations add remarkable mechanical properties to the eukaryotic biominerals and make them chemically distinct from the prokaryotic biominerals (Dove *et al.*, 2003). This being said, it should also be noted that a wide variety of mineral and mineraloid-like inclusions (oxalates, carbonates, uric acid etc.) are neither species-specific nor 'hybrid' in nature.

Biomagnetite and the death of a generalization

The above discussion of mineral formation divides the world into prokaryotic redox minerals and eukaryotic structural and functional biominerals. For example, dissimilatory iron-reducing bacteria can produce copious amounts of extracellular magnetite as a function of respiration (Lovley & Phillips, 1988; Roden & Zachara, 1996), while eukaryotic chitons can produce magnetite coatings on their teeth (Lowenstam, 1981; Lowenstam & Weiner, 1989). The eukaryotic magnetite is easily recognizable, but whether it is possible to distinguish between magnetite formed by prokaryotes and abiotically formed magnetite is not yet clear.

The division between prokaryotes and eukaryotes with regard to biominerals is not always so clear, however. Many strains of magnetotactic bacteria (no magnetotactic archaea are known) produce highly ordered, intracellular, crystalline magnetite inclusions called magnetosomes (Fig. 6). These single domain, mineralogically nearly perfect, highly magnetic crystals are often arranged in chain-like arrays (Blakemore, 1975; Bazylinski & Frankel, 2000). They provide the cells that contain them with the ability to align passively in a magnetic field, the response being that these highly motile cells move swiftly towards one magnetic pole or the other. It appears that each bacterial strain produces a characteristic magnetite pattern (size, shape and arrangement of the magnetosomes) and that specific genes (and, thus, specific proteins) are involved in the production of the magnetosomes (Schuler, 1999, 2004; Matsunaga & Okamura, 2003). Furthermore, while the function(s) of the magnetosome is not yet fully proven, nearly everyone agrees that there must be an advantage to the cells that contain the magnetosomes, and most agree that it is connected with environmental sensing and location (Frankel *et al.*, 1997).

Are there other such examples of highly structured, genetically dictated biominerals produced by bacteria? Recently, it was reported that highly crystalline forms of metal sulfides were produced by sulfate-reducing bacteria (Suzuki *et al.*, 2002). Whether these are in fact structured with a 'purpose' is not known, but their highly crystalline structure is an intriguing finding; perhaps such preordained and structured prokaryotic minerals are more prevalent than was thought (Fortin, 2004).

TIME: LINKING THE PAST WITH THE PRESENT

The timing of the emergence of various metabolic abilities is one of the most enigmatic problems in the microbial world. The evidence available to us in the ancient rock record is, at best, akin to smudged fingerprints and, at worst, like footprints in the sand after the tide has come and gone. One of the great hopes of molecular phylogenetic approaches was that they would allow one to look back in time using sequence data: using these data to estimate when major metabolic 'inventions' occurred in the past. Such 'inventions' would include not only structural innovations visible in the fossil record, but metabolic inventions like respiration and photosynthesis and other prokaryotic specialities like nitrogen fixation, denitrification and sulfur oxidation. This hope has been realized to some extent but, realistically, the methods are probably good only as far back as there is a fossil record to support them. For example, Benner *et al.* (2002) discussed the use of various types of data to understand the evolution of metabolism, concluding that, within the last 50, perhaps, 100 million years, one can probably do this with some confidence, utilizing a combination of fossil, isotope, organic geochemical and molecular evolutionary clock records to infer past patterns of evolution (in this case, the evolution of certain sugar-fermentation abilities). Thus, to some extent, the fossil record is of limited use: true (recognizable) biominerals produced by eukaryotes are visible only ~500 million years ago, when the first sponge spicules and carbonate biominerals can be seen. These processes are thus geologically young (Li *et al.*, 1998; Nealson & Rye, 2004).

Prior to the formation of these 'true' biominerals, the signatures that exist are primarily geochemical in nature. Thus, while prokaryotic minerals may not be readily recognizable by their morphologies or unique crystal structures, many can be judged to be of biological origin via the fractionation of isotopes during the 'supply' of chemical components for their formation. Kinetic fractionation occurs as a function of enzymic catalysis, leading to biological materials preferentially being composed of the 'lighter' isotopes. Thus light carbon is preferentially used by living organisms during carbon fixation and accumulates in the resulting biomass, while light sulfide is produced during sulfur or sulfate reduction and accumulates in sulfide minerals. These isotopic tracers have been of value in tracing the metabolic activities of modern organisms in both the laboratory and the field and provide a major tool for looking for indicators of metabolic activity in ancient samples. That is, it is possible, by using C and S isotopes, to see in the ancient rock record the appearance of processes that result in fractionation of these isotopes. Herein lies an important distinction that can be easily missed: while the isotopes may strongly suggest the existence of a process leading to fractionation, they cannot tell us unambiguously which process was involved and cannot be used to tell which organism or even group of organisms accounted for the fractionation. Unless one accepts this tenet, one can be easily fooled.

Carbon isotopes have been among the most valuable and widely used of the isotopes with regard to life detection and definition. Carbon fractionation occurs during its reduction (fixation) from CO_2 to organic carbon by bacteria, algae and plants and to a much greater degree (i.e. light carbon) when CO_2 is fixed into methane by methanogenic archaea. However, even for this well-known and often-used system, deciphering the isotope signatures from ancient samples is difficult because (i) these pathways are varied and unknown, (ii) subsequent diagenetic reactions are not easy to specify and (iii) the signature of the source of carbon is seldom known.

Even with these caveats, however, it seems clear from a variety of studies that biominerals can be traced to the early phases of Earth's history. Stable-isotope signatures of both carbon and sulfur suggest that metabolic activities were involved with the formation of minerals from very early times. Carbon isotope ratios ($^{13}C/^{12}C$) have been used to suggest that carbon fixation may have existed as early as 3·8 Ga (billion years) ago (Mojzsis *et al.*, 1996). While this number has been challenged, few would argue with 3·5 Ga for convincing evidence of carbon isotope signals in the ancient record. Similarly, sulfur isotopes ($^{34}S/^{33}S$) suggest that sulfur reduction of some kind was occurring 2·5 Ga and perhaps earlier (Canfield *et al.*, 2000; Shen *et al.*, 2001).

Providing definitive fossil evidence from before the time of the biomineral-producing eukaryotes is difficult for several reasons: first, because the preservation of the materials is often poor, making identification difficult, and second, because virtually none of the putative organisms seen in the samples are alive today. While they have similarities to other organisms, the nature of their behaviour and even their metabolism cannot be specified with reasonable certainty. Another discouraging development with regard to molecular evolution methods is the rampant appearance of examples in which it is now clear that the evolutionary 'clock' is neither constant (it can run at different rates) nor predictable (Doolittle *et al.*, 1996). Thus, two organisms that would be described as deeply branching and suspected of being of equal 'age' may be quite different because of differences in evolutionary clock speeds. For ancient samples, of the order of hundreds to thousands of millions of years and older, the situation gets even more uncertain.

Another difficulty that is peculiar to the prokaryotes is that the 'fossils' used to identify them are reduced to either organic geochemicals (i.e. classes of chemicals peculiar to a certain type of cell) or isotope fractionation patterns indicative of a certain type of metabolism. Both analyses have their limitations. For example, while some classes of compounds can be identified as components of cyanobacteria in the modern world, there is no way of knowing with certainty that non-cyanobacterial organisms that contained these compounds were not present before the cyanobacteria arose. Similarly, when one sees isotope fractionation, such as the appearance of light sulfur, it is tempting to invoke the appearance of sulfate reduction (and sulfate-reducing bacteria) and, in fact, this is often done (Canfield *et al.*, 2000). While the activity of sulfate-reducing bacteria would indeed explain the observed results, it is also true that sulfur can be fractionated by many other organisms, including sulfur-, polysulfide- and thiosulfate-reducers (Smock *et al.*, 1998).

As a footnote to this discussion, one must remember that much of the work with both organic biomarkers and isotope fractionations hinges on our knowledge of microbial physiology and the composition of cultured microbes. As noted above, it is estimated that less than 1 % of the microbial world has been obtained in culture (Whitman *et al.*, 1998), and therefore that we are often extrapolating from a very limited experimental base. To this end, we note in Table 2 that much progress has been made in recent years in uncovering novel organisms and novel metabolic abilities – microbes whose very existence was previously doubted or unknown. Given that methods are improving with regard to culturing microbes, one can hope that the situation will improve, but one must be cautious when trying to extrapolate backwards in time to ancient metabolisms based on such an incomplete database.

Finally, we must confront the issue of lateral (horizontal) gene transfer (Doolittle, 2002). With the advent of genome sequencing, it has become apparent that in the

Table 2. Microbial metabolic diversity

This table represents some of the metabolic abilities of prokaryotes that have been revealed since 1988.

Process	Found in	Year of discovery	Reference(s)
Dissimilatory Fe and Mn reduction	Bacteria and archaea	1988	Lovley & Phillips (1988); Myers & Nealson (1988)
Anaerobic methane oxidation consortium	Bacteria/archaea	2000	Boetius *et al.* (2000); Orphan *et al.* (2001b)
Anaerobic ammonia oxidation	Bacteria	1998	Jetten *et al.* (1998)
Anaerobic iron oxidation	Photosynthetic bacteria	1993	Ehrenreich & Widdel (1994); Widdel *et al.* (1993)
Anaerobic iron oxidation	Heterotrophic bacteria	1994	Straub *et al.* (1996)
Perchlorate reduction	Bacteria	1996	Coates *et al.* (1999)
Phosphite oxidation	Bacteria	2000	Schink & Friedrich (2000)
Dissimilatory arsenic reduction	Bacteria	1994	Ahmann *et al.* (1994)
Dissimilatory selenate reduction	Bacteria	1994	Oremland *et al.* (1994)

past there has been a large amount of mixing of genomes, such that the so-called phylogenetic 'tree of life' is in fact much more like a cross-hatched bush (Doolittle, 2002). Each of the three 'kingdoms' has obtained, and fixed into their genomes, ample information from the other two kingdoms, suggesting that major sharing of genetic information has occurred, especially among the prokaryotes, where this is still happening at rapid rates. Thus, one of the major issues in prokaryotic genomics is that of defining the gene set that characterizes a given group of microbes (as opposed to those genes that can apparently move in and out with non-lethal results).

With regard to the use of molecular methods, one of the intellectual 'traps' of this approach should be noted – namely that, for the prokaryotes, all of the organisms on which the 'phylogeny' is based are alive and evolving today. That is, while they may contain 'ancient' traits or abilities, these are surely not as they were in the past, so that the phylogenetic approach allows one to look at the most likely sequence of events, but not to place accurate times on any of the events. Thus, one could argue that we are looking at the living remnants of past microbial evolution and can have a reasonably good idea of what preceded what (within the limitations of uncertainties introduced by horizontal gene transfer), but trying to ascertain when any of these processes arose, i.e. when a given process or organism first appeared, is very difficult (if not impossible) by this method. As a precaution, one might note that almost none of the organisms seen in the fossil record of 100 million years ago are alive today. If we tried to construct those organisms simply from molecular biology alone, it would almost certainly be a resounding failure. This is the dilemma we are faced with in reconstructing prokaryotic evolution.

Kinetic biosignatures and layered communities

While the above discussion has dealt with inference of the past through reading the rock record, much biology is concerned with measurement and manipulation of the present. To this end, one can look for biosignatures connected to the chemistry of the organism and its impact on the environment (substrate removal or alteration and product deposition). Such short-term interactions define, to some extent, one of the major differences between life and non-life: specific catalysis of reactions that would other-wise occur at extremely slow rates. In the absence of life, many low-temperature geochemical reactions (i.e. less than 100 °C) proceed at rates slower than molecular diffusion, so that product accumulation and gradient formation cannot occur. Enzymic catalysis, however, leads to the consumption of reactants at rates faster than they can be supplied and the production of products at rates faster than they can diffuse away, leading to the formation of gradients that are indicative of the life forces that have produced them (Nealson & Berelson, 2003). The simple argument is that, in the absence of life, the chemical gradients (as we call them, layered microbial communities or LMCs) that are so dominant in anaerobic niches on Earth would simply not exist. For example, in the Black Sea (Fig. 7), a series of redox zones are seen, each indicative of a process that occurs in a well-defined redox zone (Nealson & Berelson, 2003). At these interfaces or layers, the chemical profiles can be used to define the microbial processes that are occurring to establish the gradients. For example, as shown in Fig. 7, the abrupt disappearance of oxygen at ~50 m is referred to as the oxygen-depletion zone, where catalysis by aerobic respiration occurs so quickly that oxygen is taken to nearly zero within a few metres. In the absence of respiration, oxygen would be nearly constant to the bottom. Similarly, the nitrate disappears a few metres below due to a process called denitrification – without this, the nitrate would almost certainly be uniformly distributed in the Black Sea, as the rates of the processes that might consume it are very slow. Below this are the zones of manganese, iron and sulfate reduction, all of which would not exist without life catalysing each process.

It should be obvious that these LMCs are inferred not from biological measurements, but from the very existence of the chemical layers and non-linear gradients (Nealson & Berelson, 2003). As will be discussed below, they are more complex than is implied above, but the gradients themselves can be used as indicators of specific catalysis and thus of life. Of interest here is that, with regard to spatial scales, we see such LMCs at scales of micrometres in biofilms to millimetres in algal mats and centimetres in lake and ocean sediments – they are a universal feature of life on Earth. In environments

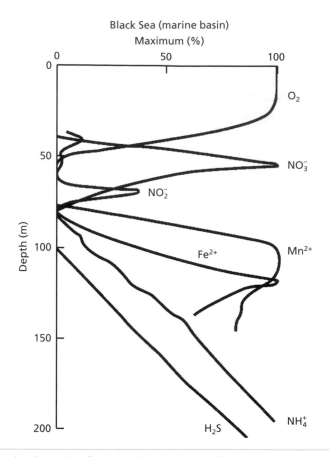

Fig. 7. Redox-related vertical profiles in the Black Sea. This profile presents a general scheme of redox interfaces seen in the Black Sea (Nealson & Berelson, 2003). The gradients here are a function of the organisms that accumulate at the interfaces – organisms that consume reactants and create products at rates sufficient to account for the gradients observed. Similar chemical gradients are used as indicators of LMCs over scales ranging from many metres, like this one, to millimetres or less in many other environments.

where rapid mixing occurs, such as the open ocean, mixed lakes and the atmosphere, such gradients are rapidly dissipated by convection and mixing, but the signals of the biological processes are nevertheless there.

EXTREME ENVIRONMENTS: WHY DO WE GO THERE?

'Extremophiles': one of the most popular buzzwords of the last decade, and the cause for great excitement among microbiologists. Finding the 'most extreme' organism with regard to any given variable (temperature, pH, UV radiation etc.) has been a sporting pleasure for many of us. However, our obsession with these physical and chemical extremes may be, to some extent, slightly misplaced. For many physical and chemical variables, the prokaryotes and eukaryotes are both able to adapt reasonably

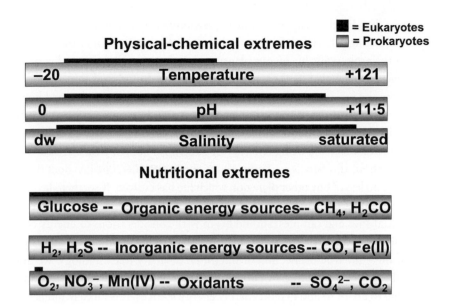

Fig. 8. Types of extremophily. This diagram shows the limits of prokaryotic and eukaryotic life for a number of variables. The top three panels are physical/chemical variables, demonstrating that the structurally simpler prokaryotes are a bit tougher and more resilient than eukaryotes, while the bottom three panels show that some types of metabolism are 'off limits' for the eukaryotes – truly extreme. dw, Distilled water.

well (Fig. 8), although the simpler and more robust prokaryotes almost always exceed the more complex eukaryotes. As seen in Fig. 8, however, the true extreme conditions might be imagined to be those of nutrition. As noted above, eukaryotes might define as 'acceptable' only those nutrients that can be converted into glucose or pyruvate, and only oxygen to respire. Thus, nearly the entire inorganic world of electron donors is 'extreme', as is the use of any electron acceptor other than oxygen. When we move into the anaerobic world, especially the anaerobic world at high temperature, we begin to see the prokaryote-dominated world that we can imagine has some relevance to the ancient Earth. It should not be surprising to see that many of us who want to think about either the ancient Earth or the possibility of life elsewhere find ourselves studying environments like the deep subsurface – places where nutritional extremophiles are abundant and, for the most part, uncharacterized.

CONCLUSIONS

Metabolic diversity has become a catchphrase for the wonders of the microbial world – wonders that may help to explain why the world is like it is through the energetic exploitation of the geosphere via the metabolism of the biosphere. A direct, but not entirely obvious, extension of this thinking is that it applies not only to understanding the evolution and distribution of life on our own planet, but to the possibility of life in abodes other than Earth: giving rise to such questions as, 'could we use this geobiological logic to distinguish a living from a dead planet?' A quick perusal of the metabolic diversity of the prokaryotic life on our planet reveals that virtually every redox-related energetic niche is occupied. If there is fuel (an electron donor) and something with which to burn the fuel (an electron acceptor), then some microbe has learned how to exploit this energetic niche. This exploitation, in its simplest sense, defines the metabolic diversity with which we work – microbes striving to make a living on whatever resources are available, and being remarkably successful in plying their trade. Given that such metabolic diversity is so dominant on our own planet, one imagines that we should expect no less of any other planet on which life has evolved. If so, then the search for life in all realms (the surface and subsurface of the Earth, as well as potential extraterrestrial sites and samples) should begin with considerations of energy types, levels and fluxes, always with an eye to finding things that should not be there in the absence of life – the true connection of microbial diversity to exobiology. We don't know whether life exists elsewhere than on our planet. However, given the remarkable ability of life to adapt to and exploit earthly energy sources, it is an easy intellectual leap to imagine that, in other abodes, with abundant energy and acceptable chemistry, some sort of life could prosper. However, assumptions about the chemistry or other details of any non-earthly life may lead one awry. What must be true is that life will evolve to take advantage of the energy sources and chemical building blocks that are available and that, to compete with the natural chemistry, it will need to invent catalysts that elevate the rates of reactions. It is this catalysis that should be the signpost of life – the geobiological indicator that something is amiss (reactant consumption, product accumulation, complex molecule generation or isotope fractionation), and the enticing possibility that life may have been responsible.

REFERENCES

Ahmann, D., Roberts, A. L., Krumholtz, L. R. & Morel, F. M. M. (1994). Microbe grows by reducing arsenic. *Nature* **371**, 750.

Bazylinski, D. A. & Frankel, R. B. (2000). Biologically controlled mineralization of magnetic iron minerals by magnetotactic bacteria. In *Environmental Microbe-Metal Interactions*, pp. 109–143. Edited by D. R. Lovley. Washington, DC: American Society for Microbiology.

Beja, O., Aravind, L., Koonin, E. V. & 9 other authors (2000). Bacterial rhodopsin: evidence for a new type of phototrophy in the sea. *Science* **289**, 1902–1906.

Benner, S. A., Caraco, M. D., Thomson, J. M. & Gaucher, E. A. (2002). Paleontological, geological, and molecular histories of life. *Science* **296**, 864–868.

Blakemore, R. (1975). Magnetotactic bacteria. *Science* **190**, 377–379.

Blankenship, R. E. (2002). *Molecular Mechanisms of Photosynthesis*. Oxford: Blackwell Science.

Boetius, A., Ravenschlag, K., Schubert, C. J. & 7 other authors (2000). A marine microbial consortium apparently mediating anaerobic oxidation of methane. *Nature* **407**, 623–626.

Canfield, D. E., Habicht, K. S. & Thamdrup, B. (2000). The archean sulfur cycle and the early history of atmospheric oxygen. *Science* **288**, 658–661.

Coates, J. D., Michaelidou, U., Bruce, R. A., O'Connor, S. M., Crespi, J. N. & Achenbach, L. A. (1999). Ubiquity and diversity of dissimilatory (per)chlorate-reducing bacteria. *Appl Environ Microbiol* **65**, 5234–5241.

de la Torre, J. R., Christianson, L. M., Beja, O., Suzuki, M. T., Karl, D. M., Heidelberg, J. & DeLong, E. F. (2003). Proteorhodopsin genes are distributed among divergent marine bacterial taxa. *Proc Natl Acad Sci U S A* **100**, 12830–12835.

Doolittle, W. F. (2002). Thinking laterally about genes (review of *Lateral DNA Transfer: Mechanisms and Consequences*, by F. Bushman). *Nature* **418**, 589–590.

Doolittle, R. F., Feng, D.-F., Tsang, S., Cho, G. & Little, E. (1996). Determining divergence times of the major kingdoms of living organisms with a protein clock. *Science* **271**, 470–477.

Dove, P. M., De Yoreo, J. J. & Weiner, S. (editors) (2003). *Biomineralization*. Reviews in Mineralogy and Geochemistry, vol. 54. Washington, DC: Mineralogical Society of America.

Ehrenreich, A. & Widdel, F. (1994). Anaerobic oxidation of ferrous iron by purple bacteria, a new type of phototrophic metabolism. *Appl Environ Microbiol* **60**, 4517–4526.

Fenchel, T. & Finlay, B. J. (1995). *Ecology and Evolution in Anoxic Worlds*. Oxford: Oxford University Press.

Fortin, D. (2004). What biogenic minerals tell us. *Science* **303**, 1618–1619.

Frankel, R. B., Bazylinski, D. A., Johnson, M. S. & Taylor, B. L. (1997). Magneto-aerotaxis in marine coccoid bacteria. *Biophys J* **73**, 994–1000.

Jetten, M. S. M., Stous, M., van de Pas-Schoonen, K. T. & 7 other authors (1998). The anaerobic oxidation of ammonium. *FEMS Microbiol Rev* **22**, 421–437.

Li, C. W., Chen, J. Y. & Hua, T. E. (1998). Precambrian sponges with cellular structures. *Science* **279**, 879–882.

Lovley, D. R. & Phillips, E. J. P. (1988). Novel mode of microbial energy metabolism: organic carbon oxidation coupled to dissimilatory reduction of iron or manganese. *Appl Environ Microbiol* **54**, 1472–1480.

Lowenstam, H. A. (1981). Minerals formed by organisms. *Science* **211**, 1126–1131.

Lowenstam, H. A. & Weiner, S. (1989). *On Biomineralization*. New York: Oxford University Press.

Margulis, L. (1981). *Symbiosis in Cell Evolution*. New York: W. H. Freeman.

Martin, W. & Muller, M. (1998). The hydrogen hypothesis for the first eukaryote. *Nature* **392**, 37–41.

Matsunaga, T. & Okamura, Y. (2003). Genes and proteins involved in bacterial magnetic particle formation. *Trends Microbiol* **11**, 536–541.

Mojzsis, S., Arrhenius, G., McKeegan, K. D., Harrison, T. M., Nutman, A. P. & Friend, C. R. L. (1996). Evidence for life on Earth before 3,800 million years ago. *Nature* **384**, 55–59.

Muller, M. (1993). The hydrogenosome. *J Gen Microbiol* **139**, 2879–2889.

Myers, C. R. & Nealson, K. H. (1988). Bacterial manganese reduction and growth with manganese oxide as the sole electron acceptor. *Science* **240**, 1319–1321.

Nealson, K. H. (1997). Sediment bacteria: who's there, what are they doing, and what's new? *Annu Rev Earth Planet Sci* **25**, 403–434.

Nealson, K. & Berelson, W. (2003). Layered microbial communities and the search for life in the universe. *Geomicrobiol J* **20**, 451–462.

Nealson, K. H. & Conrad, P. G. (1999). Life: past, present and future. *Philos Trans R Soc Lond B Biol Sci* **354**, 1923–1939.

Nealson, K. H. & Rye, R. (2004). Evolution of metabolism. In *Treatise on Geochemistry*, pp. 41–61. Edited by W. H. Schlesinger. Amsterdam: Elsevier.

Oremland, R. S., Switzer Blum, J., Culbertson, C. W., Visscher, P. T., Miller, L. G., Dowdle, P. & Strohmaier, F. E. (1994). Isolation, growth, and metabolism of an obligately anaerobic, selenate-respiring bacterium, strain SES-3. *Appl Environ Microbiol* **60**, 3011–3019.

Orphan, V. J., Hinrichs, K.-U., Ussler, W., III, Paull, C. K., Taylor, L. T., Sylva, S. P., Hayes, J. M. & DeLong, E. F. (2001a). Comparative analysis of methane-oxidizing archaea and sulfate-reducing bacteria in anoxic marine sediments. *Appl Environ Microbiol* **67**, 1922–1934.

Orphan, V. J., House, C. H., Hinrichs, K.-U., McKeegan, K. H. & DeLong, E. F. (2001b). Methane-consuming archaea revealed by directly coupled isotopic and phylogenetic analysis. *Science* **293**, 484–487.

Pace, N., Stahl, D. A., Lane, D. J. & Olsen, G. J. (1986). The analysis of natural microbial populations by ribosomal RNA sequences. In *Advances in Microbial Ecology*, pp. 1–55. Edited by K. C. Marshall. New York: Plenum.

Roden, E. E. & Zachara, J. M. (1996). Microbial reduction of crystalline iron(III) oxides: influence of oxide surface area and potential for cell growth. *Environ Sci Technol* **30**, 1618–1628.

Schink, B. & Friedrich, M. (2000). Phosphite oxidation by sulphate reduction. *Nature* **406**, 37.

Schuler, D. (1999). Formation of magnetosomes in magnetotactic bacteria. *J Mol Microbiol Biotechnol* **1**, 79–86.

Schuler, D. (2004). Molecular analysis of a subcellular compartment: the magnetosome membrane in *Magnetospirillum gryphiswaldense*. *Arch Microbiol* **181**, 1–7.

Shen, Y. A., Buick, R. & Canfield, D. E. (2001). Isotopic evidence for microbial sulphate reduction in the early Archaean era. *Nature* **410**, 77–81.

Skinner, H. C. W. & Jahren, A. H. (2004). Biomineralization. In *Biogeochemistry*, pp. 117–184. Edited by W. H. Schlesinger. Amsterdam: Elsevier.

Smock, A. M., Bottcher, M. W. & Cypionka, H. (1998). Fractionation of sulfur isotopes during thiosulfate reduction by *Desulfovibrio desulfuricans*. *Arch Microbiol* **169**, 460–463.

Stoeckenius, W. & Bogomolni, R. A. (1982). Bacteriorhodopsin and related pigments of halobacteria. *Annu Rev Biochem* **51**, 587–616.

Straub, K. L., Benz, M., Schink, B. & Widdel, F. (1996). Anaerobic, nitrate-dependent, microbial oxidation of ferrous iron. *Appl Environ Microbiol* **62**, 1458–1460.

Suzuki, Y., Kelly, S. D., Kemner, K. M. & Banfield, J. F. (2002). Nanometre-size products of uranium bioreduction. *Nature* **419**, 134–136.

Theissen, U., Hoffmeister, M., Grieshaber, M. & Martin, W. (2003). Single eubacterial origin of eukaryotic sulfide : quinone oxidoreductase, a mitochondrial enzyme conserved from the early evolution of eukaryotes during anoxic and sulfidic times. *Mol Biol Evol* **20**, 1564–1574.

Tielens, A. G. M., Rotte, C., van Hellemond, J. J. & Martin, W. (2002). Mitochondria as we don't know them. *Trends Biochem Sci* **27**, 564–572.

Venter, J. C., Remington, K., Heidelberg, J. F. & 20 other authors (2004). Environmental genome shotgun sequencing of the Sargasso Sea. *Science* **304**, 66–74.

Whitman, W. B., Coleman, D. C. & Wiebe, W. J. (1998). Prokaryotes: the unseen majority. *Proc Natl Acad Sci U S A* **95**, 6578–6583.

Whittaker, R. H. & Margulis, L. (1978). Protist classification and the kingdoms of organisms. *Biosystems* **10**, 3–18.

Widdel, F., Schnell, S., Heising, S., Ehrenreich, A., Assmus, B. & Schink, B. (1993). Ferrous iron oxidation by anoxygenic phototrophic bacteria. *Nature* **362**, 834–836.

Woese, C. R. (2004). A new biology for a new century. *Microbiol Mol Biol Rev* **68**, 173–186.

Woese, C. R., Stackebrandt, E., Weisburg, W. G. & 8 other authors (1984). The phylogeny of purple bacteria: the alpha subdivision. *Syst Appl Microbiol* **5**, 315–326.

Woese, C. R., Kandler, O. & Wheelis, M. L. (1990). Towards a natural system of organisms: proposal for the domains Archaea, Bacteria, and Eucarya. *Proc Natl Acad Sci U S A* **87**, 4576–4579.

Xiong, J., Fischer, W. M., Inoue, K., Nakahara, M. & Bauer, C. E. (2000). Molecular evidence for the early evolution of photosynthesis. *Science* **289**, 1724–1730.

Young, J. R. & Henriksen, K. (2003). Biomineralization within vesicles: the calcite of coccoliths. In *Biomineralization*, Reviews in Mineralogy and Geochemistry, vol. 54, pp. 190–215. Edited by P. M. Dove, J. J. De Yoreo & S. Weiner. Washington, DC: Mineralogical Society of America.

Zumft, W. G. (1997). Cell biology and molecular basis of denitrification. *Microbiol Mol Biol Rev* **61**, 533–616.

Biogeochemical cycling in polar, temperate and tropical coastal zones: similarities and differences

David B. Nedwell

Department of Biological Sciences, University of Essex, Colchester CO4 3SQ, UK

INTRODUCTION

This chapter will consider biogeochemical cycling in the coastal zone. This is defined as that area of estuarine and coastal, relatively shallow water where there is strong benthic–pelagic linkage and exchange between the water column and the underlying sediment. In deeper water this connection becomes increasingly tenuous as the exchange between the euphotic zone and the benthic layer declines. Longhurst *et al.* (1995) recognized the coastal boundary domain, divided into 22 provinces, as often bounded by a shelf-break front, and included coastal upwelling regions. The coastal zone generally exhibits high rates of primary production compared with the open ocean (Table 1), and there is the greatest impact from inputs from the land to the coastal sea through estuaries. Estuaries and coastal seas are highly heterotrophic systems which are net exporters of CO_2 to the atmosphere due to the mineralization and recycling of both autochthonous and allochthonous organic matter (Borges, 2005).

PHYSICO-CHEMICAL DIFFERENCES BETWEEN LATITUDINAL REGIONS

The physical–biological interactions that influence marine phytoplankton production have been reviewed by Daly & Smith (1993). Because of the spherical shape of the Earth, more solar energy falls per unit area of surface in equatorial regions than at the poles (Fig. 1a), and the incidence of light at the equator is vertical to the surface, but oblique at the poles. Furthermore, the distance radiation travels through the atmosphere is longer at the poles, thus reducing the irradiation incident at the poles compared with equatorial regions. Consequently, equatorial regions have higher energy inputs

SGM symposium 65: Micro-organisms and Earth systems – advances in geomicrobiology.
Editors G. M. Gadd, K. T. Semple & H. M. Lappin-Scott. Cambridge University Press. ISBN 0 521 86222 1 ©SGM 2005

Table 1. Areas and annual primary production rates of biogeochemical provinces based on the CZCS 1978–86 climatological data (after Longhurst *et al.*, 1995)

Geographical subset	Area ($\times 10^{-6}$ km^2)	Primary production rate (g C m^{-2} year^{-1})
Coastal domain	37·4	385
Polar domain	20·8	310
Westerlies domain	129·9	126
Trades domain	139·9	93
Arctic Ocean	1·7	645
Atlantic Ocean	74·0	199
Pacific Ocean	148·9	132
Indian Ocean	45·4	143
Southern Ocean	57·9	141
Upwelling provinces	8·4	398

Table 2. Seasonal variation in characteristics of the coastal zone in polar, temperate and tropical coastal zones

Characteristic	Polar	Temperate	Tropical
Insolation	Extremely seasonal	Moderately seasonal	Weakly seasonal
Day length	Extremely seasonal	Moderately seasonal	Weakly seasonal
Water temperature	Constant, low	Seasonally variable	Constant, high
Estuarine flushing	Highly seasonal (summer melt)	Moderately seasonal	Highly seasonal (wet/dry season rainfall)

and are hotter than polar regions. Rainfall and humidity are also extreme in the tropics, often with distinct rainy seasons which impose differences in the ecology of the coastal zones. During the tropical dry season, when there is little or no rainfall or river flow, there may be accumulation of detritus within an estuary but during the wet season heavy rainfall flushes the estuary, exporting a spike of material to the coastal zone, and 're-setting' the estuary (Eyre & Balls, 1999). For example, north Queensland tropical estuaries exhibited flushing times varying between 2 and 30 days during the dry season but 0 days in the wet, when river water essentially passed straight through the estuaries. In contrast, temperate Scottish estuaries exhibited much less seasonal variation in flushing. Pulses of nutrients (and particulate matter) leached from the catchment are therefore likely in tropical estuaries compared with temperate estuaries, whose rainfall and therefore input loads to the coastal zone are less seasonally variable. Estuaries are extremely important in attenuating the loads of nutrients passing from land to sea (see Nedwell *et al.*, 1999); in the case of nitrogen particularly by denitrification, or of phosphorus by settlement of particles to which phosphate is adsorbed. Attenuation of nutrient fluxes from land to sea is likely to be diminished if the major loads coincide with periods when freshwater flushing times within the estuary are very short, as in

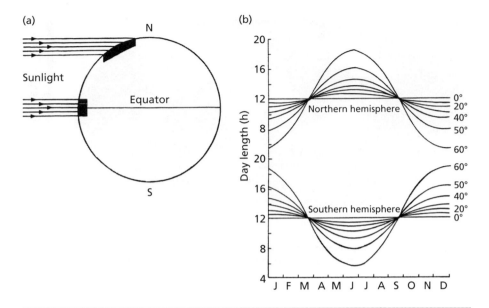

Fig. 1. (a) Owing to the spherical nature of the Earth, solar energy is concentrated over a smaller area in equatorial regions than in polar regions, and the distance travelled through the atmosphere is also shorter at the equator. Consequently, energy input per unit area is greater at the equator than at the poles. (b) Variations in day length throughout the year in relation to latitude (reproduced with permission from Osborne, 2000).

some tropical estuaries. In polar regions any terrestrial input to the coastal zone occurs only during the short summer, when ice and snow melt occurs, and is therefore also seasonally extreme.

Day length also exhibits latitudinal changes. Because of the rotation of the Earth around its axis there is little difference seasonally in the day length at the equator, whereas at the poles there is extreme change in the day length between winter and summer and intermediate seasonal changes in day length at temperate latitudes (Fig. 1b). The different environmental factors are summarized in Table 2. In essence, high-latitude polar coastal waters are stenothermal, with relatively constant low temperature (−1 to 4 °C), mid-latitude temperate waters are eurythermal, with seasonal temperature cycling between approximately 2 and 20 °C, and low-latitude tropical waters are stenothermal, with relatively constant high temperature (> 16 °C).

PRIMARY PRODUCTION

Biogeochemical cycling in the coastal zone is driven by autochthonous primary production and also by allochthonous inputs via estuaries, both of which provide the energy for subsequent chemo-organotrophic activity. To compare biogeochemical cycling in different latitudinal regions we must start by examining their primary

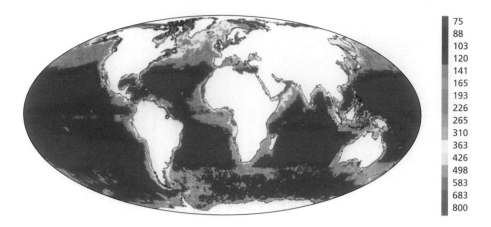

Fig. 2. Global primary production (g C m^{-2} year^{-1}) modelled from satellite estimates of chlorophyll *a*, photosynthesis–light relationships and local light environment (reproduced with permission from Longhurst *et al.*, 1995).

production. Are there consistent differences in primary production in polar, temperate or tropical coastal zones? Longhurst *et al.* (1995) used satellite radiometer data to model primary production in the oceans (Fig. 2), based on estimates of chlorophyll by the CZCS radiometer on the NIMBUS satellite from 1979 to 1986, photosynthesis–light relationships and the light environment. It is clear that the coastal zones are in general areas of high production. They estimated global oceanic primary production as 44·7–50·2 Pg C year^{-1}. The annual primary production rates for the coastal domain (Table 2) averaged 385 g C m^{-2} year^{-1}, for the Arctic Ocean 645 g C m^{-2} year^{-1} and upwelling provinces 398 g C m^{-2} year^{-1}. The annual production in the coastal domain was greater than that in all oceans except the Arctic Ocean, and the relative productivity of the ocean province was only 2·5–4·0 times that of the coastal provinces, despite the former's very much greater area. Tropical coastal provinces in the Indian and Pacific Oceans averaged 279 g C m^{-2} year^{-1}, again confirming that low-temperature polar coastal communities are as productive as those at lower latitudes and higher temperatures.

What is beyond doubt is that the seasonal pattern of primary production in the different regions varies enormously, reflecting the different seasonality imposed by insolation and day length, and in turn influencing the biogeochemical cycling that goes on. Polar regions have constantly low water temperatures (usually < 4 °C) but exhibit an extreme seasonal pattern, with short summer ice-free periods of high insolation during which pelagic primary production rapidly increases, plankton biomass quickly attains a maximum and a pulse of organic input to the coastal zone bottom sediments occurs (Fig. 3) (Nedwell *et al.*, 1993; Rysgaard *et al.*, 1999). Rysgaard *et al.* (1999) showed that there was a strong relationship in Arctic waters between the annual pelagic

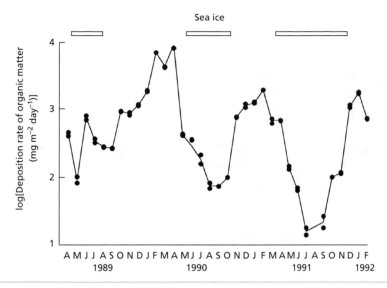

Fig. 3. Seasonal changes in the settlement rate of organic matter from the water column to coastal sediments at Signy Is., South Orkney Islands, Antarctica (reproduced with permission from Nedwell *et al.*, 1993).

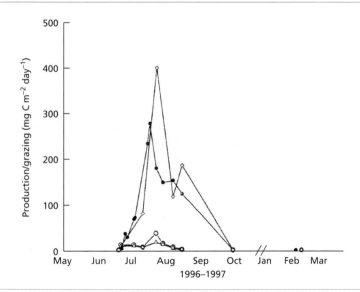

Fig. 4. Annual primary production (●) and grazing by ciliates (○), heterotrophic dinoflagellates (△) and copepods (◇) in Young Sound, northeast Greenland (reproduced with permission from Rysgaard *et al.*, 1999).

primary production and the duration of the open-water period when pelagic primary production could occur. This is because an ice layer with snow cover strongly attenuates light, and it is only after ice melt that there is significant illumination of the water. The

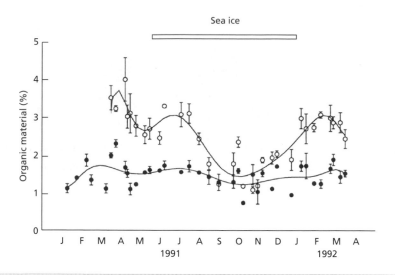

Fig. 5. Benthic organic matter content at Signy Is. Sediment depth: ○, 0–0·5 cm, ●, 1–2 cm (reproduced with permission from Nedwell *et al.*, 1993).

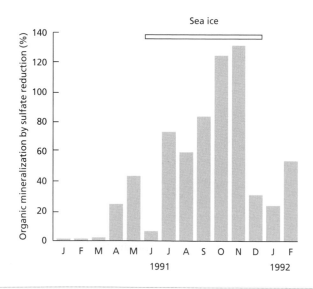

Fig. 6. Benthic anaerobic organic matter degradation driven by sulfate reduction as a percentage of total mineralization indicated by benthic O_2 uptake (reproduced with permission from Nedwell *et al.*, 1993).

seasonal cycle of pelagic primary production in Young Sound, Greenland, exhibited an extremely short but intense period of primary production (Fig. 4) and in the coastal zone around Signy Island, in the South Orkney islands, at the tip of the Antarctic Peninsula, a similar short but intense period of summer production has been described

(Nedwell *et al.*, 1993). Once sea ice breaks up, the rapidly increasing light penetrates the clear water and primary production increases exponentially, the water column seeded by ice algae from the bottom surface of the melting ice. The complete seasonal cycle of algal production at Signy Is. occurs within 2–3 months, after which the sea ice forms again, and may be as short as 4–8 weeks at higher latitudes (e.g. Rysgaard *et al.*, 1998; Karl *et al.*, 1996). Lack of turbulence beneath fast ice causes rapid settlement of organic matter from the water column and there is, in essence, a short but intense pulse of organic deposition into the coastal bottom sediments. Benthic processes such as O_2 uptake respond quickly and increase due to the input of fresh organic matter, so that there is a strong seasonal signal of benthic activity, in direct contrast to tropical coastal systems. In the few studies carried out, annual benthic organic degradation seems broadly to balance net primary production in both Arctic and Antarctic coastal systems studied so far (Nedwell *et al.*, 1993; Rysgaard *et al.*, 1998), indicating stable ecosystems. However, while primary production occurs in a short summer burst, that pulsed input of organic matter sustains benthic processes for the remainder of the year, spread out over the period when winter ice cover minimizes further organic inputs. Thus, the deposited organic matter represents a food or energy reserve for the organotrophic benthic community which tides them over the winter period with no new inputs. For example, at Signy Is. the seasonal signal of deposited organic matter could be seen only in the top 0·5 cm of sediment (Fig. 5). As the deposited organic matter was buried into the sediment by very active bioturbation, the proportion of benthic organic matter degradation which was driven by anaerobic sulfate reduction increased through the winter period, while that due to aerobic metabolism declined (Fig. 6) until, with the input of fresh organic matter in spring onto the surface of the sediment, aerobic metabolism again increased dramatically and the cycle repeated. Mincks *et al.* (2005) have argued that this pulsed input to polar benthic communities represents a 'food bank' for polar benthic detritivors, including micro-organisms, over the winter period when there were no new organic inputs, and the data from Signy Is. are consistent with this idea.

In the temperate coastal zone annual primary production still exhibits a strong seasonal signal, though less intense than that of polar regions. There are significant seasonal changes in insolation and water temperature (e.g. approximately 3–18 °C in the southern North Sea) and a seasonal cycle of spring bloom primary production occurs, subsequently limited by depleted nutrient availability, and possibly with a smaller autumn bloom if summer stratification breaks down. In estuaries and inshore waters, where there is a continuous supply of nutrients, there may be no distinct spring bloom, but an increase of primary production during spring which is sustained throughout the summer (e.g. Kocum *et al.*, 2002a, b). For example, in the north-central North Sea during 1998–99 there was a spring bloom which decreased during summer, but, in the German Bight, where there is a continued supply of nutrients from the Rhine, primary production continued in a broad peak throughout the growing season (Fig. 7). Again, there are seasonal changes in the downward fluxes of organic matter to bottom sediments, albeit less extreme than in polar regions. Upton *et al.* (1993) estimated that benthic organic matter mineralization accounted for 17–45 % of net primary production in the various regions of the southern North Sea.

In the tropics the seasonal insolation change is least and insolation remains at high values throughout the year, while water temperatures also remain relatively constant and high. Primary production is therefore less likely to be limited by insolation than at higher latitudes, but regulated by other factors such as nutrients or turbidity. The tropical regime typically reflects the relatively continuous formation of new organic matter, with little seasonal variation. In a study of coastal lagoon sediments in Fiji there were no statistically significant seasonal variations in either benthic oxygen uptake rates or rates of pelagic/benthic nutrient exchange (Sobey, 2004), reflecting a lack of seasonal change in inputs of organic matter. However, our knowledge of coastal zone biogeo-chemical processes in the tropics is based on a relatively sparse dataset compared with other regions, and requires much further study. Much of the tropical ocean has low nutrient concentrations because of the presence of a permanent thermocline which prevents vertical transport of new nutrients from deeper water. This conspires to ensure that the tropical coastal zone generally operates at lower ambient nutrient concentrations than at higher latitudes. While tropical coastal zone nutrient concen-trations may be higher than those offshore in tropical oceans, they are still usually lower than those in the coastal zone at higher latitudes. Tropical primary production therefore becomes dependent upon recycling of nutrients to sustain fresh production, and is likely to be nutrient limited. However, where nutrients become available from upwelling or from fluvial inputs in a tropical high-light environment, primary pro-duction may be intense (e.g. Robertson *et al.*, 1998).

What factors regulate primary production in the different coastal environments? Clearly, in polar regions it is light that is the principal seasonal regulator of primary production in coastal regions, at constant very low temperature, but this is not to imply that light is the only regulator. Primary production may be coregulated by a variety of interacting factors, and nutrient availability is often a significant factor despite the apparently relatively high nutrient concentrations. In the Southern Ocean the occurrence of high-nutrient, low-chlorophyll areas, where primary production is low despite the presence of relatively high concentrations of nitrate, has been attributed to lack of available iron, thus limiting photosynthesis (de Baar *et al.*, 1995; Martin *et al.*, 1990). This iron limitation apparently does not happen in the Arctic, where proximity to land masses results in higher aeolian inputs of iron and nitrate is reduced to very low

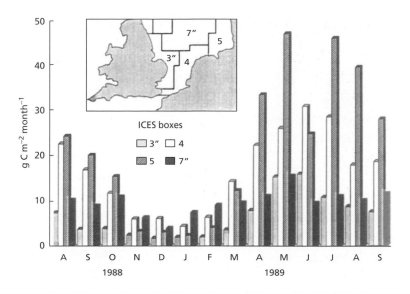

Fig. 7. Seasonal variation in pelagic primary production in different ICES regions of the North Sea (reproduced with permission from Joint & Pomroy, 1993).

concentrations during summer activity. Moreover, in coastal zones, in close proximity to land, iron supply should be higher than in the open ocean. However, low temperature may have other influences on the ability of algae to sequester nutrients at low concentrations.

My co-workers and I have shown previously (Nedwell & Rutter, 1994; Nedwell, 1999; Reay *et al.*, 1999) that low temperature apparently reduces the affinity of bacteria and algae for substrates taken up by active transport, including nitrate, phosphate and silicate. This has been attributed to stiffening of the cell membrane at low temperatures, which reduces the efficiency of transporter proteins embedded in the membrane. From the data available, there is no such inhibitory effect of low temperature upon ammonium uptake, which occurs at least partly by passive or facilitated diffusion mechanisms which do not depend upon active transport. Primary production in the Southern Ocean seems to be very dependent upon ammonium uptake, with very low f-ratios (see Reay *et al.*, 2001), and it is possible that for at least part of the summer growth season passive uptake of ammonium is sufficient to maintain the rates of primary production measured. The concentrations of dissolved inorganic nitrogen, predominantly nitrate, in Antarctic waters often remains comparatively high (> 10 µM) even during summer, but addition of nitrate still stimulates algal growth despite significant concentrations of nitrate already being present, indicating continued limitation by nitrogen even at these apparently replete concentrations.

Table 3. Rates of benthic oxygen uptake reported in polar, temperate and tropical regions

Site	Rate (mmol O_2 m^{-2} day^{-1})	Source
Polar		
E. Svalbard, 226–320 m depth	3·2–11·9	Pfannkuche & Thiel (1987)
Bering and Chukchi Seas, 19–25 m depth	8·7–19·2	Henriksen et al. (1993)
E. Svalbard, 170–240 m depth	3·9–11·2	Hulth et al. (1994)
E. Greenland, 40 m depth	17·8	Rysgaard et al. (1996)
Bering and Chukchi Seas, 30–52 m depth	7·3–25·5	Henriksen et al. (1993)
Off Newfoundland, 270 m depth	8·4	Pomeroy et al. (1991)
Signy Is., Antarctica	15–85	Nedwell et al. (1993)
Temperate		
Flax Pond, Long Is. Sound	81–137	Mackin & Swider (1989)
Aarhus Bay	27	Jorgensen & Revsbech (1985)
North Sea sediments, 25–81 m depth	5·3–27·8	Upton et al. (1993)
NW Mediterranean, 60–80 m depth	4·6–9·9	Tahey et al. (1994)
Skagerrak, 190–695 m depth	11·8–16·1	Canfield et al. (1993)
Thames estuary sediments, UK	12·2–241	Trimmer et al. (2000)
Colne estuary, UK	48–216	Dong et al. (2000)
Tropical		
Thai mangrove	17–61	Kristensen et al. (1991)
Jamaican mangrove	31–103	Nedwell et al. (1994)
Fiji lagoon sediment	0–24	Sobey (2004)
Indus delta, Pakistan	16–28	Kristensen et al. (1992)
Hinchinbrook Is., Australia	2·8–61	Alongi et al. (1999)

If affinity for nitrate (or other nutrients) is decreased at low temperatures, algae may be nitrogen-limited despite the presence of significant concentrations of nitrate because the nitrate cannot be sequestered effectively. This suggestion does not necessarily contradict the hypothesis that iron limits Southern Ocean but not Arctic Ocean primary production, as iron uptake is itself probably an active transport process and its uptake will therefore tend to be retarded at low temperature.

ORGANIC MATTER DEGRADATION AND BIOGEOCHEMICAL RECYCLING

The strong pelagic–benthic linkage in the coastal zone ensures that the bottom sediments are major sites of organic matter breakdown and biogeochemical recycling. In comparing biogeochemical cycling in the different regions we may ask a number of questions about the benthic communities which bring about this recycling and the environmental factors which control them. It is clear that primary production rates do not show any consistent latitudinal variation with temperature (see above), being as great in some polar regions as in the tropics. Similarly, is there any consistent relationship between water temperature and the biogeochemical recycling of organic matter? Inspection of reported rates of benthic oxygen uptake (Table 3), as a surrogate for benthic organic matter breakdown, again suggests that there is no consistent difference between high and low latitudes: the rates of benthic oxygen uptake in Factory Cove, Signy Is., at water temperatures near 0 °C all year were greater than those reported for many tropical sediments (Nedwell *et al.*, 1993; Glud *et al.*, 1998). However, the overlap of gross rates of O_2 uptake in the different regions does not imply that the microbial communities at high or low latitudes are operating in physiologically similar ways.

Pomeroy and co-workers (Pomeroy & Deibel, 1986; Pomeroy *et al.*, 1991) proposed the 'cold water paradigm' that 'bacterial metabolism and growth are depressed to a much greater degree than those of phytoplankton at low (<4 °C) sea water temperatures', which would tend to lead to imbalance between the production and mineralization of organic matter. They argued that less consumption of primary production by bacteria at low temperature reduced the microbial loop activity and left more organic matter for metazoan grazers. In an investigation of bacterial growth and primary production along a north–south transect in the Atlantic Hoppe *et al.* (2002) demonstrated clear gradients of primary production and bacterial growth correlated with latitudinal temperature change. The ratio of bacterial production to primary production both above and below the equator was most significantly correlated with water temperature.

In a study of growth rates of bacteria in cold oceans measured by thymidine incorporation, Rivkin *et al.* (1996) concluded that there was a weak relationship between specific growth rate and temperature, and the mean and median specific growth rates for bacteria from cold (≤4 °C) and warm (≥4 °C) regions were not significantly different, i.e. the growth rates of bacteria from cold and temperate oceans are similar at their respective environmental temperatures. However, they pointed out that the production rate is a function of both growth rate and biomass of cells in the standing crop, and the bacterial abundance in cold ocean regions can be 10-fold lower than in temperate or tropical seas.

EFFECT OF TEMPERATURE ON ABILITY TO SEQUESTER SUBSTRATES BY ACTIVE TRANSPORT

Interaction between organic substrate requirement and temperature has been reported for marine bacteria (Pomeroy *et al.*, 1991; Pomeroy & Deibel, 1986; Wiebe *et al.*, 1992, 1993); at low temperature a higher concentration of substrate is required to support a given rate of growth or respiration than at a higher environmental temperature.

This temperature–substrate concentration interaction was attributed (Nedwell & Rutter, 1994; Nedwell, 1999) to decreased affinity for substrates at temperatures below the optimum temperature for growth (T_{opt}). At temperatures below T_{opt} microorganisms become increasingly less able to sequester low concentrations of substrates from their environment effectively by active uptake, apparently because as the membrane stiffens as the temperature drops the transport proteins embedded in the membrane become less effective, and affinity for substrates declines (Nedwell, 1999). Because of the decreased affinity for substrates at low temperature, higher concentrations of substrates are necessary in the environment to counter loss of affinity and to maintain a given substrate flux into the cell. This loss of affinity below T_{opt} has been demonstrated for active uptake of both inorganic solutes such as nitrate by both algae and bacteria (Ogilvie *et al.*, 1997; Reay *et al.*, 1999) and organic substrates by chemoorganotrophic bacteria (Nedwell & Rutter, 1994). The paradigm also holds true for psychrophilic, mesophilic and thermophilic bacteria over their respective ranges of temperature (Nedwell, 1999).

For a species to be selected by their environmental temperature they must first be physiologically capable of growing at that temperature, and the membrane of each species adapts to its broad temperature range by features such as the fatty acid composition of the membrane lipids, which affects membrane fluidity (e.g. Russell, 1990, 1992, 1998). The homeoviscous model of Sinensky (1974) proposed that these membrane compositional differences adapt a species to maintain membrane fluidity, and hence biological function, over a particular range of temperature. A certain degree of flexibility of this temperature range appears to be possible by variations in the membrane and key enzymes but in essence the biokinetic range for a particular species is fixed. Within the broad temperature range for a species, however, their ability to sequester substrates from the environment appears to decline below T_{opt}. Note, however, that we would only expect an adverse effect by a low environmental temperature on affinity for substrates if the environmental temperature was below the T_{opt} for the species, so the extent to which a species population is optimally adapted to its environmental temperature would appear important. This raises the question of the extent to which species populations in different temperature environments have T_{opt} at or below their environmental temperature.

In high-temperature environments such as thermal springs the optimum temperature of the species present conforms closely to the environmental temperature, but in low-temperature, polar environments it appears that this is not the case. Morita (1975) defined psychrophiles, mesophiles and thermophiles in relation to their cardinal temperatures, with obligate psychrophiles able to grow at $0\,°C$ with $T_{opt} < 15\,°C$. A category of 'psychrotolerant' types, able to grow at $0\,°C$ but with an optimum $> 15\,°C$, was subsequently added. In reality there is probably a continuum of physiological adaptation across the temperature range. Feller & Gerday (2003) have argued that the terms stenothermal and eurythermal psychrophiles are better used, respectively, for 'obligate' psychrophiles able to grow over only a narrow temperature range or 'facultative' psychrophiles able to grow over a wide temperature range. While stenothermal psychrophiles are undoubtedly present in polar environments, the majority of organisms present seem to be eurythermal psychrotolerants with T_{opt} for growth much higher than the environmental temperature (e.g. Morita, 1975; Franzmann, 1996; Upton & Nedwell, 1989). Additionally, many reports of biogeochemical processes such as sulfate reduction (Arnosti & Jørgensen, 2003; Nedwell, 1989) also indicate optima well above environmental temperature in polar regions. Li and co-workers (Li, 1980, 1985; Li *et al.*, 1984) suggested that algal species at the equator have temperature optima for photosynthesis near to their environmental temperature, but that at higher latitudes and lower temperatures the environmental temperature is lower than the optimum for photosynthesis. If this latitudinal change is the general case then it appears that in low-temperature populations both algal and bacterial species may indeed be operating under conditions where they are less well adapted physiologically to their environmental temperature than populations at lower, warmer latitudes.

It might be predicted that it would be adaptive for an organism to operate most efficiently at its environmental temperature, so it is somewhat surprising that, while this seems to occur in the tropics, it does not seem to be true in high-latitude populations. This may mean that T_{opt} is irrelevant to selection. A new model of the effect of temperature on enzyme activity, the equilibrium model (Daniel *et al.*, 2001; Peterson *et al.*, 2004), proposes that enzyme activity at any temperature is a function of an equilibrium between an active and an inactive form of the enzyme; it is this equilibrium which leads to the lowering of enzyme activity above the T_{opt} for the enzyme. It is the inactive form of the enzyme that is irreversibly denatured at higher temperature, and the reduction in activity may be due to a reversible change of conformation at the active site. Similarly, inactivation of enzymes can occur at low temperature. The equilibrium between active and inactive forms is defined by an equilibrium constant, K_{eq}, and T_{eq} is the temperature when K_{eq} is 1 (i.e. when active and inactive forms are equal). Peterson *et al.* (2004) argue that adaptive evolution is more likely to operate through selection for T_{eq} than through T_{opt} for the enzyme or thermal stability. In polar, low-temperature populations an increase in temperature may increase the catalytic rate by the Arrhenius activation energy, but furthermore may change the equilibrium between active and inactive forms to increase the amount of active enzyme, so that the overall rate of reaction is enhanced in a compound manner as temperature rises. In that case the apparently high T_{opt} for growth of polar micro-organisms may be an artefact which does not actually reflect their degree of adaptation to their environmental temperature.

In a review of psychrophilic enzymes Feller & Gerday (2003) argue that effective enzyme activity at low temperature requires highly reactive reaction centres which are less stable than those of mesophilic enzymes, which will therefore be heat labile. Psychrophilic enzymes seem to be inactivated at temperatures much lower than those that cause unfolding of their protein structure, unlike mesophilic or thermophilic enzymes. This emphasizes the severe heat-lability of the psychrophilic reaction centre compared with those of mesophiles or thermophiles. Feller & Gerday (2003) point out that 'a mobile and flexible active site binds its substrate weakly and, indeed, most psychrophilic enzymes have higher K_m values than their mesophilic counterparts'. Furthermore, cold-active enzymes maintain reaction rates by increasing k_{cat} (the maximum enzyme reaction rate at a given temperature) at the expense of K_m, i.e. affinity as defined by K_m is likely to decline (K_m increase) at low temperatures. Feller & Gerday (2003) proposed that 'psychrophilic enzymes have reached a state that is close to the lowest possible stability... and they cannot be less stable without losing the native and active conformation'. Does this provide a rationale for why physiological adaptation by micro-organisms to low temperature is not as complete as that attained at higher environmental temperatures? The assumption is extending the model for single enzyme reactions to cell growth, but is it too fanciful to imagine that the effect of temperature upon growth may mimic its effect upon the enzyme reactions that support growth? If true, it may also explain why, at low environmental temperatures, micro-organisms are more regulated by low affinity for substrates than in warmer environments, because they are operating suboptimally in terms of temperature and/or the reaction centres of key psychrophilic enzymes (which may include transport proteins) confer a low affinity for substrates.

The implications of this interaction between affinity for substrates and environmental temperature are profound. Mincks *et al.* (2005) argue that microbial respiration and mineralization of organic matter is a product of both substrate concentration (S) and a first-order reaction rate constant (k) and that benthic community respiration (R) is therefore described by

$$R \propto \Sigma(S_i k_i)$$

where S_i and k_i are the respective rate constants and concentrations for metabolizable components (i) of the organic matter (Fig. 8a). As a given k_i decreases due to low temperature, the rate can be maintained only by an increase in S_i.

Alternatively, the rate can be described (Button, 1993) using the specific affinity, a_A, and the rate of substrate utilization is

$$R \propto \Sigma a_{A(i)} S_i X$$

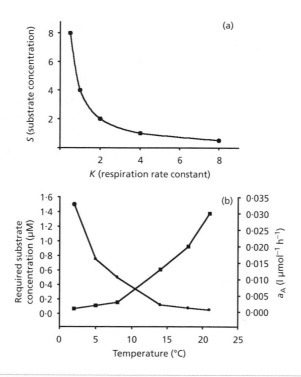

Fig. 8. (a) Conceptual comparison between effects of substrate concentration and respiration rate constant for benthic organic matter in polar and temperate sediments (after S. L. Mincks and others, personal communication). (b) Effect of temperature on specific affinity (a_A) for glucose for an Antarctic eurythermal coryneform (■) and the substrate concentration required at each temperature to maintain a constant rate of substrate utilization of 0·0015 μmol h⁻¹ (●).

where $a_{A(i)}$ and S_i are the specific affinity and concentration of substrate i, and X is the microbial biomass. As a_A decreases because of lower temperature, the rate can only be maintained by either higher substrate concentrations or a higher biomass. Fig. 8(b) shows how a_A varies with temperature for a eurythermal coryneform with a T_{opt} of 22 °C, isolated from an Antarctic lake. In order for a constant rate of substrate utilization of 0·0015 μmol h⁻¹ to be maintained as temperature decreases below T_{opt} (assuming unit biomass, X), the decreasing a_A must be countered by an increased substrate concentration, S. This leads to the conclusion that microbial communities might be regarded as operating either at high affinity (a_A) but low S, as in tropical environments, or at low affinity but high S, as in polar environments (Fig. 8a). This is analogous to fast turnover of small substrate pools in tropical systems compared with slow turnover of relatively large substrate pools in polar benthic environments (the 'food reserve' concept). Holmboe *et al.* (2001) compared anoxic decomposition in Thai mangrove sediment and temperate Wadden Sea salt-marsh sediments. They reported high C : N and C : P ratios in both sediments, but low nutrient release and faster rates of turnover of N and P by nutrient-deficient bacteria in the tropical sediment, which is consistent with the proposed model.

Presumably, temperate environments exhibit changes between the two extremes in response to the seasonal temperature cycle and seasonal selection of the populations in the microbial community. Previous studies (King & Nedwell, 1984; Sieburth, 1967) have demonstrated the selection of temperate bacterial communities by seasonal temperature change, the physiological change induced in the community lagging 2 months behind the environmental temperature change. Selection by competition is not immediate in response to seasonal temperature change, and the 2-month lag is the period during which the competition occurs, resulting in seasonal selection of better-adapted populations. This presumably results in cycling between selection of low-affinity/high-substrate-concentration populations in the spring towards higher-affinity/low-substrate-concentration populations during summer. This continual seasonal change in itself implies that adaptation to environmental temperature by microbial populations in eurythermal temperate environments with seasonally varying tempera-ture must be less complete than in stenothermal environments, and it is inevitable that physiological adaptation with respect to environmental temperature will be less optimized than in stenothermal environments. We have shown previously (Rutter & Nedwell, 1994) that, in non-steady-state temperature environments, the speed of response by a species to change may be more adaptive than their competitive ability under constant conditions. In stenothermal, constant low or high temperature environ-ments the speed of response to temperature change will not be an issue.

It would seem, therefore, that physiological adaptation in polar populations might be suboptimal relative to environmental temperature because of the inherent limitation of adaptability of enzymes to very low temperature; in temperate populations it must be suboptimal because of the time lag between selection of populations in response to seasonal environmental temperature change; but in stenothermal tropical systems populations will be physiologically better adapted, with T_{opt} near to environmental temperature.

FUNCTIONAL GROUPS

While gross rates of biogeochemical processes apparently do not differ between latitudinal regions, the microbial communities which catalyse them may differ. The process of benthic organic matter breakdown is achieved by the integrated activities of a variety of different functional groups of bacteria and archaea, together with meio- and macrofauna, including aerobes, and anaerobes such as nitrate respirers, Fe^{3+} and Mn^{4+} respirers, sulfate reducers and methanogenic archaea. Firstly, are there any differences in the relative importance of the different functional groups of micro-organisms present in the different latitudinal regions? Secondly, are there any phylogenetic differences in these functional groups in the different regions: i.e. are there any specifically low-temperature or high-temperature types?

The evidence available suggests that there are no obvious major differences in functional groups in polar or temperate sediments. Benthic oxygen uptake rates in Arctic and Antarctic coastal sediments are similar (e.g. Nedwell *et al.*, 1993; Glud *et al.*, 1998) to those of temperate sediments at temperatures 20–25 °C higher. In Young Sound, Greenland, about half of the organic input was preserved by burial in bottom sediments (Rysgaard *et al.*, 1998) while, of the organic matter mineralized to CO_2, aerobic respiration accounted for 38 %, sulfate reduction for 33 %, iron reduction for 25 % and denitrification for only 4 % (manganese reduction was unimportant in these sediments). Reduced iron (and manganese if present) will be reoxidized by O_2 in the surface, oxic layer of sediment, facilitated by bioturbation of the surface layers of sediment, but failure to measure iron and manganese reduction directly will result in the significance of anaerobic mineralization by iron and manganese being underestimated and appearing as aerobic mineralization. Similar proportions of aerobic respiration and sulfate reduction have been reported for Antarctic sediments, although iron and nitrate respiration were not measured directly (Nedwell *et al.*, 1993). Rates of sulfate reduction were similar to those reported for temperate sediments, although the optimum temperature for sulfate reduction (15 °C) was very much higher than ambient temperature.

Methanogenesis in high-sulfate marine sediments is usually low because of competitive inhibition by sulfate reducers (Nedwell, 1984). While there is some evidence that low temperature may influence carbon flow through benthic communities, inhibiting hydrogenotrophic methanogenesis and shifting towards acetoclastic methanogenesis and acetogenesis (Nozhevnikova *et al.*, 1997; Schulz & Conrad, 1996; Fey & Conrad, 2000), there is no current evidence for this occurring in high-latitude coastal marine environments.

As in most coastal zone sediments, nitrate respiration in polar sediments accounts for only a very small proportion of organic matter degradation, largely due to the relatively low concentrations of nitrate in bottom water. Of the denitrification in Young Sound, Greenland, the vast majority (93 %) was derived from coupled nitrification–denitrification (Rysgaard *et al.*, 1998) within the sediment, and not from nitrate from the water column, where nitrate concentrations were low. Rates of denitrification reported from Arctic sediments (Devol *et al.*, 1997; Glud *et al.*, 1998; Henriksen *et al.*, 1993) are similar to those reported from temperate sediments (e.g. Dong *et al.*, 2000;

Jenkins & Kemp, 1984; Lohse *et al.*, 1996; Nielsen, 1992; Rysgaard *et al.*, 1995) and give no indication that low temperature influences nitrate respiration.

While functional group activity in coastal zone biogeochemical processes between regions seems similar, is there any evidence of different phylogenetic composition of polar and temperate/tropical functional group communities bringing about these processes? Molecular studies to date have largely focused on the phylogenetically cohesive groups (sulfate reducers and methanogenic archaea) to which the 16S rRNA approach is applicable, rather than to the phylogenetically diverse aerobic and nitrate-reducing communities. Molecular studies of polar marine sediments (Purdy *et al.*, 2003; Ravenschlag *et al.*, 1999, 2001; Sahm *et al.*, 1999) have shown them to have diverse communities of sulfate reducers and methanogenic archaea, although the diversity of both the archaeal and sulfate-reducer communities in Antarctic coastal sediment was lower than in communities in temperate coastal sediments (Purdy *et al.* 2003). Only five distinct groups of archaea were present in Signy Is. coastal sediment compared with 12 from temperate estuarine sediments. It has been suggested previously (e.g. Colinvaux, 1993) that community diversity is likely to be lower in physically or chemically stressed environments than in communities which are not physically or chemically stressed but where biological competition controls diversity, and this may be the case in these Antarctic sediments.

Methanogenesis in a Signy Is. coastal sediment represented only 2 % of carbon flow, and there was an equivalently small proportion (0·1 %) of methanogenic archaeal rRNA in the total sedimentary rRNA. (In contrast, in Signy Is. freshwater lake sediment archaeal rRNA represented 34 % of the prokaryotic community.) The coastal methanogens included members of both the *Methanosarcinales*, particularly *Methano-coccoides*, which metabolize C_1 compounds, and *Methanomicrobiales*, which metabolize hydrogen. Inhibition of sulfate reducers immediately stimulated methanogenesis in the coastal sediment, which suggests that, as in temperate sediments, the methanogens are outcompeted for substrates by sulfate reducers, which dominated the terminal steps of organic matter breakdown. Of the total prokaryotic rRNA in the Signy Is. coastal sediment, 15 % represented sulfate reducers, dominated by the *Desulfotalea/Desulforhopalus* group. *Desulfotalea* are psychrophiles first isolated from Arctic sediments (Knoblauch *et al.*, 1999; Sahm *et al.*, 1999), and seem to be quantitatively important in polar sediments. They are all incomplete oxidizers of their organic substrates, as are *Desulfovibrio*, which have not been detected in either Arctic or Antarctic sediments (Sahm *et al.*, 1999); Knoblauch *et al.* (1999) suggested that the *Desulfotalea/Desulforhopalus* group is the ecological equivalent in polar sediments of *Desulfovibrio*. As these groups are incomplete oxidizers, what completes the mineralization of organic matter in low-temperature coastal sediments by oxidizing acetate to

CO_2, a step which in temperate coastal sediment is brought about by sulfate reduction (Nedwell, 1984) and which in temperate estuarine sediments was attributed to *Desulfobacter* (Purdy *et al.*, 1997, 2001)?

The acetoclastic groups *Desulfobacter* and *Desulfotomaculum* reported in temperate sediments were not detected in the Antarctic sediment. Ravenschlag *et al.* (1999) reported a large number of clones in a general bacterial clone library from Svalbard sediment that were related to S^0- or Fe^{3+}- or Mn^{4+}-reducing *Desulfuromonas*, which utilize acetate, and closely related clones were present in Signy Is. coastal sediment (Purdy *et al.*, 2003). The implication is that acetate oxidation in these low-temperature sediments may be achieved by the acetate-oxidizing, S^0-reducing *Desulfuromonas* (i.e. that S^0 reduction rather than sulfate reduction may be important in the terminal step of organic carbon mineralization at low temperature) or by acetate-utilizing *Pelobacter* or *Geobacter*.

ROLE OF COASTAL WETLANDS

Biogeochemical cycling of elements in the coastal zone may be significantly influenced by the presence of coastal wetlands, predominantly salt marshes in the temperate region and mangroves in the tropics. The two ecosystems may be regarded as ecological equivalents: which system dominates seems to be determined by temperature in the coldest month > 16 °C or in the warmest month > 24 °C. Salt marshes and mangroves are highly productive ecosystems and one of the earliest hypotheses was that of 'outwelling' (Teal, 1962); that salt marshes export energy in the form of organic matter to the coastal zone. The same concept was proposed for mangroves (Odum & Heald, 1972; Robertson & Duke, 1990), and also extended to the idea of net export of nutrients such as nitrogen to the tropical coastal zone, in which primary production tends to be nutrient limited. Subsequent research indicated that net export from both salt marshes (Nixon, 1980) and mangroves varied enormously depending upon local physical conditions and the productivity of the biological community. Twilley *et al.* (1992) suggested that export from mangroves declined with increasing latitude as their productivity and litter production decreased; export from both mangroves and salt marshes has been related (e.g. Wolanski *et al.*, 1992) to the strength of the local tidal influence, with high-amplitude tides and strong directional water flow in estuaries resulting in increased tidal export from estuarine salt marshes or mangroves. Export of nitrogen to the coastal zone can undoubtedly occur from some mangroves, predominantly as particulate or dissolved organic nitrogen (e.g. Rivera-Monroy *et al.*, 1995), but they are usually sinks for inorganic nitrogen. However, the zone of influence of the mangrove or salt marsh in the adjacent coastal zone seems to be limited. Rodelli *et al.* (1984) measured $\delta^{13}C$ values in mangrove tissue, algae and consumer organisms and could trace the signal from mangrove detritus in the adjacent coastal zone in

Malaysia only within about 2 km of the mangrove; similar restricted zones of influence have been reported for Australian (Alongi, 1990), African (Hemminga *et al.*, 1994; Marguillier *et al.*, 1997), Indian (Bouillon *et al.*, 2002) and South American (Jennerjahn & Ittekkot, 1999, 2002) mangroves. The direct impact of mangrove-derived organic matter therefore seems to be of only limited extent in the adjacent coastal zone.

While the net export of either energy or nutrients from salt marshes or mangroves to the coastal zone may not be significant on an annual basis, in the temperate region there may be significant seasonal differences in their exchange and impact on the coastal zone. Thus, a UK salt marsh was balanced overall with respect to its annual nitrogen exchange with tidal water but there was net export of dissolved and particulate organic nitrogen and a smaller amount of ammonium during summer, when coastal sea-water nitrogen concentrations are low and the coastal zone phytoplankton are nutrient-depleted, but consistent removal of nitrate from tidal water by the salt-marsh sediments (Azni & Nedwell, 1986). This ability means that salt marshes, mangroves and estuarine sediments are highly efficient processors of nutrients in coastal sea water and buffer the loads of nutrients in the coastal zone, particularly when ambient inorganic nitrogen, usually nitrate, concentrations are increased by anthropogenic inputs (Corredor & Morell, 1994; Nedwell, 1975; Nedwell *et al.*, 1999). It has been estimated that estuarine and coastal sediments remove up to 50 % of the nitrate load through temperate estuaries (Nedwell, 1975; Seitzinger,1988; Nixon *et al.*, 1996). Jickells *et al.* (2000) argued that the pristine Humber estuary, with its extensive wetland and salt-marsh area, was likely to have been an effective sink for both N and P, but removal of > 90 % of these wetlands has reduced this capacity. Nonetheless, the removal of anthropogenically increased N and P loads through estuaries by estuarine sediments, including salt marshes and mangroves, remains an important 'buffer' to ameliorate the impact of these loads on the coastal zone.

The coastal ocean, estuaries and coastal wetlands may also be important in the global carbon budget. In an extensive review of ocean–atmosphere CO_2 fluxes in the coastal ocean Borges (2005) has argued that water–air fluxes of CO_2 and fluxes of dissolved inorganic carbon (DIC) from mangroves and salt marshes are important terms in their carbon budgets which have so far been ignored. Although the DIC fluxes do not seem to have significant effect on the DIC and pCO_2 values in adjacent coastal waters, their air–water CO_2 exchange, when scaled up, seems to have a significant influence upon global CO_2 budgets. If estuaries and coastal wetlands are not taken into consideration, the coastal ocean appears to be a sink for atmospheric CO_2 ($-1\cdot17$ mol C m^{-2} year^{-1}), and the estimated uptake of CO_2 by the global ocean is $-1\cdot93$ Pg C year^{-1}. If estuaries and coastal wetlands, with their high heterotrophic activity, are included in the scale-up, the coastal ocean behaves as a net source of CO_2 ($0\cdot38$ mol C year^{-1}) and the CO_2

uptake by the total global ocean decreases to -1.44 Pg year^{-1}. There are also interesting latitudinal differences. At high latitudes and in the tropics and subtropics the model predicts that, including estuaries and coastal wetlands, the coastal ocean is a net source of CO_2 to the atmosphere, but at temperate latitudes it remains a moderate sink for atmospheric CO_2. Much work remains to be done to constrain these global estimates adequately, but the significance of the coastal zone processes seems clear.

ACKNOWLEDGEMENTS

I would particularly like to thank my colleagues Kevin Purdy and Martin Embley for their productive, enjoyable, extensive and continued collaboration on the application of molecular techniques to coastal zone microbial ecology over an extended period and to the Natural Environment Research Council, UK, for their support in a number of research grants for this work. Thanks also to the British Antarctic Survey, particularly Cynan Ellis-Evans, for their collaboration and logistic support for the Antarctic research at Signy Island. I also acknowledge the helpful comments of Professor Michael Danson on adaptation of extremophile enzymes to high or low temperature.

REFERENCES

Alongi, D. M. (1990). Abundances of benthic microfauna in relation to outwelling of mangrove detritus in a tropical coastal region. *Mar Ecol Prog Ser* **63**, 53–63.

Alongi, D. M., Tirendi, F., Dixon, P., Trott, L. A. & Brunskill, G. J. (1999). Mineralization of organic matter in intertidal sediments of a tropical semi-enclosed delta. *Estuar Coastal Shelf Sci* **48**, 451–467.

Arnosti, C. & Jørgensen, B. B. (2003). High activity and low temperature optima of extracellular enzymes in Arctic sediments: implications for carbon cycling by heterotrophic microbial communities. *Mar Ecol Prog Ser* **249**, 15–24.

Azni, S. A. & Nedwell, D. B. (1986). The nitrogen cycle of an East Coast, U.K. saltmarsh. II. Nitrogen fixation, nitrification, denitrification, tidal exchange. *Estuar Coastal Shelf Sci* **22**, 689–704.

Borges, A. V. (2005). Do we have enough pieces of the jigsaw to integrate CO_2 fluxes in the coastal ocean? *Estuaries* **28**, 3–27.

Bouillon, S., Raman, A. V., Dauby, P. & Dehairs, F. (2002). Carbon and nitrogen stable isotope ratios of subtidal benthic invertebrates in an estuarine mangrove ecosystem (Andhra Pradesh, India). *Estuar Coastal Shelf Sci* **54**, 901–913.

Button, D. K. (1993). Nutrient-limited microbial growth kinetics: overview and recent advances. *Antonie van Leeuwenhoek* **63**, 225–235.

Canfield, D. E., Jørgensen, B. B., Fossing, H. & 7 other authors (1993). Pathways of organic carbon oxidation in three continental margin sediments. *Mar Geol* **113**, 27–40.

Colinvaux, P. (1993). *Ecology 2*. New York: Wiley.

Corredor, J. E. & Morell, J. M. (1994). Nitrate depuration of secondary sewage effluents in mangrove sediments. *Estuaries* **17**, 295–300.

Daly, K. L. & Smith, W. O., Jr (1993). Physical-biological interactions influencing marine plankton production. *Annu Rev Ecol Syst* **24**, 555–585.

Daniel, R. M., Danson, M. J. & Eisenthal, R. (2001). The temperature optima of enzymes: a new perspective on an old phenomenon. *Trends Biochem Sci* **26**, 223–225.

de Baar, H. J. W., de Jong, J. T. M., Bakker, D. C. E., Löscher, B. M., Veth, C., Bathmann, U. & Smetacek, V. (1995). Importance of iron for plankton blooms and carbon dioxide drawdown in the Southern Ocean. *Nature* **373**, 412–415.

Devol, A. H., Codispoti, L. A. & Christensen, J. P. (1997). Summer and winter denitrification rates in western Arctic shelf sediments. *Continent Shelf Res* **17**, 1029–1033.

Dong, L. F., Thornton, D. C. O., Nedwell, D. B. & Underwood, G. J. C. (2000). Denitrification in sediments of the River Colne estuary, England. *Mar Ecol Prog Ser* **203**, 109–122.

Eyre, B. & Balls, P. (1999). A comparative study of nutrient behavior along the salinity gradient of tropical and temperate estuaries. *Estuaries* **22**, 313–326.

Feller, G. & Gerday, C. (2003). Psychrophilic enzymes: hot topics in cold adaptation. *Nat Rev Microbiol* **1**, 200–208.

Fey, A. & Conrad, R. (2000). Effect of temperature on carbon and electron flow and on the archaeal community in methanogenic rice field soil. *Appl Environ Microbiol* **66**, 4790–4797.

Franzmann, P. D. (1996). Examination of Antarctic prokaryote diversity through molecular comparisons. *Biodivers Conserv* **5**, 1295–1305.

Glud, R. N., Holby, O., Hoffmann, F. & Canfield, D. E. (1998). Benthic mineralization and exchange in Arctic sediments (Svalbard, Norway). *Mar Ecol Prog Ser* **173**, 237–251.

Hemminga, M. A., Slim, F. J., Kazungu, J., Gaanssen, G. M., Nieuwenhuize, J. & Kruyt, N. M. (1994). Carbon outwellings from a mangrove forest with adjacent seagrass beds and coral reefs (Gazi Bay, Kenya). *Mar Ecol Prog Ser* **106**, 291–301.

Henriksen, K., Blackburn, T. H., Lomstein, B. A. & McRoy, C. P. (1993). Rates of nitrification, distribution of nitrifying bacteria and inorganic N fluxes in northern Bering-Chukchi shelf sediments. *Continent Shelf Res* **13**, 629–651.

Holmboe, N., Kristensen, E. & Andersen, F. Ø. (2001). Anoxic decomposition in sediments from a tropical mangrove forest and the temperate Wadden Sea: implications of N and P addition experiments. *Estuar Coastal Shelf Sci* **53**, 125–140.

Hoppe, H.-G., Gocke, K., Koppe, R. & Begler, S. (2002). Bacterial growth and primary production along a north-south transect of the Atlantic Ocean. *Nature* **416**, 168–171.

Hulth, S., Blackburn, T. H. & Hall, P. O. J. (1994). Arctic sediments (Svalbard): consumption and microdistribution of oxygen. *Mar Chem* **46**, 293–316.

Jenkins, M. C. & Kemp, W. M. (1984). The coupling of nitrification and denitrification in two estuarine sediments. *Limnol Oceanogr* **29**, 609–619.

Jennerjahn, T. C. & Ittekkot, V. (1999). Changes in organic matter from surface waters to continental slope sediments off the São Francisco River, eastern Brazil. *Mar Geol* **161**, 129–140.

Jennerjahn, T. C. & Ittekkot, V. (2002). Relevance of mangroves for the production and deposition of organic matter along tropical continental margins. *Naturwissenschaften* **89**, 23–30.

Jickells, T., Andrews, J., Samways, G. & 7 other authors (2000). Nutrient fluxes through the Humber estuary – past, present and future. *Ambio* **29**, 130–135.

Joint, I. & Pomroy, A. (1993). Phytoplankton biomass and production in the southern North Sea. *Mar Ecol Prog Ser* **99**, 169–182.

Jørgensen, B. B. & Revsbech, N. P. (1985). Diffusive boundary layer and the oxygen uptake of sediments and detritus. *Limnol Oceanogr* **30**, 111–122.

Karl, D. M., Christian, J. R., Dore, J. E. & Letelier, R. M. (1996). Microbial oceanography in the region west of the Antarctic Peninsula: microbial dynamics, nitrogen cycle and carbon flux. In *Foundation for Ecological Research West of the Antarctic Peninsula*, Antarctic Research Series vol. 70, pp. 303–332. Washington, DC: American Geophysical Union.

King, D. & Nedwell, D. B. (1984). Changes in the nitrate-reducing community of an anaerobic saltmarsh sediment in response to seasonal selection by temperature. *J Gen Microbiol* **130**, 2935–2941.

Knoblauch, C., Sahm, K. & Jørgensen, B. B. (1999). Psychrophilic sulfate-reducing bacteria isolated from permanently cold Arctic marine sediments: description of *Desulfofrigus oceanense* gen. nov., sp. nov., *Desulfofrigus fragile* sp. nov., *Desulfofaba gelida* gen. nov., sp. nov., *Desulfotalea psychrophila* gen. nov., sp. nov. and *Desulfotalea arctica* sp. nov. *Int J Syst Bacteriol* **49**, 1631–1643.

Kocum, E., Underwood, G. J. C. & Nedwell, D. B. (2002a). Simultaneous measurement of phytoplanktonic primary production, nutrient and light availability along a turbid, eutrophic UK east coast estuary (the Colne Estuary). *Mar Ecol Prog Ser* **231**, 1–12.

Kocum, E., Nedwell, D. B. & Underwood, G. J. C. (2002b). Regulation of phytoplankton primary production along a hypernutrified estuary. *Mar Ecol Prog Ser* **231**, 13–22.

Kristensen, E., Holmer, M. & Bussarawit, N. (1991). Benthic metabolism and sulphate reduction in a southern Asian mangrove swamp. *Mar Ecol Prog Ser* **73**, 93–103.

Kristensen, E., Devol, A., Ahmed, S. & Saleem, M. (1992). Preliminary study of benthic metabolism and sulfate reduction in a mangrove swamp of the Indus Delta, Pakistan. *Mar Ecol Prog Ser* **90**, 287–297.

Li, W. K. W. (1980). Temperature adaptation in phytoplankton: cellular and photosynthetic characteristics. In *Primary Productivity in the Sea*, Environmental Science Research series, vol. 19, pp. 259–279. Edited by P. G. Falkowski. New York: Plenum.

Li, W. K. W. (1985). Photosynthetic response to temperature of marine phytoplankton along a latitudinal gradient (16° N to 74° N). *Deep Sea Res* **32**, 1381–1391.

Li, W. K. W., Smith, J. C. & Platt, T. (1984). Temperature responses of photosynthetic capacity and carboxylase activity in Arctic marine phytoplankton. *Mar Ecol Prog Ser* **17**, 237–243.

Lohse, L., Kloosterhuis, T., van Raaphorst, W. & Helder, W. (1996). Denitrification rates as measured by the isotope pairing method and by the acetylene inhibition technique in continental shelf sediments of the North Sea. *Mar Ecol Prog Ser* **132**, 169–179.

Longhurst, A., Sathyendranath, S., Platt, T. & Caverhill, C. (1995). An estimate of global primary production in the ocean from satellite radiometer data. *J Plankton Res* **17**, 1245–1271.

Mackin, J. E. & Swider, K. T. (1989). Organic matter decomposition pathways and oxygen consumption in coastal marine sediments. *J Mar Res* **47**, 681–716.

Marguillier, S., van der Velde, G., Dehairs, F., Hemminga, M. A. & Rajagopal, S. (1997). Trophic relationships in an interlinked mangrove-seagrass ecosystem as traced by $\delta^{13}C$ and $\delta^{15}N$. *Mar Ecol Prog Ser* **151**, 115–121.

Martin, J. H., Gordon, R. M. & Fitzwater, S. E. (1990). Iron in Antarctic waters. *Nature* **345**, 156–158.

Mincks, S. L., Smith, C. R. & DeMaster, D. J. (2005). Persistence of labile organic matter and microbial biomass in Antarctic shelf sediments: evidence of a sediment 'food bank'. *Mar Ecol Prog Ser* (in press).

Morita, R. Y. (1975). Psychrophilic bacteria. *Bacteriol Rev* **39**, 144–167.

Nedwell, D. B. (1975). Inorganic nitrogen metabolism in a eutrophicated tropical mangrove estuary. *Water Res* **9**, 221–231.

Nedwell, D. B. (1984). The input and mineralisation of organic carbon in anaerobic aquatic sediments. In *Advances in Microbial Ecology*, pp. 93–131. Edited by K. C. Marshall. London: Plenum.

Nedwell, D. B. (1989). Benthic microbial activity in an Antarctic coastal sediment at Signy Island, South Orkney Islands. *Estuar Coastal Shelf Sci* **28**, 507–516.

Nedwell, D. B. (1999). Effect of low temperature on microbial growth: lowered affinity for substrates limits growth at low temperature. *FEMS Microbiol Ecol* **30**, 101–111.

Nedwell, D. B. & Rutter, M. (1994). Influence of temperature on growth rate and competition between two psychrotolerant Antarctic bacteria: low temperature diminishes affinity for substrate uptake. *Appl Environ Microbiol* **60**, 1984–1992.

Nedwell, D. B., Walker, T. R., Ellis-Evans, J. C. & Clarke, A. (1993). Measurements of seasonal rates and annual budgets of organic carbon fluxes in an Antarctic coastal environment at Signy Island, South Orkney Islands, suggest a broad balance between production and decomposition. *Appl Environ Microbiol* **59**, 3989–3995.

Nedwell, D. B., Blackburn, T. H. & Wiebe, W. J. (1994). Dynamic nature of the turnover of organic carbon, nitrogen and sulphur in the sediments of a Jamaican mangrove forest. *Mar Ecol Prog Ser* **110**, 223–231.

Nedwell, D. B., Jickells, T., Trimmer, M. T. & Sanders, R. (1999). Nutrients in estuaries. In *Estuaries*, Advances in Ecological Research vol. 29, pp. 43–92. Edited by D. B. Nedwell & D. Raffaelli. London: Harcourt.

Nielsen, L. P. (1992). Denitrification in sediment determined from nitrogen isotope pairing. *FEMS Microbiol Ecol* **86**, 357–362.

Nixon, S. W. (1980). Between coastal marshes and coastal waters – a review of twenty years of speculation and research on the role of saltmarshes in estuarine productivity and water chemistry. In *Estuarine and Wetland Processes with Emphasis upon Modelling*, pp. 437–525. Edited by P. Hamilton & K. B. Macdonald. New York: Plenum.

Nixon, S. W., Ammerman, J. W., Atkinson, L. P. & 13 other authors (1996). The fate of nitrogen and phosphorus at the land-sea margin of the North Atlantic Ocean. *Biogeochemistry* **35**, 141–180.

Nozhevnikova, A. N., Holliger, C., Ammann, A. & Zehnder, A. J. B. (1997). Methanogenesis in sediments from deep lakes at different temperatures (2–70 °C). *Water Sci Technol* **36** (6–7), 57–64.

Odum, W. E. & Heald, E. J. (1972). Trophic analyses of an estuarine mangrove community. *Bull Mar Sci* **22**, 671–738.

Ogilvie, B. G., Rutter, M. & Nedwell, D. B. (1997). Selection by temperature of nitrate-reducing bacteria from estuarine sediments: species composition and competition for nitrate. *FEMS Microbiol Ecol* **23**, 11–22.

Osborne, P. L. (2000). *Tropical Ecosystems and Ecological Concepts*. Cambridge: Cambridge University Press.

Peterson, M. E., Eisenthal, R., Danson, M. J., Spence, A. & Daniel, R. M. (2004). A new intrinsic thermal parameter for enzymes reveals true temperature optima. *J Biol Chem* **279**, 20717–20722.

Pfannkuche, O. & Thiel, H. (1987). Meiobenthic stocks and benthic activity on the NE-Svalbard Shelf and in the Nansen Basin. *Polar Biol* **7**, 253–266.

Pomeroy, L. R. & Deibel, D. (1986). Temperature regulation of bacterial activity during the spring bloom in Newfoundland coastal waters. *Science* **233**, 359–361.

Pomeroy, L. R., Wiebe, W. J., Deibel, D., Thompson, R. J., Rowe, G. T. & Pakulski, J. D. **(1991).** Bacterial responses to temperature and substrate concentration during the Newfoundland spring bloom. *Mar Ecol Prog Ser* **75**, 143–159.

Purdy, K. J., Nedwell, D. B., Embley, T. M. & Takii, S. **(1997).** Use of 16S rRNA-targeted oligonucleotide probes to investigate the occurrence and selection of sulfate-reducing bacteria in response to nutrient addition to sediment slurry microcosms from a Japanese estuary. *FEMS Microbiol Ecol* **24**, 221–234.

Purdy, K. J., Nedwell, D. B., Embley, T. M. & Takii, S. **(2001).** Using 16S rRNA-targeted oligonucleotide probes to investigate the distribution of sulphate-reducing bacteria in estuarine sediments. *FEMS Microbiol Ecol* **36**, 165–168.

Purdy, K. J., Nedwell, D. B. & Embley, T. M. **(2003).** Analysis of the sulfate-reducing bacterial and methanogenic archaeal populations in contrasting Antarctic sediments. *Appl Environ Microbiol* **69**, 3181–3191.

Ravenschlag, K., Sahm, K., Pernthaler, J. & Amman, R. **(1999).** High bacterial diversity in permanently cold marine sediments. *Appl Environ Microbiol* **65**, 3982–3989.

Ravenschlag, K., Sahm, K. & Amman, R. **(2001).** Quantitative molecular analysis of the microbial community in marine Arctic sediments (Svalbard). *Appl Environ Microbiol* **67**, 387–395.

Reay, D. S., Nedwell, D. B., Priddle, J. & Ellis-Evans, C. J. **(1999).** Temperature dependence of inorganic nitrogen uptake: reduced affinity for nitrate at suboptimal temperatures in both algae and bacteria. *Appl Environ Microbiol* **65**, 2577–2584.

Reay, D. S., Priddle, J., Nedwell, D. B., Whitehouse, M. J., Ellis-Evans, J. C., Deubert, C. & Connelly, D. **(2001).** Regulation by low temperature of phytoplankton growth and nutrient uptake in the Southern Ocean. *Mar Ecol Prog Ser* **219**, 51–64.

Rivera-Monroy, H., Day, J. W., Twilley, R. W., Vera-Herrera, F. & Coronado-Molina, C. **(1995).** Flux of nitrogen and sediment in a fringe mangrove forest in Terminos Lagoon, Mexico. *Estuar Coastal Shelf Sci* **40**, 139–160.

Rivkin, R. B., Anderson, M. R. & Lajzerowicz, C. **(1996).** Microbial processes in cold oceans. I. Relationship between temperature and bacterial growth rate. *Aquat Microb Ecol* **10**, 243–254.

Robertson, A. I. & Duke, N. C. **(1990).** Mangrove fish communities in tropical Queensland, Australia: spatial and temporal patterns in densities, biomass and community structure. *Mar Biol* **104**, 369–379.

Robertson, A. I., Dixon, P. & Alongi, D. M. **(1998).** The influence of fluvial discharge on pelagic production in the Gulf of Papua, Northern Coral Sea. *Estuar Coastal Shelf Sci* **46**, 319–331.

Rodelli, M. R., Gearing, J. N., Gearing, P. J., Marshall, N. & Sasekumar, A. **(1984).** Stable isotope ratio as a tracer of mangrove carbon in Malaysian ecosystems. *Oecologia* **61**, 326–333.

Russell, N. J. **(1990).** Cold adaptation of microorganisms. *Philos Trans R Soc Lond B Biol Sci* **326**, 595–611.

Russell, N. J. **(1992).** Psychrophilic bacteria. In *Molecular Biology and Biotechnology of Extremophiles*, pp. 203–224. Edited by R. A. Herbert & R. S. Sharp. Glasgow: Blackie.

Russell, N. J. **(1998).** Molecular adaptations in psychrophilic bacteria: potential for biotechnological applications. *Adv Biochem Eng Biotechnol* **61**, 1–21.

Rutter, M. & Nedwell, D. B. **(1994).** Influence of changing temperature on growth rate and competition between two psychrotolerant Antarctic bacteria: competition and survival in non-steady-state temperature environments. *Appl Environ Microbiol* **60**, 1993–2002.

Rysgaard, S., Christensen, P. B. & Nielsen, L. P. (1995). Seasonal variation in nitrification and denitrification in estuarine sediment colonized by benthic microalgae and bioturbating infauna. *Mar Ecol Prog Ser* **126**, 111–121.

Rysgaard, S., Finster, K. & Dahlgaard, H. (1996). Primary production, nutrient dynamics and mineralisation in a northeastern Greenland fjord during the summer thaw. *Polar Biol* **16**, 497–506.

Rysgaard, S., Thamdrup, B., Risgaard-Petersen, N., Fossing, H., Berg, P., Christensen, P. B. & Dalsgaard, T. (1998). Seasonal carbon and nutrient mineralization in a high-Arctic coastal marine sediment, Young Sound, Northeast Greenland. *Mar Ecol Prog Ser* **175**, 261–276.

Rysgaard, S., Nielsen, T. G. & Hansen, B. W. (1999). Seasonal variation in nutrients, pelagic primary production and grazing in a high-Arctic coastal marine ecosystem, Young Sound, Northeast Greenland. *Mar Ecol Prog Ser* **179**, 13–25.

Sahm, K., Knoblauch, C. & Amann, R. (1999). Phylogenetic affiliation and quantification of psychrophilic sulfate-reducing isolates in marine Arctic sediments. *Appl Environ Microbiol* **65**, 3976–3981.

Schulz, S. & Conrad, R. (1996). Influence of temperature on pathways to methane production in the permanently cold profundal sediment of Lake Constance. *FEMS Microbiol Ecol* **20**, 1–14.

Seitzinger, S. P. (1988). Denitrification in freshwater and coastal marine ecosystems: ecological and geochemical significance. *Limnol Oceanogr* **33**, 702–724.

Sieburth, J. M. (1967). Seasonal selection of estuarine bacteria by water temperature. *J Exp Mar Biol Ecol* **1**, 98–121.

Sinensky, M. (1974). Homeoviscous adaptation – a homeostatic process that regulates the viscosity of membrane lipids in *Escherichia coli*. *Proc Natl Acad Sci U S A* **71**, 522–525.

Sobey, N. S. (2004). *The role of bottom sediments in nitrogen cycling in a tropical lagoon.* PhD thesis, University of Essex, Colchester, UK.

Tahey, T. M., Duinevald, G. C. A., Berghuis, E. M. & Helder, W. (1994). Relation between sediment-water fluxes of oxygen and silicate and faunal abundance at continental shelf, slope and deep water stations in northwest Mediterranean. *Mar Ecol Prog Ser* **104**, 119–130.

Teal, J. M. (1962). Energy flow in the salt marsh ecosystem of Georgia. *Ecology* **43**, 614–624.

Trimmer, M., Nedwell, D. B., Sivyer, D. B. & Malcolm, S. J. (2000). Seasonal benthic organic matter mineralisation measured by oxygen uptake and denitrification along a transect of the inner and outer River Thames estuary, UK. *Mar Ecol Prog Ser* **197**, 103–119.

Twilley, R. R., Chen, R. H. & Hargis, T. (1992). Carbon sinks in mangroves and their implications to carbon budget of tropical coastal ecosystems. *Water Air Soil Pollut* **64**, 265–288.

Upton, A. C. & Nedwell, D. B. (1989). Temperature responses of bacteria isolated from different Antarctic environments. In *University Research in Antarctica: Proceedings of British Antarctic Survey Special Topic Award Scheme Symposium*, pp. 97–101. Edited by R. B. Heywood. Cambridge: British Antarctic Survey.

Upton, A. C., Nedwell, D. B., Parkes, R. J. & Harvey, S. M. (1993). Seasonal benthic microbial activity in the southern North Sea: oxygen uptake and sulphate reduction. *Mar Ecol Prog Ser* **101**, 273–281.

Wiebe, W. J., Sheldon, W. M., Jr & Pomeroy, L. R. (1992). Bacterial growth in the cold:

evidence for an enhanced substrate requirement. *Appl Environ Microbiol* **58**, 359–364.

Wiebe, W. J., Sheldon, W. M., Jr & Pomeroy, L. R. (1993). Evidence for an enhanced substrate requirement by marine mesophilic bacterial isolates at minimal growth temperatures. *Microb Ecol* **25**, 151–159.

Wolanski, E., Mazda, Y. & Ridd, P. (1992). Mangrove hydrodynamics. In *Tropical Mangrove Ecosystems*, pp. 43–62. Edited by A. I. Robertson & D. M. Alongi. Washington, DC: American Geophysical Union.

Fungal roles and function in rock, mineral and soil transformations

Geoffrey M. Gadd, Marina Fomina and Euan P. Burford

Division of Environmental and Applied Biology, Biological Sciences Institute,
School of Life Sciences, University of Dundee, Dundee DD1 4HN, Scotland, UK

INTRODUCTION

The most important perceived environmental roles of fungi are as decomposer organisms, plant pathogens and symbionts (mycorrhizas, lichens), and in the maintenance of soil structure through their filamentous growth habit and production of exopolymers. However, a broader appreciation of fungi as agents of biogeochemical change is lacking and, apart from obvious connections with the carbon cycle, they are frequently neglected within broader microbiological and geochemical research contexts. While the profound geochemical activities of bacteria and archaea receive considerable attention, especially in relation to carbon-limited and/or anaerobic environments (see elsewhere in this volume), in aerobic environments fungi are of great importance, especially when considering rock surfaces, soil and the plant root–soil interface (Gadd, 2005a). For example, mycorrhizal fungi are associated with ~ 80 % of plant species and are involved in major mineral transformations and redistributions of inorganic nutrients, e.g. essential metals and phosphate, as well as carbon flow. Free-living fungi have major roles in the decomposition of plant and other organic materials, including xenobiotics, as well as mineral solubilization (Gadd, 2004). Lichens (a symbiosis between an alga or cyanobacterium and a fungus) are one of the commonest members of the microbial consortia inhabiting exposed subaerial rock substrates, and play fundamental roles in early stages of rock colonization and mineral soil formation. Fungi are also major biodeterioration agents of stone, wood, plaster, cement and other building materials, and it is now realized that they are important components of rock-inhabiting microbial communities, with significant roles in mineral dissolution and secondary mineral formation (Burford *et al.*, 2003a, b). The objective of this chapter is to outline

SGM symposium 65: Micro-organisms and Earth systems – advances in geomicrobiology.
Editors G. M. Gadd, K. T. Semple & H. M. Lappin-Scott. Cambridge University Press. ISBN 0 521 86222 1 ©SGM 2005

important fungal roles and function in rock, mineral and soil transformations and to emphasize the importance of fungi as agents of geological change.

WEATHERING PROCESSES AND THE INFLUENCE OF MICROBES

The composition of the Earth's lithosphere, biosphere, hydrosphere and atmosphere is influenced by weathering processes (Ferris et al., 1994; Banfield et al., 1999; Vaughn et al., 2002). Rock substrates and their mineral constituents are weathered through physical (mechanical), chemical and biological mechanisms; the relative significance of each process varying widely depending on environmental and other conditions. Near-surface weathering of rocks and minerals, which occurs in subaerial (i.e. situated, formed or occurring on or immediately adjacent to the surface of the Earth) and sub-soil (i.e. not exposed to the open air) environments, will often involve an interaction between all three mechanisms (White et al., 1992), and it is the biological component of the overall process that provides much microbiological interest. At or near the Earth's surface, interaction between minerals, metals and non-metallic species in an aqueous fluid nearly always involves the presence of microbes or their metabolites (Banfield & Nealson, 1998). Mineral replacement reactions in rocks mainly occur by dissolution–re-precipitation processes (e.g. cation exchange, chemical weathering, leaching and diagenesis), where one mineral or mineral assemblage is replaced by a more stable assemblage (Putnis, 2002). Micro-organisms can influence this by mineral dissolution, biomineralization and alteration of mineral surface chemistry and reactivity (Hochella, 2002). Mineral dissolution can also be inhibited by extracellular microbial poly-saccharides that passivate reactive centres on minerals (Welch & Vandevivere, 1994; Welch et al., 1999). In contrast, mineral dissolution may be accelerated by microbially mediated pH changes and many other changes in solution chemistry. Excretion of organic ligands and siderophores not only enhances nutrient acquisition by micro-organisms but can markedly affect mineral composition and dissolution reactions (Grote & Krumbein, 1992; Maurice et al., 1995; Hersman et al., 1995; Stone, 1997; Gadd, 1999; Kraemer et al., 1999; Liermann et al., 2000; Sayer & Gadd, 2001). In addition, the formation of secondary minerals (biogenic crystalline precipitates) can occur through metabolism-independent and metabolism-dependent processes, often being influenced by abiotic and biotic factors such as environmental pH, microbial cell density and the composition of cell walls (Ferris et al., 1987; Gadd, 1990, 1993; Arnott, 1995; Thompson & Ferris, 1990; Douglas & Beveridge, 1998; Fortin et al., 1998; Sterflinger, 2000; Verrecchia, 2000). The term 'biomineralization' refers to biologically induced mineralization where an organism modifies the local microenvironment, creating conditions that promote chemical precipitation of extracellular mineral phases (Hamilton, 2003). Secondary minerals can form by direct nucleation on cellular macromolecules, e.g. melanin and chitin in fungal cell walls (Gadd, 1990, 1993; Fortin

& Beveridge, 1997; Beveridge *et al.*, 1997). However, indirect precipitation of secondary minerals also frequently occurs as a result of microbially mediated changes in solution conditions (Fortin & Beveridge, 1997; Banfield *et al.*, 2000). Regardless of the mechanism, lithospheric weathering of rocks and minerals can result in the mobilization and redistribution of essential nutrients (e.g. P, S) and metals (e.g. Na, K, Mg, Ca, Mn, Fe, Cu, Zn, Co and Ni) required for plant and microbial growth. In addition, non-essential metals (e.g. Cs, Al, Cd, Hg and Pb) may also be mobilized from mineral and soil pools (Gadd, 1993, 2001a, b; Morley *et al.*, 1996).

Bioweathering can be defined as the erosion, decay and decomposition of rocks and minerals mediated by living organisms. Micro-organisms, as well as animals and plants, can weather rock aggregates through biomechanical and biochemical attack on component minerals (Goudie, 1996; Adeyemi & Gadd, 2005). Filamentous micro-organisms, plant roots and burrowing animals can physically affect rocks and enhance splitting and fractionation: the disruptive (hydraulic) pressure of growing roots and hyphae is important here (Sterflinger, 2000; Money, 2001). However, biochemical actions of organisms are believed to be more significant processes than mechanical degradation (Sterflinger, 2000; Etienne, 2002). Microbes, e.g. bacteria, algae and fungi, and plants can mediate chemical weathering of rocks and minerals through excretion of for example H^+, organic acids and other metabolites, while respiratory CO_2 can lead to carbonic acid attack on mineral surfaces (Johnstone & Vestal, 1993; Ehrlich, 1998; Sterflinger, 2000; Gadd & Sayer, 2000). Biochemical weathering of rocks can result in changes in the microtopography of minerals through pitting and etching, mineral displacement reactions and even complete dissolution of mineral grains (Ehrlich, 1998; Kumar & Kumar, 1999; Adeyemi & Gadd, 2005).

MICROBES IN ROCK AND MINERAL HABITATS

Micro-organisms occur in rock and building stone in a variety of microhabitats, being classed as epilithic, hypolithic, endolithic, chasmolithic, cryptoendolithic and euendolithic organisms (Fig. 1) (Gerrath *et al.*, 1995, 2000; May, 2003). Epiliths are common under humid conditions and occur on the surface of rocks and building stone. Hypolithic micro-organisms are often found under and attached to pebbles, particularly in hot and cold deserts. Endoliths inhabit the rock subsurface and may form distinct masses or brightly coloured layers. Endolithic micro-organisms can occur as chasmoliths that grow in pre-existing cracks and fissures within rock, often being visible from the rock surface. Conversely, cryptoendoliths grow inside cavities and among crystal grains and cannot be observed from the rock surface. Euendolithic microbes are a specialized group of cryptoendoliths that are capable of actively penetrating (boring) into submerged rock (Ehrlich, 1998; Gerrath *et al.*, 1995).

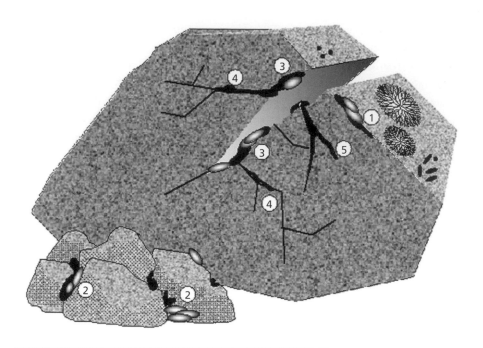

Fig. 1. Terminology for rock-dwelling micro-organisms. 1, Epilithic micro-organisms, occurring on the surface of rocks and building stone; 2, hypolithic micro-organisms, found under and attached to pebbles; 3–5, endolithic micro-organisms, inhabiting the rock subsurface, which include chasmoliths (3), which grow in pre-existing cracks and fissures within the rock, cryptoendoliths (4), which grow inside cavities and among crystal grains, and euendoliths (5), which are a specialized group of cryptoendoliths that are capable of actively penetrating (boring) into rock. Organisms shown can represent various microbial groups including bacteria, cyanobacteria, microalgae, fungi and lichens.

Micro-organisms play a fundamental role in mineral transformations in the natural environment, most notably in the formation of mineral soils from rock and the cycling of elements (May, 2003; Gadd, 2005a). It is not surprising therefore that a wide variety of micro-organisms, including bacteria, algae and fungi, inhabit rocks and stone-work of buildings and historic monuments (Ehrlich, 2002; Burford *et al.*, 2003a; Gleeson *et al.*, 2005). Exposed surfaces are not necessarily conducive to microbial growth, as a result of moisture deficit, exposure to UV solar radiation and limited availability of nutrients. However, complex interactions between microbes and the mineral substrate are frequently observed, often to some distance into the mineral (May, 2003). The rock micro-environment is subject to diurnal and seasonal changes in for example temperature and moisture as well as in available nutrients (Gorbushina & Krumbein, 2000; Roldan *et al.*, 2002). Nutrients may accumulate as a result of water interactions, wind-blown dust particles, animal faeces and death and degradation of living organisms and be utilized by microbes. Mineral grains within the host-rock may also serve as a source of metals essential for microbial growth. The transfer of biological material (e.g. fungal spores and other reproductive structures) from external sources may also play a role in the colonization of subaerial environments by microbes. Physical properties (e.g. porosity) and elemental composition of the host rock (e.g. C, P, K, S and metal content) may govern initial establishment, growth and survival of microbial communities (Gleeson *et al.*, 2005). Thus, colonization of rock substrates by microbes and the development of a microbial consortium is likely to be influenced by physical and chemical properties and interactions based on environmental (e.g. macro-/micro-climate) and biological factors resulting in and influencing ecological succession at the micro-scale.

MICROBIAL PROCESSES INFLUENCED BY MINERALS

Many important microbial processes can be influenced by minerals, including energy generation, nutrient acquisition, cell adhesion and biofilm formation (Hochella, 2002; see elsewhere in this volume). Micro-organisms can also acquire essential nutrients for microbial growth from mineral surfaces, which effectively concentrate these vital nutrients far above surrounding environmental levels, e.g. C, N, P, Fe and various organic compounds (Vaughn *et al.*, 2002). Some environmental contaminants, which may be concentrated on mineral surfaces by various sorption reactions, can be displaced by similar microbial processes (Kraemer *et al.*, 1999). In addition, it is likely that potentially toxic metals, released from minerals as a result of physico-chemical and biological processes, will have an effect on microbial communities (Gadd, 2005b). Mineral surface properties (e.g. microtopography, surface composition, surface charge and hydrophobicity) also play an integral role in microbial attachment and detachment. They are therefore critical in biofilm formation and the ecology of microbial populations in and on mineral substrates (Wolfaardt *et al.*, 1994; Fredrickson *et al.*, 1995; Bennett *et al.*, 1996; Rogers *et al.*, 1998).

FUNGI IN THE TERRESTRIAL ENVIRONMENT

Fungi are ubiquitous components of terrestrial microbial communities, with soil being regarded as their most characteristic habitat. Subaerial rock surfaces can be considered an inhospitable habitat for fungal (and other microbial) growth due to their high degree of insolation, desiccation and limited availability of nutrients (Gorbushina & Krumbein, 2000). Micro-organisms that thrive under these extreme conditions have been termed 'poikilotrophic', i.e. able to deal with varying micro-climatic conditions such as light, salinity, pH and moisture. Microbial biofilms on and in rocks are believed to be major factors in rock decay and also in the formation of various patinas, films, varnishes, crusts and stromatolites in rock substrates (Gorbushina & Krumbein, 2000).

Fungi have been reported from a wide range of rock types including limestone, soapstone, marble, granite, sandstone, andesite, basalt, gneiss, dolerite, amphibolite and quartz, from a variety of environments (Staley *et al.*, 1982; Gorbushina *et al.*, 1993;

Sterflinger, 2000; Verrecchia, 2000; Burford *et al.*, 2003a, b). It is likely that they are ubiquitous components of the microflora of all rocks and building stone, throughout a wide range of geographical and climatic zones. Despite the apparent inhospitality of the rock environment, the presence of organic and inorganic residues on mineral surfaces or within cracks is thought to encourage proliferation of fungi and other microbes. Waste products of algae and bacteria, dead cells, decaying plant material, dust particles, aerosols and animal faeces can all act as nutrient sources for fungi (Sterflinger, 2000). Some extremophilic fungi are especially evolved to survive and exploit microhabitats on and within mineral substrata, occurring within the lichen symbiosis or as free-living microcolonial fungi (Gorbushina *et al.*, 1993; Bogomolova *et al.*, 1998; Sterflinger, 2000). Microcolonial fungi include those black fungi that occur as spherical clusters of tightly packed thick-pigmented-walled cells or hyphae (Gorbushina *et al.*, 1993; Bogomolova *et al.*, 1998). As well as these, other filamentous fungi, including zygomycetes, ascomycetes and basidiomycetes, often occur on rock surfaces (epiliths) and in cracks, fissures and pores (endoliths). Certain fungi may also actively 'burrow' into rock substrates (cryptoendoliths). In addition, many deutero-mycetes (Fungi Imperfecti), which only exhibit asexual reproduction, are commonly found in mineral substrates (Kumar & Kumar, 1999; Sterflinger, 2000; Verrecchia, 2000).

In soil, fungi generally comprise the largest pool of biomass (including other microbes and invertebrates) and this, combined with their filamentous growth habit, ensures that fungus–mineral interactions are an integral component of biogeochemical processes in the soil (Gadd, 1993, 1999, 2000a). They occur as free-living filamentous forms, plant symbionts, unicellular yeasts and animal and plant pathogens, and play an important role in carbon cycling and other biogeochemical cycles (Gadd & Sayer, 2000; Gadd, 2005a). Their ability to translocate nutrients through the mycelial network also provides significant environmental advantages (Fomina *et al.*, 2003; Jacobs *et al.*, 2004). Mycorrhizal fungi in particular are one of the most important ecological groups of soil fungi in terms of mineral weathering and dissolution of insoluble minerals (Paris *et al.*, 1995; Jongmans *et al.*, 1997; Lundström *et al.*, 2000; Hoffland *et al.*, 2002; Martino *et al.*, 2003; Fomina *et al.*, 2004, 2005c).

MECHANISMS OF ROCK WEATHERING BY FUNGI

Biomechanical deterioration

Fungi are an important component of lithobiotic communities (an association of micro-organisms forming a biofilm at the mineral–microbe interface), where they interact with the lithic substrate, both geophysically and geochemically (de los Rios *et al.*, 2002; Burford *et al.*, 2003a, b). Biomechanical deterioration of rocks may occur through hyphal penetration (e.g. into decayed limestone) and by tunnelling into otherwise intact mineral material (e.g. along crystal planes in calcitic and dolomitic rocks) (Kumar & Kumar, 1999; Sterflinger, 2000). Fungal hyphae can also exploit grain boundaries, cleavages and cracks to gain access to mineral surfaces (Adeyemi & Gadd, 2005). In lichens, cleavage-bound mineral fragments as small as 5 μm in diameter can accumulate within the lower thallus (Banfield *et al.*, 1999). An important feature of hyphal growth is spatial exploration of the environment to locate and exploit new substrates (Boswell *et al.*, 2002, 2003; Jacobs *et al.*, 2002a). This is facilitated by a range of tropic responses that determine the direction of hyphal growth. Among the tropisms that may occur, thigmotropism or contact guidance is a well-known property of fungi that grow on and within solid substrates (Watts *et al.*, 1998). Hyphal growth can often be influenced by grooves, ridges and pores in solid substrate and is more prevalent in weakened mineral surfaces. Chemotropism, and other nutritional responses, may also be important in stress avoidance, such as that raised by toxic metals (Fomina *et al.*, 2000a; Gadd *et al.*, 2001).

The process of invasive hyphal growth due to turgor pressure inside hyphae allows fungi to acquire nutrients from many solid materials (Money, 2001). Highly pressurized hyphae can penetrate tougher substrates than those with lower pressures, and fungi that naturally invade hard materials generate extraordinarily high pressures (Money, 1999; Money & Howard, 1996). Melanin has also been implicated in the penetrative ability of plant-pathogenic fungi, as this black pigment facilitates the development of infection structures (Wheeler & Bell, 1988; Money & Howard, 1996). Rock-dwelling fungi are often melanized (Gorbushina *et al.*, 1993; Sterflinger, 2000).

Biochemical deterioration

Biochemical actions of fungi on rocks are believed to be more important than mechanical degradation (Kumar & Kumar, 1999). Fungi can solubilize minerals and metal compounds through several mechanisms, including acidolysis, complexolysis, redoxolysis and by mycelial metal accumulation (Burgstaller & Schinner, 1993). So-called 'heterotrophic leaching' by fungi primarily involves the first two mechanisms and occurs as a result of several processes, including proton efflux via the plasma membrane H^+-ATPase and/or maintenance of charge balance during nutrient uptake, the production of siderophores [for Fe(III) mobilization] or as a result of respiratory CO_2 production. In many fungal strains, however, an important leaching mechanism occurs through the production of organic acids (e.g. oxalic and citric acid) (Adams *et al.*, 1992; Gadd, 1999, 2001a; Sayer & Gadd, 2001; Jarosz-Wilkolazka & Gadd, 2003; Fomina *et al.*, 2005a). In addition, fungi excrete many other metabolites with metal-complexing properties, e.g. amino acids and phenolic compounds (Manley & Evans, 1986; Müller *et al.*, 1995). Fungal carboxylic acids can play a significant role in the chemical attack of mineral surfaces, since the production of organic acids provides a source of protons for solubilization and metal-chelating anions which complex metal cations (Müller *et al.*, 1995; Gadd, 1999, 2001a; Sayer & Gadd, 2001; Fomina *et al.*, 2004). In one study on the effect of microscopic fungi on the mobility of copper, nickel and zinc compounds in polluted Al–Fe–humus podzols, it was found that the mobility of all studied metals increased under the impact of fungi and was predetermined mostly by the decomposition of soil organic matter (Bespalova *et al.*, 2002).

FUNGAL DETERIORATION OF ROCK AND BUILDING STONE

Microbial attack on minerals may be specific and may depend on the groups of micro-organisms involved, e.g. some lichen hyphae overgrew augite and mica but avoided quartz (Aristovskaya, 1980). Substrate acidification by a free-living and different ecto-mycorrhizal species varied between species as well as in relation to the different minerals. *Mycena galopus* and *Cortinarius glaucopus* produced the highest acidity per unit biomass density, with higher substrate acidification resulting during growth on tricalcium phosphate (Rosling *et al.*, 2004).

In podzols, quartz and kaolin are usually overgrown by fungi and algae, with abundant fungal hyphae also being associated with apatite particles (Aristovskaya, 1980). It seems that alkaline (basic) rocks are generally more susceptible to fungal attack than acidic rocks (Eckhardt, 1985; Kumar & Kumar, 1999). Along with other organisms, fungi are believed to contribute to the weathering of silicate-bearing rocks, e.g. mica and orthoclase, and iron- and manganese-bearing minerals, e.g. biotite, olivine and pyroxene (Kumar & Kumar, 1999). Callot *et al.* (1987) showed that siderophore-producing fungi were able to pit and etch microfractures in samples of olivine and glasses under laboratory conditions. A polycarboxylate siderophore, rhizoferrin, showed the ability to bind Cr(III), Fe(III) and Al(III) (Pillichshammer *et al.*, 1995). Fungi can also deteriorate natural glass and man-made antique and medieval glass (Krumbein *et al.*, 1991). Degradation of aluminosilicates and silicates is believed to occur as a result of the production of organic acids, inorganic acids, alkalis and complexing agents (Rossi & Ehrlich, 1990). It is also likely that CO_2 released during fungal respiration can enhance silicate degradation by carbonic acid attack (Sterflinger, 2000). *Aspergillus niger* can degrade olivine, dunite, serpentine, muscovite, feldspar, spodumene, kaolin and nepheline. *Penicillium expansum* can degrade basalt, while *Penicillium simplicissimum* and *Scopulariopsis brevicaulis* both release aluminium from aluminosilicates (Mehta *et al.*, 1979; Rossi, 1979; Sterflinger, 2000). *Piloderma* was able to extract K and/or Mg from biotite, microcline and chlorite to satisfy nutritional requirements. When grown under low-K conditions, the organism showed fibrillar growths, hyphal swellings and hyphae devoid of ornamentation, possibly indicating nutrient deficiency. Differences were found in growth rates, morphologies and the Mg content of hyphae grown with chlorite and biotite, suggesting that Mg was limiting to normal growth.

Energy-dispersive X-ray analysis indicated that *Piloderma* extracted significantly more K from biotite than from microcline. The high Ca and O content of hyphal ornamentation mainly resulted from calcium oxalate crystals (Glowa *et al.*, 2003).

In podzol E horizons under European coniferous forests, the weathering of horn-blendes, feldspars and granitic bedrock has been attributed to oxalic, citric, succinic, formic and malic acid excretion by saprotrophic and mycorrhizal hyphae. Ectomycorrhizal fungi could form micropores (3–10 μm) in weatherable minerals and hyphal tips could produce micro- to millimolar concentrations of these organic acids (Jongmans *et al.*, 1997; van Breemen *et al.*, 2000). In order to quantify the contribution of mineral tunnelling to the weathering of feldspars and ecosystem influx of Ca and K, surface soils of 11 podzols were studied by Smits *et al.* (2005). Tunnels were observed only in soils older than 1650 years, with the contribution of tunnelling to mineral weathering in the upper mineral soil being less than 1 %. Feldspar tunnelling corresponded to an average ecosystem influx of 0·4 g ha^{-1} year^{-1} for K and 0·2 g ha^{-1} year^{-1} for Ca over 5000 years of soil development. These data indicate that the contribution of tunnelling to weathering is more important in older soils, but remains low (Smits *et al.*, 2005).

Fungal weathering of limestone, sandstone and marble is also known to occur (Kumar & Kumar, 1999; Ehrlich, 2002). In hot and cold deserts and semi-arid regions, clump-like colonies of epi- and endolithic darkly pigmented microcolonial fungi are common inhabitants of limestone, sandstone, marble and granite, as well as other rock types (Staley *et al.*, 1982; Sterflinger, 2000; Gorbushina *et al.*, 1993). Analysis of desert rock samples has shown colonies or single cells in connection with pitting and etching patterns, suggesting acid attack of the mineral surface, possibly a result of organic acid or carbonic acid production (Sterflinger, 2000). Microcolonial fungi have also been shown to be common inhabitants of biogenic oxalate crusts on granitic rocks (Blazquez *et al.*, 1997).

Acidolysis, complexolysis and metal accumulation were involved in solubilization of zinc phosphate and pyromorphite by a selection of soil fungi representing ericoid and ectomycorrhizal plant symbionts and an endophytic/entomopathogenic fungus, *Beauveria caledonica*. Acidolysis (protonation) was found to be the major mechanism of both zinc phosphate and pyromorphite dissolution for most of the fungi examined and, in general, the more metal-tolerant fungal strains yielded more biomass, acidified the medium more and dissolved more of the metal mineral than less-tolerant strains. However, *Beauveria caledonica* excreted a substantial amount of oxalic acid (up to 0·8 mM) in the presence of pyromorphite that coincided with a dramatic increase in lead mobilization, providing a clear example of complexolysis (Fig. 2) (Fomina *et al.*, 2004, 2005a).

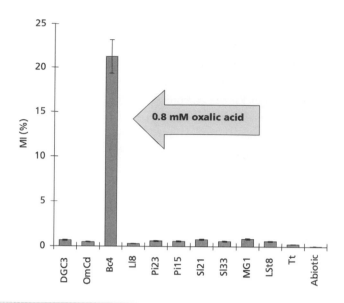

Fig. 2. Importance of oxalic acid in fungal solubilization of pyromorphite. The mobilization index (MI), which represents the ratio of solubilized lead to the initial amount of pyromorphite, is shown for several fungal strains and an abiotic control. Bars represent SEM. Fungal cultures are identified as: DGC3, *Hymenoscyphus ericae* DGC3(UZ); OmCd, *Oidiodendron maius* Cd; Bc4, *Beauveria caledonica* 4; Ll8, *Laccaria laccata* 8; Pi23, *Paxillus involutus* 23; Pi15, *P. involutus* 15; Sl21, *Suillus luteus* 21; Sl33, *S. luteus* 33; MG1, *Suillus bovinus* MG1; LSt8, *S. bovinus* LSt8; Tt, *Telephora terrestris* (adapted from Fomina *et al.*, 2004).

Calcium carbonate ($CaCO_3$) and calcium magnesium carbonate [$CaMg(CO_3)_2$] occur extensively on the Earth's surface as limestone and dolomite (Ehrlich, 2002). Near-surface calcretes and dolocretes cover as much as 13 % of the total land surface and are an important reservoir of carbon, accounting for almost 80 % of the total HCO_3^-, CO_3^{2-} and CO_2 in the Earth's lithosphere (Ehrlich, 2002; Goudie, 1996). Numerous micro-organisms, including bacteria and fungi, have been isolated from natural limestone formations. Cryptoendolithic (i.e. actively penetrating the rock matrix to several millimetres in depth) and chasmolithic or endolithic (i.e. living in hollows, cracks and fissures) fungi are known to occur in limestone. The production of organic acids is believed to play a major role in degradation of limestone (Ehrlich, 2002).

The chemical basis for carbonate weathering is the instability of carbonates in acid solution:

$$CaCO_3 + H^+ \leftrightarrow Ca^{2+} + HCO_3^- \qquad \text{(equation 1)}$$

$$HCO_3^- + H^+ \leftrightarrow H_2CO_3 \qquad \text{(equation 2)}$$

$$H_2CO_3 \leftrightarrow H_2O + CO_2 \qquad \text{(equation 3)}$$

Since $Ca(HCO_3)_2$ is very soluble compared with $CaCO_3$, the $CaCO_3$ dissolves even in weakly acidic solutions. In strong acid solutions, the $CaCO_3$ dissolves more rapidly as carbonate is lost from the solution as CO_2. Any organism capable of producing acidic metabolites is capable of dissolving carbonates and even the production of respiratory CO_2 during respiration can have the same effect:

$$CO_2 + H_2O \leftrightarrow H_2CO_3 \hspace{5cm} \text{(equation 4)}$$

$$H_2CO_3 + CaCO_3 \leftrightarrow Ca^{2+} + 2HCO_3^- \hspace{4cm} \text{(equation 5)}$$

Degradation of sandstone by fungi is also well documented and this has been attributed to the production of for example acetic, oxalic, citric, formic, fumaric, glyoxylic, gluconic, succinic and tartaric acids (Gómez-Alarcón *et al.*, 1994; Hirsch *et al.*, 1995; Sterflinger, 2000). Fungi can also attack rock surfaces through redox attack of mineral constituents such as Mn and Fe (Timonin *et al.*, 1972; Grote & Krumbein, 1992; de la Torre & Gómez-Alarcón, 1994).

METAL BINDING AND ACCUMULATION BY FUNGI

Fungal biomass provides a sink for metals, by either (i) metal biosorption to biomass (cell walls, pigments and extracellular polysaccharides), (ii) intracellular accumulation and sequestration or (iii) precipitation of metal compounds onto hyphae. As well as immobilizing metals, the reduction in external free metal activity may shift the equilibrium so that more metal is released into the soil solution (Gadd, 1993, 2000b; Sterflinger, 2000). Ectomycorrhizal fungi can release elements from apatite and wood ash and accumulate them in the mycelia. *Suillus granulatus* contained 3–15 times more K (3 mg g^{-1}) and showed large calcium-rich crystals on rhizomorphs when grown with apatite, while *Paxillus involutus* had the largest amount of Ca (2–7 mg g^{-1}). Wood ash addition increased the amounts of Ti, Mn and Pb in rhizomorphs (Wallander *et al.*, 2003).

Fungi are effective biosorbents for a variety of metals, including Ni, Zn, Ag, Cu, Cd and Pb (Gadd, 1990, 1993). Metal binding by fungi can occur through metabolism-dependent or metabolism-independent binding of ions onto cell walls and other external surfaces and can be an important passive process in both living and dead fungal biomass (Gadd, 1990, 1993; Sterflinger, 2000). Metal-binding capacity can be influenced by external pH, with binding capacity decreasing at low pH for metals such as Cu, Zn and Cd (de Rome & Gadd, 1987). Cell density also effects binding capacity, with lower cell densities allowing a higher specific metal uptake per unit of biomass (Gadd, 1993). The presence of melanin in fungal cell walls also strongly influences biosorptive capacity (Gadd & Mowll, 1995; Manoli *et al.*, 1997). Metal localization was investigated in the lichenized ascomycete *Trapelia involuta* growing on a range of uraniferous minerals, including metazeunerite [$Cu(UO_2)_2(AsO_4)_2 . 8H_2O$], meta-

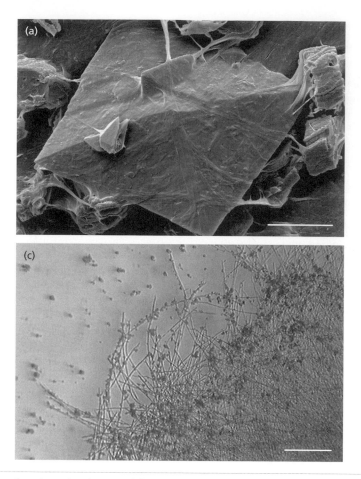

Fig. 3. Formation of metal oxalate crystals by *Beauveria caledonica*. (a) Dry-mode environmental scanning electron microscope (ESEM) image of calcium oxalate (weddelite and whewellite) crystals produced on agar containing calcium carbonate (M. Fomina and G. M. Gadd, unpublished); (b) wet-mode ESEM image of zinc oxalate dihydrate formed by the mycelium grown on zinc phosphate and

torbernite [$Cu(UO_2)_2(PO_4)_2 \cdot 8H_2O$], autunite [$Ca(UO_2)_2(PO_4)_2 \cdot 10H_2O$] and uranium-enriched iron oxide and hydroxide minerals. The highest U, Fe and Cu concentrations occurred in the outer parts of melanized apothecia, indicating that metal biosorption by melanin-like pigments was likely to be responsible for the observed metal fixation (Purvis *et al.*, 2004).

MYCOGENIC MINERAL FORMATION

Fungi have been shown to precipitate a number of inorganic and organic mineral compounds, e.g. oxalates, carbonates and oxides (Arnott, 1995; Verrecchia, 2000; Gadd, 2000a; Grote & Krumbein, 1992). This process may be important in soil, as precipitation of carbonates, phosphates and hydroxides increases soil aggregation.

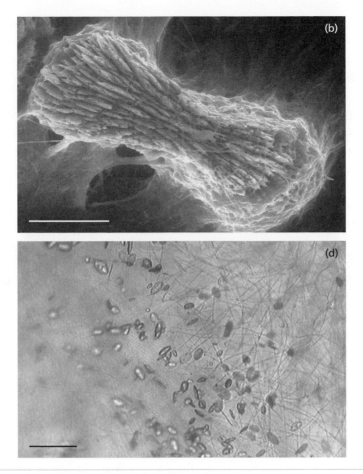

covered with a mucilaginous hyphal network; (c) light microscopy images of zinc phosphate particles adsorbed by the fungal mycelium after 7 days of growth at 25 °C in static liquid medium; (d) crystals of zinc oxalate dihydrate in the same mycelium after 10 days of growth (adapted from Fomina *et al.*, 2005a). Bars, 20 μm (a, c, d) and 50 μm (b).

Cations like Si^{4+}, Fe^{3+}, Al^{3+} and Ca^{2+} (that may be released through dissolutive mechanisms) stimulate precipitation of such compounds, which act as bonding agents for soil particles. Roots and hyphae can enmesh particles together, alter alignment and release organic metabolites that assist aggregate stability (Bronick & Lal, 2005).

Oxalate precipitation

Calcium oxalate is the most common form of oxalate associated with soils and leaf litter, occurring as the dihydrate (weddelite) or the more stable monohydrate (whewellite). Calcium oxalate crystals are commonly associated with free-living, pathogenic and plant-symbiotic fungi and are formed by the reprecipitation of solubilized calcium as calcium oxalate (Fig. 3) (Arnott, 1995). Fungal-derived calcium oxalate can exhibit a variety of crystalline forms (tetragonal, bipyramidal, plate-like, rhombohedral or needles). The formation of calcium oxalate by fungi is important geochemically, since it acts as a reservoir for calcium but also influences phosphate availability (Gadd, 1993, 1999; Jacobs *et al.*, 2002a, b).

Fungi can produce other metal oxalates with a variety of different metals and metal-bearing minerals, e.g. Cd, Co, Cu, Mn, Pb, Sr and Zn (Fig. 3) (Gadd, 2000a, b; Jarosz-Wilkołazka & Gadd, 2003; Burford *et al.*, 2003b; Fomina *et al.*, 2005a). Formation of metal oxalates may provide a mechanism whereby fungi can tolerate environments containing potentially high concentrations of toxic metals. A similar mechanism occurs in lichens growing on copper-sulfide-bearing rocks, where precipitation of copper oxalate occurs within the thallus (Arnott, 1995; Easton, 1997).

Lichens are one of the most common members of the microbial consortia inhabiting exposed subaerial rock substrates and building stone. Oxalic acid excretion by lichens can result in the dissolution of insoluble carbonates and silicates and formation of water-soluble and -insoluble oxalates (Jones & Wilson, 1985; May *et al.*, 1993; Edwards *et al.*, 1991; Verrecchia, 2000). Oxalates can be formed with a variety of metal ions wherever lichens grow (Purvis, 1996). In particular, there has been concern over the deteriorative effects of biologically formed oxalic acid on architecturally important buildings, monuments and frescoes (Nimis *et al.*, 1992). Conversely, however, several studies have suggested that lichen cover protects certain rock surfaces, acting as an 'umbrella' and reducing the erosion of for example slightly soluble calcium sulfates (Mottershead & Lucas, 2000).

Carbonate precipitation: calcified fungal filaments in limestone and calcareous soils

Physico-chemical processes are known to play a significant role in calcrete formation and development, although it is now recognized that micro-organisms, particularly bacteria and algae, may also play a crucial and even dominant role in calcrete transformation (Goudie, 1996). In limestone, fungi and lichens are generally considered to be important agents of carbonate mineral deterioration. A less studied area of fungal action, however, is their influence on carbonate precipitation (Goudie, 1996; Sterflinger, 2000). It is already known that many near-surface limestones (calcretes), calcic and petrocalcic horizons in soils are often secondarily cemented with calcite ($CaCO_3$) and whewellite (calcium oxalate monohydrate, $CaC_2O_4 \cdot H_2O$) (Verrecchia, 2000). Although this phenomenon has partly been attributed to physico-chemical processes, the presence of calcified fungal filaments in limestone and calcareous soils from a range of localities confirms that fungi may play a prominent role in secondary calcite precipitation. Fungal filaments mineralized with calcite, together with whewellite, have been reported in limestone and calcareous soils from a range of localities (Kahle, 1977; Klappa, 1979; Calvet, 1982; Callot *et al.*, 1985a, b; Verrecchia *et al.*, 1993; Monger & Adams, 1996; Bruand & Duval, 1999; Verrecchia, 2000). It has also been proposed that calcium oxalate can be degraded to calcium carbonate, e.g. in semi-arid environments, where such a process may again act in the cementation of pre-existing limestones (Verrecchia *et al.*, 1990).

Calcite formation by fungi may occur through indirect processes via the fungal excretion of oxalic acid and the precipitation of calcium oxalate (Verrecchia *et al.*, 1990; Gadd, 1999; Verrecchia, 2000). For example, oxalic acid excretion and the formation of calcium oxalate results in the dissolution of the internal pore walls of the limestone matrix, so that the solution becomes enriched in free carbonate. During passage of the solution through the pore walls, calcium carbonate recrystallizes as a result of a decrease in CO_2, and this contributes to hardening of the material. Biodegradation of oxalate as a result of microbial activity can also lead to transformation into carbonate, resulting in precipitation of calcite in the pore interior, leading to closure of the pore system and hardening of the chalky parent material. During decomposition of fungal hyphae, calcite crystals can act as sites of further secondary calcite precipitation (Verrecchia, 2000). Manoli *et al.* (1997) demonstrated that chitin, the major component of fungal cell walls, is a substrate on which calcite will readily nucleate and subsequently grow.

Reduction or oxidation of metals and metalloids

Many fungi precipitate reduced forms of metals and metalloids in and around fungal hyphae: for example, Ag(I) reduction to elemental silver [Ag(0)], selenate [Se(VI)] and selenite [Se(IV)] to elemental selenium [Se(0)] and tellurite [Te(IV)] to elemental tellurium [Te(0)]. Reduction of Hg(II) to volatile Hg(0) can also be mediated by fungi (Gadd, 1993, 2000a, b). An *Aspergillus* sp., P37, was able to grow at arsenate concentrations of 0·2 M (more than 20-fold higher than that withstood by *Escherichia coli*, *Saccharomyces cerevisiae* and *Aspergillus nidulans*), and it was suggested that increased arsenate reduction contributed to the hypertolerant phenotype of this fungus (Canovas *et al.*, 2003a, b).

Desert varnish, an oxidized metal layer (patina) a few millimetres thick found on rocks and in soils of arid and semi-arid regions, is believed to be of fungal and bacterial origin. *Lichenothelia* spp. can oxidize manganese and iron in metal-bearing minerals, such as siderite ($FeCO_3$) and rhodochrosite ($MnCO_3$), and precipitate them as oxides (Grote & Krumbein, 1992). Similar oxidation of Fe(II) and Mn(II) by fungi leads to the formation of dark patinas on glass surfaces (Erkhardt, 1985). A Mn-depositing fungus, identified as an *Acremonium*-like hyphomycete, was isolated from a variety of laboratory and natural locations including Mn(III,IV)-oxide-coated stream-bed pebbles. A proposed role for a laccase-like multicopper oxidase was postulated, analogous to the Mn(II)-oxidizing factors found in certain bacteria (Miyata *et al.*, 2004).

Other minerals associated with fungal communities

A wide range of minerals may be deposited under conditions that deviate strongly from 'normal' pressure/temperature diagrams of precipitation and stability and have been found in association with poikilophilic communities on rock surfaces (Gorbushina *et al.*, 2002). Some evaporite minerals (gypsum, $CaSO_4 . 2H_2O$) and iron oxides (magnetite, Fe_3O_4) also display distinctive morphologies indicative of the presence of microbial communities (Gorbushina *et al.*, 2002). Another biogenic mineral (tepius) has been identified in association with a lichen carpet that covers high mountain ranges in Venezuela (Gorbushina *et al.*, 2002). Forsterite (Mg_2SiO_4), the magnesium member of the olivine [$(Mg,Fe)_2SiO_4$] mineral solid solution series, is known to occur only in volcanic rocks, meteorites and metamorphosized carbonates (e.g. skarn deposits). The presence of forsterite in surficial deposits on rock surfaces can therefore be considered to be a possible biosignature for former or extant life (Gorbushina *et al.*, 2002).

FUNGAL–CLAY INTERACTIONS

Clay mineral formation and impact on soil properties

Silicon dioxide, when combined with oxides of Mg, Al, Ca and Fe, forms the silicate minerals in rocks and soil (Bergna, 1994). These high-temperature minerals are unstable in the biosphere and break down readily to form clays. Micro-organisms play a fundamental role in the dissolution of silicate structure in rock weathering, and therefore in the genesis of clay minerals, and soil and sediment formation (Banfield *et al.*, 1999). In fact, the presence of clay minerals can be a typical symptom of biogeochemically weathered rocks (Barker & Banfield, 1996, 1998; Rodriguez Navarro *et al.*, 1997). For example, in lichen weathering of silicate minerals, Ca, K, Fe clay minerals and nano-crystalline aluminous Fe oxyhydroxides were mixed with fungal organic polymers (Barker & Banfield, 1998), while biotite was interpenetrated by fungal hyphae growing along cleavages and partially converted to vermiculite (Barker & Banfield, 1996). Other studies have shown that transformation of mica and chlorite to 2:1 expandable clays was predominant in the ectomycorrhizosphere compared with non-ectomycorrhizosphere soils, likely to be a result of the high production of organic acids and direct extraction of K^+ and Mg^{2+} by fungal hyphae (Arocena *et al.*, 1999).

Soil, which can be considered to be a biologically active loose mass of weathered rock fragments mixed with organic matter, is the ultimate product of rock weathering, i.e. the interaction between the biota, climate and rocks. Clay minerals are generally present in soil in larger amounts than organic matter and, because of their ion-exchange capacity, charge and adsorption powers, they perform a significant buffering function in mineral soils (Ehrlich, 2002) and are important reservoirs of cations and organic molecules (Wild, 1993; Li & Li, 2000; Dinelli & Tateo, 2001; Dong *et al.*, 2001; Krumhansl *et al.*, 2001).

Biological effects of clay minerals

Fungi are in close contact with clay minerals in soils and sediments. Numerous studies have shown that interactions of micro-organisms with solid adsorbents lead to increases in biomass, growth rate and the production of enzymes and metabolites (Stotzky, 1966, 2000; Martin *et al.*, 1976; Fletcher, 1987; Marshall, 1988; Claus & Filip, 1990; Lee & Stotzky, 1999; Lotareva & Prozorov, 2000; Lunsdorf *et al.*, 2000; Demaneche *et al.*, 2001; Fomina & Gadd, 2003). Some clays may stimulate or inhibit fungal metabolism (Fomina *et al.*, 2000b; Fomina & Gadd, 2003). Stimulatory effects may arise from the abilities of different clays to serve as (i) pH buffers, (ii) a source of metal cationic nutrients, (iii) specific adsorbents of metabolic inhibitors, other nutrients and growth stimulators and (iv) modifiers of the microbial microenvironment because of their physico-chemical properties such as surface area and adsorptive capacity (Stotzky, 1966; Marshall, 1988; Martin *et al.*, 1976; Fletcher, 1987). It has also been shown that clay minerals (bentonite, palygorskite and kaolinite) influence the size, shape and structure of mycelial pellets in liquid medium (Fig. 4) (Fomina & Gadd, 2002).

Fungal–clay mineral interactions in soil aggregation

Fungal–clay mineral interactions play an important role in soil evolution. An examination of soil clay aggregation by saprophytic (*Rhizoctonia solani* and *Hyalodendron* sp.) and mycorrhizal (*Hymenoscyphus ericae* and *Hebeloma* sp.) fungi supported the hypothesis that fungal hyphae bring mineral particles and organic materials together to form stable microaggregates at least < 2 μm in size and enmesh such microaggregates into stable aggregates of > 50 μm in diameter (Tisdall *et al.*, 1997). Fungi not only entrap soil particles in their hyphae but take part in polysaccharide aggregation as well (Dorioz *et al.*, 1993; Martens & Frankenberger, 1992; Schlecht-Pietsch *et al.*, 1994; Puget *et al.*, 1999; Chantigny *et al.*, 1997). Only a few studies have been carried out on the sorption properties of mixtures of clay minerals (montmorillonite, kaolinite) and microbial biomass (algae, fungi): interactions between clay minerals and fungi alter the adsorptive properties of both clays and hyphae (Garnham *et al.*, 1991; Kadoshnikov *et al.*, 1995; Morley & Gadd, 1995; Fomina & Gadd, 2002, 2003).

Clay and silicate weathering by fungi

Fungi and bacteria play an important role in the mobilization of silica and silicates (Ehrlich, 2002). Their action is mainly indirect, either through the production of

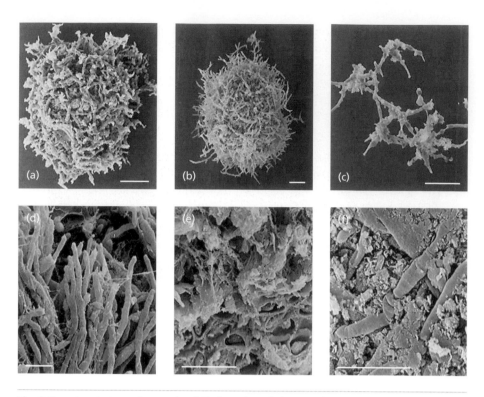

Fig. 4. Scanning electron micrographs of *Cladosporium cladosporioides* grown in the presence of different clay minerals. (a)–(c) Mycelial pellets resulting from growth in medium containing 0·5 % (w/v) (a) and 5 % (w/v) (b) palygorskite or 0·5 % (w/v) kaolinite (c). (d)–(f) Internal structure of the central zone of fractured pellets grown in clay-free control medium (d) and in the presence of 0·5 % (w/v) (e) and 5 % (w/v) (f) bentonite. Bars, 100 μm (a, b, c) and 10 μm (d, e, f). Reproduced from Fomina & Gadd (2002).

chelates or the production of acids (mineral or organic) or, as for certain bacteria, the production of ammonia or amines. Fungi isolated from weathered rock surfaces (*Botrytis*, *Mucor*, *Penicillium* and *Trichoderma* spp.) were shown to be able to solubilize calcium, magnesium and zinc silicates (Webley *et al.*, 1963). Mobilization of silicate from clay minerals by *Aspergillus niger* was found to be a result of oxalic acid excretion (Henderson & Duff, 1963). The majority of fungal strains belonging to the genera *Aspergillus*, *Paecilomyces*, *Penicillium*, *Scopulariopsis* and *Trichoderma* could leach iron in submerged culture from a China clay sample (Mandal *et al.*, 2002). Large amounts of oxalic, citric and gluconic acids were produced by *Penicillium frequentans* in liquid culture. This caused extensive deterioration of clay silicates, as well as micas and feldspars, from sandstone and granite as a result of organic salts formation such as calcium, magnesium and ferric oxalates and calcium citrates (de la Torre *et al.*, 1993). The oxalate-excreting fungus *Hysterangium crassum* also weathered clay minerals *in situ* (Cromack *et al.*, 1979).

Fig. 5. Deterioration of a concrete block that was exposed to fungal weathering for 2 years in an experimental microcosm (M. Fomina and G. M. Gadd, unpublished). Cracking is evident, as well as an extensive hyphal net over the surface. Bar, 200 μm.

MINERAL MYCOTRANSFORMATIONS AND ENVIRONMENTAL BIOTECHNOLOGY

Concrete biodeterioration and radioactive waste disposal

Fungi play an important role in the deterioration of concrete (Fig. 5) (Perfettini *et al.*, 1991; Nica *et al.*, 2000), with complexolysis suggested as the main mechanism of calcium mobilization (Gu *et al.*, 1998). This ability of fungi (and other microbes) to degrade concrete and other structural materials has implications for underground storage of nuclear waste. In high-level nuclear waste disposal, the bentonite buffer around the copper canisters is considered to be a hostile environment for most microbes because of the combination of radiation, heat and low water availability, but discrete microbial species can cope with each of these constraints (Pedersen, 1999). Endolithic, indigenous micro-organisms are capable of surviving gamma-irradiation doses simulating the near-field environment surrounding waste canisters (Pitonzo *et al.*, 1999). In 1997–1998, extensive fungal growth was observed on the walls and other building constructions in the inner part of the 'Shelter' built on the fourth unit of the Chernobyl nuclear power plant damaged in 1986 (Zhdanova *et al.*, 2000). It was discovered that low-level gamma-radiation did not affect spore germination, but led to directed growth of fungal tips towards the radiation source (so-called positive radiotropism) (Zhdanova *et al.*, 2001).

Bioremediation

Some of the processes detailed previously have the potential for treatment of contaminated land (Gadd, 2000a, b, 2001a, b, 2002, 2004; Hochella, 2002; Fomina *et al.*, 2005b). Solubilization processes provide a route for removal of metals from soil matrices, whereas immobilization processes enable metals to be transformed into insoluble, chemically more inert forms. Living or dead fungal biomass and fungal metabolites have been used to remove metal or metalloid species, compounds and particulates and organometal(loid) compounds from solution by biosorption. Fungi with Cr(VI)-reducing activity may have potential for treatment of Cr-polluted soils (Cervantes *et al.*, 2001). Fungal–clay complex biomineral sorbents may combine the sorptive advantages of the individual counterparts, i.e. the high density of metal-binding sites per unit area and high sorption affinity and capacity of fungal biomass, and the high surface area per unit weight, mechanical strength and efficient sorption at high metal concentrations of the clay minerals (Fomina & Gadd, 2003). There has also been the use of extracellular ligands excreted by fungi, especially from *Aspergillus* and *Penicillium* spp., to leach metals such as Zn, Cu, Ni and Co from a variety of solid materials, including low-grade mineral ores (Brandl, 2001). Mycorrhizal associations may also be used with plants for metal clean-up in the general area of phytoremediation. Phytoextraction involves the use of plants to remove toxic metals from soil by accumulation in above-ground parts. Mycorrhiza may enhance phytoextraction directly or indirectly by increasing plant biomass, and some studies have shown increased plant accumulation of metals, especially when inoculated with mycorrhiza isolated from metalliferous environments. However, this is a simplistic hypothesis and many complicating factors affect successful exploitation (Meharg, 2003). Several other studies have shown reduced plant metal uptake (Tullio *et al.*, 2003). Arbuscular mycorrhiza can depress translocation of zinc to shoots of host plants in soils moderately polluted with zinc, with binding of metals in mycorrhizal structures and immobilization of metals in the mycorrhizosphere contributing to these effects (Christie *et al.*, 2004). Ectomycorrhizal fungi persistently fixed Cd(II) and Pb(II), and formed an efficient biological barrier that reduced movement of these metals in birch tissues (Krupa & Kozdrój, 2004). Such mycorrhizal metal immobilization around plant roots, including biomineral formation, may also assist soil remediation and revegetation. The development of stress-tolerant plant–mycorrhizal associations may therefore be a promising new strategy for phytoremediation and soil amelioration (Schutzendubel & Polle, 2002). Because of the symbiosis with ericoid mycorrhizal fungi, ericaceous plants are able to grow in highly polluted environments, where metal ions can reach toxic levels in the soil substrate (Perotto *et al.*, 2002; Martino *et al.*, 2003). Ericoid mycorrhizal fungal endophytes, and sometimes their plant hosts, can evolve toxic-metal resistance which enables ericoid mycorrhizal plants to colonize polluted soil. This seems to be a major factor in the success of ericoid mycorrhizal taxa in a range of harsh environments (Cairney & Meharg, 2003).

CONCLUSIONS

It is clear that fungi have important biogeochemical roles in the biosphere. Symbiotic mycorrhizal fungi are responsible for major mineral transformations and redistribution of inorganic nutrients, for example essential metals and phosphate, as well as carbon flow, while free-living fungi have major roles in the decomposition of plant and other organic materials, including xenobiotics, as well as for example phosphate solubilization. Fungi are dominant members of the soil microflora, especially in acidic environments, and may operate over a wider pH range than most heterotrophic bacteria. Fungi are also major agents of biodeterioration of stone, wood, plaster, cement and other building materials, and are important components, including lichens, of rock-inhabiting microbial communities, with significant roles in mineral dissolution and secondary mineral formation. It is timely to draw attention to 'geomycology' within the umbrella of geomicrobiology and to the interdisciplinary approach that is necessary to further understanding of the important roles that all micro-organisms play in the biogeochemical cycling of elements, the chemical and biological mechanisms that are involved and their environmental and biotechnological significance.

ACKNOWLEDGEMENTS

Some of the authors' research described within was funded by the BBSRC/BIRE programme (94/BRE13640) and CLRC Daresbury SRS (SRS grant 40107), which is gratefully acknowledged. E. P. B. gratefully acknowledges receipt of an NERC post-graduate research studentship. Thanks are also due to Mr Martin Kierans (Centre for High Resolution Imaging and Processing, School of Life Sciences, University of Dundee, Scotland) for assistance with scanning electron microscopy.

REFERENCES

Adams, J. B., Palmer, F. & Staley, J. T. (1992). Rock weathering in deserts – mobilization and concentration of ferric iron by microorganisms. *Geomicrobiol J* **10**, 99–114.

Adeyemi, A. O. & Gadd, G. M. (2005). Fungal degradation of calcium-, lead- and silicon-bearing minerals. *Biometals* **18**, 269–281.

Aristovskaya, T. V. (1980). *Microbiology of the Processes of Soil Formation.* Leningrad: Nauka USSR (in Russian).

Arnott, H. J. (1995). Calcium oxalate in fungi. In *Calcium Oxalate in Biological Systems*, pp. 73–111. Edited by S. R. Khan. Boca Raton, FL: CRC Press.

Arocena, J. M., Glowa, K. R., Massicotte, H. B. & Lavkulich, L. (1999). Chemical and mineral composition of ectomycorrhizosphere soils of subalpine fir (*Abies lasiocarpa* (Hook.) Nutt.) in the Ae horizon of a luvisol. *Can J Soil Sci* **79**, 25–35.

Banfield, J. F. & Nealson, K. H. (editors) (1998). *Geomicrobiology: Interactions between Microbes and Minerals.* Reviews in Mineralogy, vol. 35. Washington, DC: Mineralogical Society of America.

Banfield, J. P., Barker, W. W., Welch, S. A. & Taunton, A. (1999). Biological impact on mineral dissolution: application of the lichen model to understanding mineral weathering in the rhizosphere. *Proc Natl Acad Sci U S A*, **96**, 3404–3411.

Banfield, J. F., Welch, S. A., Zhang, H., Ebert, T. T. & Penn, R. L. (2000). Aggregation-based crystal growth and microstructure development in natural iron oxyhydroxide biomineralization products. *Science* **289**, 751–754.

Barker, W. W. & Banfield, J. F. (1996). Biologically versus inorganically mediated weathering reactions: relationships between minerals and extracellular microbial polymers in lithobiotic communities. *Chem Geol* **132**, 55–69.

Barker, W. W. & Banfield, J. F. (1998). Zones of chemical and physical interaction at interfaces between microbial communities and minerals: a model. *Geomicrobiol J* **15**, 223–244.

Bennett, P. C., Hiebert, F. K. & Choi, W. J. (1996). Microbial colonization and weathering of silicates in petroleum-contaminated groundwater. *Chem Geol* **132**, 45–53.

Bergna, H. E. (1994). Colloid chemistry of silica – an overview. In *The Colloid Chemistry of Silica*, Advances in Chemistry Series vol. 234, pp. 1–47. Edited by H. E. Bergna. Washington, DC: American Chemical Society.

Bespalova, A. Yu., Marfenina, O. E. & Motuzova, G. V. (2002). The effect of microscopic fungi on the mobility of copper, nickel, and zinc in polluted Al–Fe-humus podzols of the Kola Peninsula. *Eurasian Soil Sci* **35**, 945–950.

Beveridge, T. J., Pouwels, P. H., Sara, M. & 22 other authors (1997). Function of S-layers. *FEMS Microbiol Rev* **20**, 99–149.

Blazquez, F., Garcia-Vallez, M., Krumbein, W. E., Sterflinger, K. & Vendrell-Saz, M. (1997). Microstromatolithic deposits on granitic monuments: development and decay. *Eur J Mineral* **9**, 889–901.

Bogomolova, E. V., Vlasov, D. Yu. & Panina, L. K. (1998). On the nature of the microcolonial morphology of epilithic black yeasts *Phaeococcomyces* de Hoog. *Dokl Ross Akad Sci* **363**, 707–709.

Boswell, G. P., Jacobs, H., Davidson, F. A., Gadd, G. M. & Ritz, K. (2002). Functional consequences of nutrient translocation in mycelial fungi. *J Theor Biol* **217**, 459–477.

Boswell, G. P., Jacobs, H., Davidson, F. A., Gadd, G. M. & Ritz, K. (2003). Growth and function of fungal mycelia in heterogeneous environments. *Bull Math Biol* **65**, 447–477.

Brandl, H. (2001). Heterotrophic leaching. In *Fungi in Bioremediation*, pp. 383–423. Edited by G. M. Gadd. Cambridge: Cambridge University Press.

Bronick, C. J. & Lal, R. (2005). Soil structure and management: a review. *Geoderma* **124**, 3–22.

Bruand, A. & Duval, O. (1999). Calcified fungal filaments in the petrocalcic horizon of eutrochrepts in Beauce, France. *Soil Sci Soc Am J* **63**, 164–169.

Burford, E. P., Kierans, M. & Gadd, G. M. (2003a). Geomycology: fungal growth in mineral substrata. *Mycologist* **17**, 98–107.

Burford, E. P., Fomina, M. & Gadd, G. M. (2003b). Fungal involvement in bioweathering and biotransformation of rocks and minerals. *Mineral Mag* **67**, 1127–1155.

Burgstaller, W. & Schinner, F. (1993). Leaching of metals with fungi. *J Biotechnol* **27**, 91–116.

Cairney, J. W. G. & Meharg, A. A. (2003). Ericoid mycorrhiza: a partnership that exploits harsh edaphic conditions. *Eur J Soil Sci* **54**, 735–740.

Callot, G., Mousain, D. & Plassard, C. (1985a). Concentrations de carbonate de calcium sur les parois des hyphes mycéliens. *Agronomie* **5**, 143–150 (in French).

Callot, G., Guyon, A. & Mousain, D. (1985b). Inter-relation entre aiguilles de calcite et hyphes mycéliens. *Agronomie* 5, 209–216 (in French).

Callot, G., Maurette, M., Pottier, L. & Dubois, A. (1987). Biogenic etching of micro-fractures in amorphous and crystalline silicates. *Nature* 328, 147–149.

Calvet, F. (1982). Constructive micrite envelope developed in vadose continental environment in pleistocene eoliantes of Mallorca (Spain). *Acta Geol Hisp* 17, 169–178.

Canovas, D., Duran, C., Rodriguez, N., Amils, R. & de Lorenzo, V. (2003a). Testing the limits of biological tolerance to arsenic in a fungus isolated from the River Tinto. *Environ Microbiol* 5, 133–138.

Canovas, D., Mukhopadhyay, R., Rosen, B. P. & de Lorenzo, V. (2003b). Arsenate transport and reduction in the hyper-tolerant fungus *Aspergillus* sp. P37. *Environ Microbiol* 5, 1087–1093.

Cervantes, C., Campos-Garcia, J., Devars, S., Gutiérrez-Corona, F. G., Loza-Tavera, H., Torres-Guzmán, J. C. & Moreno-Sánchez, R. (2001). Interactions of chromium with microorganisms and plants. *FEMS Microbiol Rev* 25, 335–347.

Chantigny, M. H., Angers, D. A., Prevost, D., Vezina, L. P. & Chalifour, F. P. (1997). Soil aggregation and fungal and bacterial biomass under annual and perennial cropping systems. *Soil Sci Soc Am J* 61, 262–267.

Christie, P., Li, X. L. & Chen, B. D. (2004). Arbuscular mycorrhiza can depress translocation of zinc to shoots of host plants in soils moderately polluted with zinc. *Plant Soil* 261, 209–217.

Claus, H. & Filip, Z. (1990). Effects of clay and other solids on the activity of phenolozidases produced by some fungi and actinomycetes. *Soil Biol Biochem* 22, 483–488.

Cromack, K., Jr, Sollins, P., Graustein, W. C., Speidel, K., Todd, A. W., Spycher, G., Li, C. Y. & Todd, R. L. (1979). Calcium oxalate accumulation and soil weathering in mats of the hypogeous fungus *Hysterangium crassum. Soil Biol Biochem* 11, 463–468.

de la Torre, M. A. & Gómez-Alarcón, G. (1994). Manganese and iron oxidation by fungi isolated from building stone. *Microb Ecol* 27, 177–188.

de la Torre, M. A., Gómez-Alarcón, G., Vizcaino, C. & Garcia, M. T. (1993). Biochemical mechanisms of stone alteration carried out by filamentous fungi living in monuments. *Biogeochemistry* 19, 129–147.

de los Rios, A., Wierzchos, J. & Ascaso, C. (2002). Microhabitats and chemical micro-environments under saxicolous lichens growing on granite. *Microb Ecol* 43, 181–188.

Demaneche, S., Jocteur-Monrozier, L., Quiquampoix, H. & Simonet, P. (2001). Evaluation of biological and physical protection against nuclease degradation of clay-bound plasmid DNA. *Appl Environ Microbiol* 67, 293–299.

de Rome, L. & Gadd, G. M. (1987). Copper adsorption by *Rhizopus arrhizus, Cladosporium resinae,* and *Penicillium italicum. Appl Microbiol Biotechnol* 26, 84–90.

Dinelli, E. & Tateo, F. (2001). Sheet silicates as effective carriers of heavy metals in the ophiolitic mine area of Vigonzano (northern Italy). *Mineral Mag* 65, 121–132.

Dong, W. M., Wang, X. K., Bian, X. Y., Wang, A. X., Du, J. Z. & Tao, Z. Y. (2001). Comparative study on the sorption/desorption of radioeuropium on alumina, bentonite and red earth: effects of pH, ionic strength, fulvic acid, and iron oxide in red earth. *Appl Radiat Isotopes* 54, 603–610.

Dorioz, J. M., Robert, M. & Chenu, C. (1993). The role of roots, fungi and bacteria on clay particles organization. An experimental approach. *Geoderma* 56, 179–194.

Douglas, S. & Beveridge, T. J. (1998). Mineral formation by bacteria in natural microbial communities. *FEMS Microbiol Ecol* 26, 79–88.

Easton, R. M. (1997). Lichen-rock-mineral interactions: an overview. In *Biological-Mineral-ogical Interactions*, vol. 21, pp. 209–239. Edited by J. M. McIntosh & L. A. Groat. Mineralogical Association of Canada Short Course Series. Ottawa: Mineralogical Association of Canada.

Eckhardt, F. E. W. (1985). Solubilisation, transport, and deposition of mineral cations by microorganisms-efficient rock-weathering agents. In *The Chemistry of Weathering*, NATO Advanced Study Institutes Series, Series C, vol. 149, pp 161–173. Edited by J. I. Drever. Dordrecht: Reidel.

Edwards, H. G. M., Farwell, D. W. & Seaward, M. R. D. (1991). Raman spectra of oxalates in lichen encrustations in Renaissance frescoes. *Spectrochim Acta A Mol Spectrosc* **47**, 1531–1539.

Ehrlich, H. L. (1998). Geomicrobiology: its significance for geology. *Earth Sci Rev* **45**, 45–60.

Ehrlich, H. L. (2002). *Geomicrobiology*. New York: Marcel Dekker.

Etienne, S. (2002). The role of biological weathering in periglacial areas: a study of weathering rinds in south Iceland. *Geomorphology* **47**, 75–86.

Ferris, F. G., Fyfe, W. S. & Beveridge, T. J. (1987). Bacteria as nucleation sites for authigenic minerals in a metal-contaminated lake sediment. *Chem Geol* **63**, 225–232.

Ferris, F. G., Wiese, R. G. & Fyfe, W. S. (1994). Precipitation of carbonate minerals by microorganisms: implications for silicate weathering and the global carbon dioxide budget. *Geomicrobiol J* **12**, 1–13.

Fletcher, M. (1987). How do bacteria attach to solid surfaces? *Microbiol Sci* **4**, 133–136.

Fomina, M. & Gadd, G. M. (2002). Influence of clay minerals on the morphology of fungal pellets. *Mycol Res* **106**, 107–117.

Fomina, M. & Gadd, G. M. (2003). Metal sorption by biomass of melanin-producing fungi grown in clay-containing medium. *J Chem Technol Biotechnol* **78**, 23–34.

Fomina, M., Ritz, K. & Gadd, G. M. (2000a). Negative fungal chemotropism to toxic metals. *FEMS Microbiol Lett* **193**, 207–211.

Fomina, M. A., Kadoshnikov, V. M., Gromosova, E. N. & Podgorsky, V. S. (2000b). The role of clay minerals during fungal growth under extreme conditions. In *Proceedings of International III International Seminar 'Mineralogy and Life: Biomineral Homologies'*, pp. 146–148. Syktyvkar, Russia: Geoprint.

Fomina, M., Ritz, K. & Gadd, G. M. (2003). Nutritional influence on the ability of fungal mycelia to penetrate toxic metal-containing domains. *Mycol Res* **107**, 861–871.

Fomina, M. A., Alexander, I. J., Hillier, S. & Gadd, G. M. (2004). Zinc phosphate and pyromorphite solubilization by soil plant-symbiotic fungi. *Geomicrobiol J* **21**, 351–366.

Fomina, M., Hillier, S., Charnock, J. M., Melville, K., Alexander, I. J. & Gadd, G. M. (2005a). Role of oxalic acid overexcretion in transformations of toxic metal minerals by *Beauveria caledonica*. *Appl Environ Microbiol* **71**, 371–381.

Fomina, M., Burford, E. P. & Gadd, G. M. (2005b). Toxic metals and fungal communities. In *The Fungal Community: its Organization and Role in the Ecosystem*, pp. 733–758. Edited by J. Dighton, P. Oudemans & J. White. Boca Raton, FL: CRC Press.

Fomina, M. A., Alexander, I. J., Colpaert, J. V. & Gadd, G. M. (2005c). Solubilization of toxic metal minerals and metal tolerance of mycorrhizal fungi. *Soil Biol Biochem* **37**, 851–866.

Fortin, D. & Beveridge, T. J. (1997). Role of the bacterium *Thiobacillus* in the formation of silicates in acidic mine tailings. *Chem Geol* **141**, 235–250.

Fortin, D., Ferris, F. G. & Scott, S. D. (1998). Formation of Fe-silicates and Fe-oxides on bacterial surfaces in samples collected near hydrothermal vents of Southern Explorer Ridge in Northeast Pacific Ocean. *Am Mineral* **83**, 1399–1408.

Fredrickson, J. K., McKinley, J. P., Nierzwicki-Bauer, S. A., White, D. C., Ringelberg, D. B., Rawson, S. A., Shu-Mei, L., Brockman, F. J. & Bjornstad, B. N. (1995). Microbial community structure and biogeochemistry of Miocene subsurface sediments: implications for long-term microbial survival. *Mol Ecol* **4**, 619–626.

Gadd, G. M. (1990). Fungi and yeasts for metal accumulation. In *Microbial Mineral Recovery*, pp. 249–275. Edited by H. L. Ehrlich & C. Brierley. New York: McGraw-Hill.

Gadd, G. M. (1993). Interactions of fungi with toxic metals. *New Phytol* **124**, 25–60.

Gadd, G. M. (1999). Fungal production of citric and oxalic acid: importance in metal speciation, physiology and biogeochemical processes. *Adv Microb Physiol* **41**, 47–92.

Gadd, G. M. (2000a). Heavy metal pollutants: environmental and biotechnological aspects. In *Encyclopedia of Microbiology*, pp. 607–617. Edited by J. Lederberg. New York: Academic Press.

Gadd, G. M. (2000b). Bioremedial potential of microbial mechanisms of metal mobilization and immobilization. *Curr Opin Biotechnol* **11**, 271–279.

Gadd, G. M. (editor) (2001a). *Fungi in Bioremediation*. Cambridge: Cambridge University Press.

Gadd, G. M. (2001b). Accumulation and transformation of metals by microorganisms. In *Biotechnology, a Multi-volume Comprehensive Treatise*, vol. 10, *Special Processes*, pp. 225–264. Edited by H.-J. Rehm, G. Reed, A. Puhler & P. Stadler. Weinheim: Wiley-VCH.

Gadd, G. M. (2002). Interactions between microorganisms and metals/radionuclides: the basis of bioremediation. In *Interactions of Microorganisms with Radionuclides*, pp. 179–203. Edited by M. J. Keith-Roach & F.R. Livens. Amsterdam: Elsevier.

Gadd, G. M. (2004). Mycotransformation of organic and inorganic substrates. *Mycologist* **18**, 60–70.

Gadd, G. M. (editor) (2005a). *Fungi in Biogeochemical Cycles*. Cambridge: Cambridge University Press (in press).

Gadd, G. M. (2005b). Microorganisms in toxic metal polluted soils. In *Microorganisms in Soils: Roles in Genesis and Functions*, pp. 325–356. Edited by F. Buscot & A. Varma. Berlin: Springer.

Gadd, G. M. & Mowll, J. L. (1995). Copper uptake by yeast-like cells, hyphae and chlamydospores of *Aureobasidium pullulans*. *Exp Mycol* **9**, 230–240.

Gadd, G. M. & Sayer, J. A. (2000). Fungal transformations of metals and metalloids. In *Environmental Microbe-Metal Interactions*, pp. 237–256. Edited by D. R. Lovley. Washington, DC: American Society for Microbiology.

Gadd, G. M., Ramsay, L., Crawford, J. W. & Ritz, K. (2001). Nutritional influence on fungal colony growth and biomass distribution in response to toxic metals. *FEMS Microbiol Lett* **204**, 311–316.

Garnham, G. W., Codd, G. A. & Gadd, G. M. (1991). Uptake of cobalt and cesium by microalgal- and cyanobacterial-clay mixtures. *Microb Ecol* **25**, 71–82.

Gerrath, J. F., Gerrath, J. A. & Larson, D. W. (1995). A preliminary account of endolithic algae of limestone cliffs of the Niagara Escarpment. *Can J Bot* **73**, 788–793.

Gerrath, J. F., Gerrath, J. A., Matthes, U. & Larson, D. W. (2000). Endolithic algae and cyanobacteria from cliffs of the Niagara Escarpment, Ontario, Canada. *Can J Bot* **78**, 807–815.

Gleeson, D. B., Clipson, N. J. W., Melville, K., Gadd, G. M. & McDermott, F. P. (2005).

Mineralogical control of fungal community structure in a weathered pegmatitic granite. *Microb Ecol* (in press).

Glowa, K. R., Arocena, J. M. & Massicotte, H. B. (2003). Extraction of potassium and/or magnesium from selected soil minerals by *Piloderma*. *Geomicrobiol J* **20**, 99–111.

Gómez-Alarcón, G., Muñoz, M. L. & Flores, M. (1994). Excretion of organic acids by fungal strains isolated from decayed limestone. *Int Biodeterior Biodegrad* **34**, 169–180.

Gorbushina, A. A. & Krumbein, W. E. (2000). Subaerial microbial mats and their effects on soil and rock. In *Microbial Sediments*, pp. 161–169. Edited by R. E. Riding & S. M. Awramik. Berlin: Springer.

Gorbushina, A. A., Krumbein, W. E., Hamann, R., Panina, L., Soucharjevsky, S. & Wollenzien, U. (1993). On the role of black fungi in colour change and biodeterioration of antique marbles. *Geomicrobiol J* **11**, 205–221.

Gorbushina, A. A., Lyalikova, N. N., Vlasov, D. Y. & Khizhnyak, T. V. (2002). Microbial communities on the monuments of Moscow and St. Petersburg: biodiversity and trophic relations. *Microbiology* (English translation of *Mikrobiologiia*) **71**, 350–356.

Goudie, A. S. (1996). Organic agency in calcrete development. *J Arid Environ* **32**, 103–110.

Grote, G. & Krumbein, W. E. (1992). Microbial precipitation of manganese by bacteria and fungi from desert rock and rock varnish. *Geomicrobiol J* **10**, 49–57.

Gu, J.-D., Ford, T. E., Berke, N. S. & Mitchell, R. (1998). Biodeterioration of concrete by the fungus *Fusarium*. *Int Biodeterior Biodegrad* **41**, 101–109.

Hamilton, W. A. (2003). Microbially influenced corrosion as a model system for the study of metal microbe interactions: a unifying electron transfer hypothesis. *Biofouling* **19**, 65–76.

Henderson, M. E. K. & Duff, R. B. (1963). The release of metallic and silicate ions from minerals, rocks and soils by fungal activity. *J Soil Sci* **14**, 236–246.

Hersman, L., Lloyd, T. & Sposito, G. (1995). Siderophore-promoted dissolution of hematite. *Geochim Cosmochim Acta* **59**, 3327–3330.

Hirsch, P., Eckhardt, F. E. W. & Palmer, R. J., Jr (1995). Fungi active in weathering rock and stone monuments. *Can J Bot* **73**, 1384–1390.

Hochella, M. F. (2002). Sustaining Earth; thoughts on the present and future roles in mineralogy in environmental science. *Mineral Mag* **66**, 627–652.

Hoffland, E., Giesler, R., Jongmans, T. & van Breemen, N. (2002). Increasing feldspar tunnelling by fungi across a north Sweden podzol chronosequence. *Ecosystems* **5**, 11–22.

Jacobs, H., Boswell, G. P., Ritz, K., Davidson, F. A. & Gadd, G. M. (2002a). Solubilization of calcium phosphate as a consequence of carbon translocation by *Rhizoctonia solani*. *FEMS Microbiol Ecol* **40**, 65–71.

Jacobs, H., Boswell, G. P., Harper, F. A., Ritz, K., Davidson, F. A. & Gadd, G. M. (2002b). Solubilization of metal phosphates by *Rhizoctonia solani*. *Mycol Res* **106**, 1468–1479.

Jacobs, H., Boswell, G. P., Scrimgeour, C. M., Davidson, F. A., Gadd, G. M. & Ritz, K. (2004). Translocation of carbon by *Rhizoctonia solani* in nutritionally-heterogeneous environments. *Mycol Res* **108**, 453–462.

Jarosz-Wilkołazka, A. & Gadd, G. M. (2003). Oxalate production by wood-rotting fungi growing in toxic metal-amended medium. *Chemosphere* **52**, 541–547.

Johnstone, C. G. & Vestal, J. R. (1993). Biogeochemistry of oxalate in the antarctic cryptoendolithic lichen-dominated community. *Microb Ecol* **25**, 305–319.

Jones, D. & Wilson, M. J. (1985). Chemical activities of lichens on mineral surfaces – a review. *Int Biodeterior* **21**, 99–104.

Jongmans, A. G., Van Breemen, N., Lungstrom, U. & 7 other authors (1997). Rock-eating fungi. *Nature* **389**, 682–683.

Kadoshnikov, V. M., Zlobenko, B. P., Zhdanova, N. N. & Redchitz, T. I. (1995). Studies of application of micromycetes and clay composition for decontamination of building materials. In *Mixed Wastes and Environmental Restoration – Working Towards a Cleaner Environment*, WM'95, pp. 61–63. Tucson, AZ: WM Symposia.

Kahle, C. F. (1977). Origin of subaerial Holocene calcareous crusts: role of algae, fungi and sparmicristisation. *Sedimentology* **24**, 413–435.

Klappa, C. F. (1979). Calcified filaments in Quaternary calcretes: organo-mineral interactions in the subaerial vadose environment. *J Sediment Petrol* **49**, 955–968.

Kraemer, S. M., Cheah, S.-F., Zapf, R., Xu, J., Raymond, K. N. & Sposito, G. (1999). Effect of hydroxamate siderophores on Fe release and Pb(II) adsorption by goethite. *Geochim Cosmochim Acta* **63**, 3003–3008.

Krumbein, W. E., Urzi, C. & Gehrmann, C. (1991). On the biocorrosion and bio-deterioration of antique and medieval glass. *Geomicrobiol J* **9**, 139–160.

Krumhansl, J. L., Brady, P. V. & Anderson, H. L. (2001). Reactive barriers for ^{137}Cs retention. *J Contam Hydrol* **47**, 233–240.

Krupa, P. & Kozdrój, J. (2004). Accumulation of heavy metals by ectomycorrhizal fungi colonizing birch trees growing in an industrial desert soil. *World J Microbiol Biotechnol* **20**, 427–430.

Kumar, R. & Kumar, A. V. (1999). *Biodeterioration of Stone in Tropical Environments: an Overview.* Los Angeles: Getty Conservation Institute.

Lee, G.-H. & Stotzky, G. (1999). Transformation and survival of donor, recipient, and transformants of *Bacillus subtilis* in vitro and in soil. *Soil Biol Biochem* **31**, 1499–1508.

Li, L. Y. & Li, R. S. (2000). The role of clay minerals and the effect of H^+ ions on removal of heavy metal (Pb^{2+}) from contaminated soils. *Can Geotechnol J* **37**, 296–307.

Liermann, L. J., Kalinowski, B. E., Brantley, S. L. & Ferry, J. G. (2000). Role of bacterial siderophores in dissolution of hornblende. *Geochim Cosmochim Acta* **64**, 587–602.

Lotareva, O. V. & Prozorov, A. A. (2000). Effect of the clay minerals montmorillonite and kaolinite on the generic transformation of competent *Bacillus subtilis* cells. *Microbiology* (English translation of *Mikrobiologiia*) **69**, 571–574.

Lundström, U. S., van Breemen, N. & Bain, D. (2000). The podzolization process. A review. *Geoderma* **94**, 91–107.

Lunsdorf, H., Erb, R. W., Abraham, W. R. & Timmis, K. N. (2000). 'Clay hutches': a novel interaction between bacteria and clay minerals. *Environ Microbiol* **2**, 161–168.

Mandal, S. K., Roy, A. & Banerjee, P. C. (2002). Iron leaching from china clay by fungal strains. *Trans Indian Inst Metals* **55**, 1–7.

Manley, E. & Evans, L. (1986). Dissolution of feldspars by low-molecular-weight aliphatic and aromatic acids. *Soil Sci* **141**, 106–112.

Manoli, F., Koutsopoulos, E. & Dalas, E. (1997). Crystallization of calcite on chitin. *J Crystal Growth* **182**, 116–124.

Marshall, K. C. (1988). Adhesion and growth of bacteria at surfaces in oligotrophic habitats. *Can J Microbiol* **34**, 593–606.

Martens, D. A. & Frankenberger, W. T. (1992). Decomposition of bacterial polymers in soil and their influence on soil structure. *Biol Fertil Soils* **13**, 65–73.

Martin, J. P., Filip, Z. & Haider, K. (1976). Effect of montmorollonite and humate on growth and metabolic activity of some actinomycetes. *Soil Biol Biochem* **8**, 409–413.

Martino, E., Perotto, S., Parsons, R. & Gadd, G. M. (2003). Solubilization of insoluble inorganic zinc compounds by ericoid mycorrhizal fungi derived from heavy metal polluted sites. *Soil Biol Biochem* **35**, 133–141.

Maurice, P. A., Hochella, M. F., Parks, G. A., Sposito, G. & Schwertmann, U. (1995). Evolution of hematite surface microtopography upon dissolution by simple organic acids. *Clays Clay Miner* **43**, 29–38.

May, E. (2003). Microbes on building stone for good or bad? *Culture* **24**, 4–8.

May, E., Lewis, F. J., Periera, S., Taylor, S., Seaward, M. R. D. & Allsopp, D. (1993). Microbial deterioration of building stone – a review. *Biodeterior Abstr* **7**, 109–112.

Meharg, A. A. (2003). The mechanistic basis of interactions between mycorrhizal associations and toxic metal cations. *Mycol Res* **107**, 1253–1265.

Mehta, A. P., Torma, A. E. & Murr, L. E. (1979). Effect of environmental parameters on the efficiency of biodegradation of basalt rock by fungi. *Biotechnol Bioeng* **21**, 875–885.

Miyata, N., Tani, Y., Iwahori, K. & Soma, M. (2004). Enzymatic formation of manganese oxides by an *Acremonium*-like hyphomycete fungus, strain KR21-2. *FEMS Microbiol Ecol* **47**, 101–109.

Money, N. P. (1999). Biophysics: fungus punches its way in. *Nature* **401**, 332–333.

Money, N. P. (2001). Biomechanics of invasive hyphal growth. In *The Mycota* VIII, pp. 3–17. Edited by K. Esser. Berlin: Springer.

Money, N. P. & Howard, R. J. (1996). Confirmation of a link between fungal pigmentation, turgor pressure, and pathogenicity using a new method of turgor measurement. *Fungal Genet Biol* **20**, 217–227.

Monger, C. H. & Adams, H. P. (1996). Micromorphology of calcite-silica deposits, Yucca Mountain, Nevada. *Soil Sci Soc Am J* **60**, 519–530.

Morley, G. F. & Gadd, G. M. (1995). Sorption of toxic metals by fungi and clay minerals. *Mycol Res* **99**, 1429–1438.

Morley, G., Sayer, J., Wilkinson, S., Gharieb, M. & Gadd, G. M. (1996). Fungal sequestration, solubilization and transformation of toxic metals. In *Fungi and Environmental Change*, pp. 235–256. Edited by J. C. Frankland, N. Magan & G. M. Gadd. Cambridge: Cambridge University Press.

Mottershead, D. & Lucas, G. (2000). The role of lichens in inhibiting erosion of a soluble rock. *Lichenologist* **32**, 601–609.

Müller, B., Burgstaller, W., Strasser, H., Zanella, A. & Schinner, F. (1995). Leaching of zinc from an industrial filter dust with *Penicillium*, *Pseudomonas* and *Coryne-bacterium*: citric acid is the leaching agent rather than amino acids. *J Ind Microbiol* **14**, 208–212.

Nica, D., Davis, J. L., Kirby, L., Zuo, G. & Roberts, D. J. (2000). Isolation and characterization of microorganisms involved in the biodeterioration of concrete in sewers. *Int Biodeterior Biodegrad* **46**, 61–68.

Nimis, P. L., Pinna, D. & Salvadori, O. (1992). *Licheni e Conservazione dei Monumenti.* Bologna: Cooperativa Libraria Universitaria Ediatrice Bologna (in Italian).

Paris, F., Bonnaud, P., Ranger, J. & Lapeyrie, F. (1995). *In vitro* weathering of phlogopite by ectomycorrhizal fungi. *Plant Soil* **177**, 191–201.

Pedersen, K. (1999). Subterranean microorganisms and radioactive waste disposal in Sweden. *Eng Geol* **52**, 163–176.

Perfettini, J. V., Revertegat, E. & Langomazino, N. (1991). Evaluation of cement degradation by the metabolic activities of two fungal strains. *Experientia* **47**, 527–533.

Perotto, S., Girlanda, M. & Martino, E. (2002). Ericoid mycorrhizal fungi: some new perspectives on old acquaintances. *Plant Soil* **244**, 41–53.

Pillichshammer, M., Pumpel, T., Poder, R., Eller, K., Klima, J. & Schinner, F. (1995). Biosorption of chromium to fungi. *Biometals* **8**, 117–121.

Pitonzo, B. J., Amy, P. S. & Rudin, M. (1999). Effect of gamma radiation on native endolithic microorganisms from a radioactive waste deposit site. *Radiat Res* **152**, 64–70.

Puget, P., Angers, D. A. & Chenu, C. (1999). Nature of carbohydrates associated with water-stable aggregates of two cultivated soils. *Soil Biol Biochem* **31**, 55–63.

Purvis, O. W. (1996). Interactions of lichens with metals. *Sci Prog* **79**, 283–309.

Purvis, O. W., Bailey, E. H., McLean, J., Kasama, T. & Williamson, B. J. (2004). Uranium biosorption by the lichen *Trapelia involuta* at a uranium mine. *Geomicrobiol J* **21**, 159–167.

Putnis, A. (2002). Mineral replacement reactions; from macroscopic observations to microscopic mechanisms. *Mineral Mag* **66**, 689–708.

Rodriguez Navarro, C., Sebastian, E. & Rodriguez Gallego, M. (1997). An urban model for dolomite precipitation: authigenic dolomite on weathered building stones. *Sediment Geol* **109**, 1–11.

Rogers, J. R., Bennett, P. C. & Choi, W. J. (1998). Feldspars as a source of nutrients for microorganisms. *Am Mineral* **83**, 1532–1540.

Roldan, M., Clavero, E. & Hernandez-Marine, M. (2002). Biofilm structure of cyanobacteria in catacombs. *Coalition* **5**, 6–8.

Rosling, A., Lindahl, B. D., Taylor, A. F. S. & Finlay, R. D. (2004). Mycelial growth and substrate acidification of ectomycorrhizal fungi in response to different minerals. *FEMS Microbiol Ecol* **47**, 31–37.

Rossi, G. (1979). Potassium recovery through leucite bioleaching: possibilities and limitations. In *Metallurgical Applications of Bacterial Leaching and Related Phenomena*, pp. 279–319. Edited by L. E. Murr, A. E. Torma & J. E. Brierley. New York: Academic Press.

Rossi, G. & Ehrlich, H. L. (1990). Other bioleaching processes. In *Microbial Mineral Recovery*, pp. 149–170. Edited by H. L. Ehrlich & C. L. Brierley. New York: McGraw-Hill.

Sayer, J. A. & Gadd, G. M. (2001). Binding of cobalt and zinc by organic acids and culture filtrates of *Aspergillus niger* grown in the absence or presence of insoluble cobalt or zinc phosphate. *Mycol Res* **105**, 1261–1267.

Schlecht-Pietsch, S., Wagner, U. & Anderson, T. H. (1994). Changes in composition of soil polysaccharides and aggregate stability after carbon amendments to different textured soils. *Appl Soil Ecol* **1**, 145–154.

Schutzendubel, A. & Polle, A. (2002). Plant responses to abiotic stresses: heavy metal-induced oxidative stress and protection by mycorrhization. *J Exp Bot* **53**, 1351–1365.

Smits, M. M., Hoffland, E., Jongmans, A. G. & van Breemen, N. (2005). Contribution of mineral tunnelling to total feldspar weathering. *Geoderma* **125**, 59–69.

Staley, J. T., Palmer, F. & Adams, J. B. (1982). Microcolonial fungi: common inhabitants on desert rocks. *Science* **215**, 1093–1095.

Sterflinger, K. (2000). Fungi as geologic agents. *Geomicrobiol J* **17**, 97–124.

Stone, A. T. (1997). Reactions of extracellular organic ligands with dissolved metal ions and mineral surface. In *Geomicrobiology: Interactions between Microbes and Minerals*, Reviews in Mineralogy, vol. 35, pp. 309–341. Edited by J. F. Banfield & K. H. Nealson. Washington, DC: Mineralogical Society of America.

Stotzky, G. (1966). Influence of clay minerals on microorganisms. II. Effect of various clay species, homoionic clays, and other particles on bacteria. *Can J Microbiol* **12**, 831–848.

Stotzky, G. (2000). Persistence and biological activity in soil of insecticidal proteins from *Bacillus thuringiensis* and of bacterial DNA bound on clays and humic acids. *J Environ Qual* **29**, 691–705.

Thompson, J. B. & Ferris, F. G. (1990). Cyanobacterial precipitation of gypsum, calcite, and magnesite from natural alkaline lake water. *Geology* **18**, 995–998.

Timonin, M. I., Illman, W. I. & Hartgerink, T. (1972). Oxidation of manganous salts of manganese by soil fungi. *Can J Bot* **18**, 793–799.

Tisdall, J. M., Smith, S. E. & Rengasamy, P. (1997). Aggregation of soil by fungal hyphae. *Aust J Soil Res* **35**, 55–60.

Tullio, M., Pierandrei, F., Salerno, A. & Rea, E. (2003). Tolerance to cadmium of vesicular arbuscular mycorrhizae spores isolated from a cadmium-polluted and unpolluted soil. *Biol Fertil Soils* **37**, 211–214.

van Breemen, N., Lundström, U. S. & Jongmans, A. G. (2000). Do plants drive podzolization via rock-eating mycorrhizal fungi? *Geoderma* **94**, 163–171.

Vaughn, D. J., Pattrick, R. A. D. & Wogelius, R. A. (2002). Minerals, metals and molecules; ore and environmental mineralogy in the new millennium. *Mineral Mag* **66**, 653–676.

Verrecchia, E. P. (2000). Fungi and sediments. In *Microbial Sediments*, pp. 69–75. Edited by R. E. Riding & S. M. Awramik. Berlin: Springer.

Verrecchia, E. P., Dumont, J. L. & Rolko, K. E. (1990). Do fungi building limestones exist in semi-arid regions? *Naturwissenschaften* **77**, 584–586.

Verrecchia, E. P., Dumont, J. L. & Verrecchia, K. E. (1993). Role of calcium-oxalate biomineralization by fungi in the formation of calcretes – a case-study from Nazareth, Israel. *J Sediment Petrol* **63**, 1000–1006.

Wallander, H., Mahmood, S., Hagerberg, D., Johansson, L. & Pallon, J. (2003). Elemental composition of ectomycorrhizal mycelia identified by PCR-RFLP analysis and grown in contact with apatite or wood ash in forest soil. *FEMS Microbiol Ecol* **44**, 57–65.

Watts, H. J., Very, A. A., Perera, T. H. S., Davies, J. M. & Gow, N. A. R. (1998). Thigmotropism and stretch-activated channels in the pathogenic fungus *Candida albicans*. *Microbiology* **144**, 689–695.

Webley, D. M., Henderson, M. E. F. & Taylor, I. F. (1963). The microbiology of rocks and weathered stones. *J Soil Sci* **14**, 102–112.

Welch, S. A. & Vandevivere, P. (1994). Effect of microbial and other naturally occurring polymers on mineral dissolution. *Geomicrobiol J* **12**, 227–238.

Welch, S. A., Barker, W. W. & Banfield, J. F. (1999). Microbial extracellular polysaccharides and plagioclase dissolution. *Geochim Cosmochim Acta* **63**, 1405–1419.

Wheeler, M. H. & Bell, A. A. (1988). Melanins and their importance in pathogenic fungi. *Curr Topics Med Mycol* **2**, 338–385.

White, I. D., Mottershead, D. N. & Harrison, S. J. (1992). *Environmental Systems: an Introductory Text*. London: Chapman & Hall.

Wild, A. (1993). *Soils and the Environment: an Introduction*. Cambridge: Cambridge University Press.

Wolfaardt, G. M., Lawrence, J. R., Robarts, R. D., Caldwell, S. J. & Caldwell, D. E. (1994). Multicellular organization in a degradative biofilm community. *Appl Environ Microbiol* **60**, 434–446.

Zhdanova, N. N., Zakharchenko, V. A., Vember, V. V. & Nakonechnaya, L. T. (2000). Fungi from Chernobyl: mycobiota of the inner regions of the containment structures of the damaged nuclear reactor. *Mycol Res* **104**, 1421–1426.

Zhdanova, N., Fomina, M., Redchitz, T. & Olsson, S. (2001). Chernobyl effects: growth characteristics of soil fungi *Cladosprorium cladosporioides* (Fresen) de Vries with and without positive radiotropism. *Pol J Ecol* **49**, 309–318.

The deep intraterrestrial biosphere

Karsten Pedersen

Deep Biosphere Laboratory, Department of Cell & Molecular Biology, Göteborg University, Box 462, SE-405 30 Göteborg, Sweden

INTRODUCTION

Exploration of the microbial world started slowly about 350 years ago, when van Leeuwenhoek and his contemporaries first focused their microscopes on extremely small living things. It is only during the last 20 years, however, that exploration of the world of intraterrestrial microbes has gathered momentum. Previously, it had generally been assumed that persistent life could not exist deep underground, out of reach of the sun and a photosynthetic ecosystem base. In the mid-1980s, scientists started to drill deep holes, from hundreds to a thousand metres deep, in both hard and sedimentary bedrock, and up came microbes in numbers equivalent to those found in many surface ecosystems. The world of intraterrestrial microbes had been discovered.

Intraterrestrial ecosystems have been reviewed elsewhere and the content of those reports need not be repeated here (Ghiorse & Wilson, 1988; Pedersen, 1993a, 2000; Bachofen, 1997; Bachofen *et al.*, 1998; Fredrickson & Fletcher, 2001; Amend & Teske, 2005). Instead, this chapter will focus on characteristics that distinguish the intraterrestrial from the terrestrial world.

Most ecosystem environments have specific, distinguishing characteristics. The environments of intraterrestrial microbial ecosystems occupy a special position, differing substantially in many respects from those of most surface-based ecosystems. In many ways, underground ecosystems must be approached quite differently from the way in which those on the surface would be approached. The intraterrestrial world is huge and diverse and this chapter can only provide a brief overview of the following matters:

SGM symposium 65: Micro-organisms and Earth systems – advances in geomicrobiology.
Editors G. M. Gadd, K. T. Semple & H. M. Lappin-Scott. Cambridge University Press. ISBN 0 521 86222 1 ©SGM 2005

- Where do intraterrestrial environments begin?
- Variability of intraterrestrial environments.
- Strategies for exploring intraterrestrial environments.
- Organisms living in intraterrestrial ecosystems.
- Range of biomass in various intraterrestrial environments.
- Intraterrestrial species diversity.
- Energy sources for intraterrestrial life.
- Activity of intraterrestrial life forms.

Several different terms are used to refer to environments under the ground and seafloor surface. Though it is difficult for one word to encompass all such environments, this paper generally uses the term 'intraterrestrial'. Other common terms are 'subseafloor', 'subsurface' and 'underground'.

WHERE DO INTRATERRESTRIAL ENVIRONMENTS BEGIN?

There is no consensus as to the answer to this question, and various scientists would probably suggest different answers. The view adopted in this chapter is that the intraterrestrial world begins where contact with surface ecosystems is lost. This lies beneath the soil and root zones, beneath the groundwater table, beneath sediment and crust surfaces. A long time must have passed since the last surface contact, 'a long time' implying at least several decades and often hundreds of years or more. So, it is not depth per se that defines an intraterrestrial ecosystem, but rather the duration of isolation from the surface.

VARIABILITY OF INTRATERRESTRIAL ENVIRONMENTS

Many different minerals and rock types constitute our planet. These rock types lie along a continuum, ranging from very hard rocks (such as granites and basalts) through sedimentary rocks (such as sandstones) to fairly soft sediments not yet defined as rock. Just two types will be defined and discussed here: (i) those that are too tight to allow the presence of microbes elsewhere than in fractures, i.e. hard rocks, and (ii) those that are porous enough to allow microbes inside the rock, i.e. sedimentary rocks.

When formed, hard rocks generally experience temperatures that greatly exceed the limit for life and are therefore sterile after formation. They cool off and fracture with time and the fractures can become colonized by microbes. Sedimentary rocks form slowly, generally on the seafloor; over geological time scales, plate tectonics may eventually move these rocks up from the sea to constitute land. Sedimentary rocks that do not exceed, when formed, the temperature limit for life (currently defined as 113 °C) can harbour microbes borne by the sediment particles during the sedimentation process as well as microbes that arrived later.

Intraterrestrial environments can be described as basically solid but containing some groundwater in fractures and pores. In this environment, huge surface areas are exposed to groundwater; solid dissolution and precipitation processes are consequently very important. At the opposite extreme, aquatic environments, such as sea or lake environments, consist of water with tiny amounts of dispersed solids; these solids are thus of very limited importance for sea and lake water composition.

Mixing processes also differ between intraterrestrial and surface aquatic environments. In seas and lakes, the water is comparatively homogeneous and well mixed by wind and currents (although there certainly are layered environments such as meromictic lakes). Underground, in particular in fractured rocks, there are many different water types that mix at fracture crossroads. Each water type has it own flow-path history and has met with different minerals and organisms on its way to a mixing point. Understanding mixing is crucial when dealing with groundwater. Any groundwater can be characterized by its age relative to when it left the ground surface and also by its origin, usually defined as an end member of a mixed groundwater. It may seem a simple matter to clarify the age and origin of a groundwater sample, but it is not. Different dating methods, for example, analysing the amounts of tritium originating from nuclear bomb tests or of ^{14}C or ^{37}Cl, commonly determine the same groundwater to be of different ages. This is because a single groundwater sample comprises a mixture of waters of various ages and origins. For example, rainwater penetrating deep underground in the Scandinavian region, mixing with water of glacial meltwater origin, would produce a groundwater containing two age signatures: those of recent and of 10 000-year-old water. It would not be obvious whether any microbes detected in this groundwater were originally borne by the rainwater or the glacial meltwater. Understanding the origins of the various component waters of a groundwater is a challenge, but can be approached by statistical methods. By statistical analysis of extensive datasets of groundwater chemistry, end members have been identified for Scandinavian groundwater (Laaksoharju et al., 1999). Models have been built that guide scientists as to how different end members contribute to the groundwater being analysed.

The results indicate that, as well as a continuum of types of solids in the intraterrestrial environment, as explained above, there is also a continuum of groundwater types. Every sample brought up from underground is more or less different from all other such samples. This is because groundwater systems are non-homogeneous. Comparing a surface aquatic environment, such as a sea, with most intraterrestrial environments results in contrasts. Spatially, the sea varies little within a local area, while a groundwater environment can vary greatly over very short distances. However, when it comes to temporal variation, the reverse holds true: the sea has marked diurnal and seasonal cycles, while the intraterrestrial environment shows no change over days or seasons. Observation of intraterrestrial environments may need to continue for many years before significant changes can be resolved.

STRATEGIES FOR EXPLORING INTRATERRESTRIAL ENVIRONMENTS

Sampling intraterrestrial micro-organisms requires a borehole or a tunnel. The preferred method of scientific drilling is the core-drilling technique, which allows the retrieved cores to be analysed for mineralogy and microbiology. The borehole can later be explored in several different ways.

To sample groundwater from water-conducting fractures or aquifers, part of the borehole can be packed off and sampled. Fig. 1 depicts several possible sampling techniques. A fairly simple sampling technique is to pump water though tubing (Fig. 1a); however, in the case of deep boreholes, approaching a kilometre or more in depth, it make take considerable time to pump out the water, and much of the microbiology and some geochemistry may change in the process. Such changes are induced by the pressure drop and change in temperature. A second option is to lower samplers and open them at the sampling depth (Fig. 1b, c); in this way, the pressure of the sample can be kept at the pressure of the sampling depth. This strategy has been applied in sampling deep Fennoscandian groundwaters (Haveman et al., 1999; Haveman & Pedersen, 2002). If there is access to a tunnel or a mine beneath the surface of the groundwater table, groundwater can be obtained from boreholes without pumping (Fig. 1d). Deep South African goldmines have allowed scientists to access the deep intraterrestrial biosphere via boreholes drilled to depths as great as 3·5 km (Onstott et al., 1998). Sampling of fractures that deliver groundwater to a tunnel does not require drilling (Fig. 1e). This approach was used by Ekendahl et al. (2003) when sampling for fungi in deep groundwater. Finally, it may be possible to work in situ by installing a laboratory near the microbes in a tunnel and working under in situ conditions. One example of an underground laboratory can be found in the Swedish Äspö Hard Rock Laboratory (Fig. 2). There, three water-bearing fractures have been intersected by drilling and then packed off. Groundwater from those fractures is circulated through flow cells in the laboratory and then back to the fracture again, maintaining the pressure, temperature and chemistry exactly at in situ levels. These laboratory flow cells can be regarded as artificial extensions of the fractures.

The superdeep well SG-3 (12 262 m deep) in the Pechenga-Zapolyarny area of the Kola Peninsula, Russia, is currently the deepest drilled hole in the world (Butler, 1994). Drilling a superdeep borehole becomes increasingly difficult with increasing depth, and success depends on the qualities of the geological formation drilled, the quality of equipment used and the skills of the drilling personnel. Other superdeep boreholes

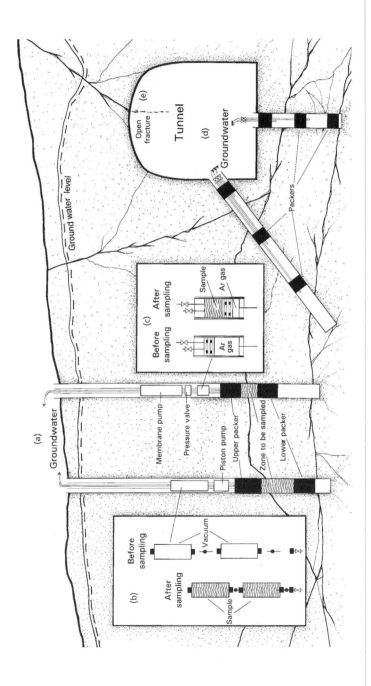

Fig. 1. Access to aquifer material and groundwater occurs via the drilling of boreholes from the ground surface or tunnels. After retrieval of drill core material, the boreholes are packed off in one or several sections, each of which isolates one or more specific aquifers. (a) Down-hole pumps of various types force groundwater from the aquifer to the ground surface for subsampling. (b) The 'BAT' borehole sampler, which can be opened and closed from the surface, is designed for gas sampling. (c) The 'PAVE' borehole sampler, which can be opened and closed from the surface, is designed for gas and microbiological sampling and allows one or more sample vessels to be used simultaneously. (d) Tunnel boreholes do not require pumps when the tunnel lies below the groundwater table. Aquifers can be packed off and connected to sampling devices in the tunnel via pressure-resistant tubes. It is important, however, to understand the potential danger and technical problems connected with the high hydrostatic pressure which occurs at depth and increases by approximately 1 atmosphere per 10 m depth penetration. (e) Open fractures in tunnels can be sampled directly and represent groundwater with minimal disturbance, except for the pressure decrease due to the tunnel.

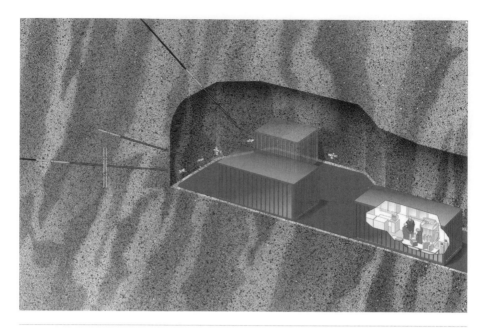

Fig. 2. Artist's impression of the MICROBE underground laboratory at the 450 m level in the Äspö hard rock tunnel. The laboratory (to the right) is situated in a steel container and connected to three discrete fractures in the rock matrix. Tubing connects the systems in the laboratory with the groundwater in the aquifers under *in situ* pressure, chemistry and temperature.

include several drilled as part of the German Continental Deep Drilling Program (KTB) into the crystalline rock of the Bavarian Black Forest (Schwarzwald) basement in Central Europe (Butler, 1994); the deepest of the six wells drilled reached a depth of 9100 m, where an *in situ* temperature of 265 °C was recorded. One of these KTB wells was searched for hyperthermophiles at a depth of 4000 m; culturable micro-organisms could not be demonstrated. Another very deep borehole was drilled in Gravenberg, Sweden, in a search for deep earth gases (Gold, 1992). It reached a depth of 6800 m and thermophilic bacteria were successfully enriched and isolated from a depth of 5278 m, where the temperature was 65–75 °C (Szewzyk *et al.*, 1994). The temperature data obtained from these boreholes indicate that the temperature increase with depth varies depending on the geographical location of the borehole. In most of Scandinavia, the increase is very slow, ranging between 1 and 1·5 °C per 100 m; elsewhere the gradient is much steeper, with the steepest temperature gradients obviously being found in hot-spring areas.

ORGANISMS THAT LIVE IN INTRATERRESTRIAL ENVIRONMENTS

Generally, representatives of most major groups of surface-living microbes should be able to live underground – with some obvious exceptions. The underground and

Table 1. Range of total counts in various intraterrestrial environments

Values are means of data from various reviews: Ghiorse & Wilson (1988), Pedersen (1993a, 2000), Bachofen (1997), Bachofen et al. (1998) and Fredrickson & Fletcher (2001). gdw, Gram dry weight.

Environment	Total counts		
	Unattached (ml^{-1})	Attached (cm^{-2})	Attached (gdw^{-1})
Hard rock fractures	10^3–10^6	10^5–10^7	–
Sedimentary rock	10^3–10^6	–	10^4–10^8
Subseafloor sediments	–	–	10^5–10^8

subseafloor world is dark, water-filled and usually anaerobic with reducing conditions. The boundary between anaerobic and aerobic environments commonly follows the groundwater table. Consequently, strictly and facultatively anaerobic archaea, bacteria, phages, fungi and protozoa should be able to live an active life there, but not strict aerobes, pathogens, photosynthetic organisms or symbionts of plants, animals or insects.

RANGE OF BIOMASS IN VARIOUS INTRATERRESTRIAL ENVIRONMENTS

Many microbes live in a planktonic state, but even more tend to live in an attached state (Characklis & Marshall, 1990). As the intraterrestrial world is a world of solids with some interpenetrating water, there is always a nearby surface on which to attach. Experiments suggest that underground ecosystems in fractured rock are dominated by attached microbes making up biofilms (Pedersen, 2001), but it remains to be demonstrated that microbial biofilms exist on fracture surfaces in deep environments. It is easier to obtain a reasonably undisturbed groundwater sample from a borehole than it is to obtain a drill core sample of an undisturbed intersected fracture. The data concerning intraterrestrial organisms, in particular in fractured rock, are thus biased towards planktonic microbes found in groundwater samples.

Most total counts in hard rock refer to the *number* of unattached microbes; in sediments, however, the data generally refer to the *weight* and include both attached and unattached cells. Typically, the numbers found in fractures and sediments are relatively equal to those found in clean surface waters and sediments, although the range is large (Table 1). A hundred million cells may sound like a lot, but bear in mind that this refers to cell *numbers* and not weight. In any case, the total intraterrestrial biomass is very impressive: the calculations of Whitman *et al.* (1998) suggest that the intraterrestrial biomass may nearly equal the surface biomass. Whether these calculations are off by 50 % or so is immaterial, as there will still be a tremendous amount of life dwelling deep under our feet.

INTRATERRESTRIAL SPECIES DIVERSITY

The cloning and sequencing of the 16S/18S rRNA genes of microbes living in their natural environments has revealed a genetic diversity exceeding any earlier estimates (Pace, 1997). This methodology has been applied to microbes from various underground sites and has generally revealed great species diversity. The case of the Fennoscandian hard-rock aquifers will be used to demonstrate typical results for hard rock. A total of 385 clones from two sites were sequenced (Ekendahl *et al.*, 1994; Pedersen *et al.*, 1996a, 1997); 122 unique sequences were found, each representing a possible species that was not recorded in international databases at the time the analysis was conducted. Therefore, on average, approximately one-third of the sequenced clones represented unique species. These studies clearly have not exhausted the sequences, as new sequences were found in nearly every additional sample repetition. This molecular work indicates that the deep, hard-rock groundwater environments studied are inhabited by diverse microbial populations, consistent with the great variability of hydrogeochemical conditions. Similar results were obtained with DNA sequences of micro-organisms from the alkaline springs of Maqarin in Jordan (Pedersen *et al.*, 2004), natural nuclear reactors in Oklo, Gabon (Pedersen *et al.*, 1996b), and in nuclear-waste buffer material (Stroes-Gascoyne *et al.*, 1997).

The molecular work described above has provided a good insight into the phylogenetic diversity of hard-rock aquifer micro-organisms but does not reveal species-specific information unless 100 % identity of the full 16S rRNA gene sequence of a known and described micro-organism is obtained. The huge diversity of the microbial world makes the probability of such a hit very small – none of the 122 specific sequences mentioned above had 100 % identities with described species. Even if a 100 % identity is obtained, there may yet be strain-specific differences in some characteristics which are not revealed by the 16S rRNA gene sequence information (Fuhrman & Campbell, 1998). If species information is required, time-consuming methods in systematic microbiology must be applied to a pure culture; known genera or species can be identified through these methods.

Sequencing 16S/18S rRNA genes is a fairly blunt tool for obtaining an understanding of microbial activity and metabolic diversity. Therefore, our laboratory has been applying cultivation methods for a long time, concurrent with molecular methods. The most probable numbers of a range of physiologically different micro-organisms have been analysed (Haveman *et al.*, 1999; Haveman & Pedersen, 2002); these micro-organisms are iron-reducing bacteria, manganese-reducing bacteria, sulfate-reducing bacteria, autotrophic methanogens, heterotrophic methanogens, autotrophic acetogens, heterotrophic acetogens and methanotrophic bacteria. All those groups have been found in hard-rock groundwater. In addition, unicellular eukaryotes, in particular fungi, have

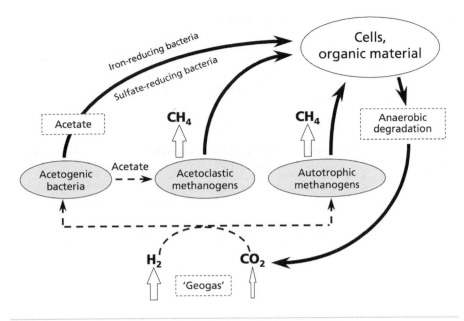

Fig. 3. The deep hydrogen-driven biosphere hypothesis, illustrated by its carbon cycle. Under relevant temperature and water-availability conditions, intraterrestrial micro-organisms are capable of performing a life cycle that is independent of solar-driven ecosystems. Hydrogen and carbon dioxide from the deep crust of the Earth are used as energy and carbon sources.

been detected (Ekendahl *et al.*, 2003). Phages are probably present but remain to be detected.

ENERGY SOURCES FOR INTRATERRESTRIAL LIFE

Where do all these organisms get their energy? Hydrogen may be an important electron and energy source and carbon dioxide an important carbon source in deep subsurface ecosystems. Hydrogen, methane and carbon dioxide have been found in micromolar concentrations at all sites tested for these gases (Sherwood Lollar *et al.*, 1993a, b; Pedersen, 2001). A model has been proposed (Fig. 3) of a hydrogen-driven biosphere in deep igneous rock aquifers of the Fennoscandian Shield (Pedersen, 1993b, 2000). The organisms at the base of this ecosystem are assumed to be autotrophic acetogens capable of reacting hydrogen with carbon dioxide to produce acetate, autotrophic methanogens that produce methane from hydrogen and carbon dioxide and aceto-clastic methanogens that produce methane from the acetate product of the autotrophic acetogens. This hydrogen-driven, intraterrestrial biosphere should conceptually be possible anywhere underground or under the seafloor where hydrogen and carbon dioxide are available. The cycle produces acetate; this is a very versatile source of energy and carbon for anaerobic microbes, such as iron- and sulfate-reducing bacteria and acetoclastic methanogens. All components needed for this life cycle have been found

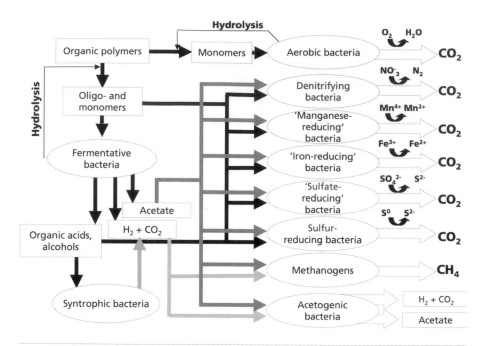

Fig. 4. The degradation of organic carbon can occur via a number of different metabolic pathways, characterized by the principal electron acceptor in the carbon oxidation reaction. A range of significant groundwater compounds are formed or consumed during this process. The degradation follows a typical redox ladder pattern.

in the deep igneous rock aquifers and the expected microbial activities have been demonstrated (Pedersen, 2001), so the model is supported by the qualitative data obtained so far. Quantitative data are currently being obtained at the underground laboratory (Fig. 2).

ACTIVITY OF INTRATERRESTRIAL LIFE FORMS

How much do intraterrestrial organisms depend on their environment and to what extent can they influence their environment? The organisms may have little to do with hard rock formation but they certainly take part in the formation of less violently formed, sedimentary rock types. Once the rock is formed, there are still many processes that can be influenced by microbes and their activities. Precipitation and dissolution, weathering of rocks and cycling of nitrogen, carbon, phosphorus and many metals can be influenced by microbes to various degrees. Such processes generally require *active* microbes, and gaining an understanding of activity generally requires *in situ* and/or laboratory cultivation of the organisms of interest. Radiotracers can be used (Pedersen & Ekendahl, 1992a, b; Ekendahl & Pedersen, 1994), and there is also the possibility of analysing stable isotope fractionations using substrates added by nature (Des Marais, 1999).

Another approach to exploring intraterrestrial micro-organisms is to study and interpret geochemical data. For example, why is there no oxygen in deep groundwater? Oxygen is the preferred electron acceptor of most microbes and its source is photosynthesis, a surface-based process. Microbes tend to degrade organic carbon down the redox ladder (Fig. 4), and oxygen is the first electron acceptor to be utilized. As rainwater percolates down through the ground to become groundwater, it carries organic carbon and oxygen. Microbes degrade the organic substances along the ladder in an ongoing process (Banwart *et al.*, 1996). Oxygen thus does not penetrate very far underground (although there will be spots where this happens), so the intraterrestrial world is thus anaerobic. The intriguing thing about this situation is that the underground environment is obviously kept anaerobic by microbes. Life has changed the surface of our planet dramatically by filling it with oxygen via photosynthesis, and it seems likely that life is also responsible for keeping oxygen out of the intraterrestrial biosphere. The intraterrestrial biosphere thus reflects times on our planet when photosynthesis had not yet had an effect.

REFERENCES

Amend, J. P. & Teske, A. (2005). Expanding frontiers in deep subsurface microbiology. *Palaeogeogr Palaeoclimatol Palaeoecol* **219**, 131–155.

Bachofen, R. (editor) (1997). Proceedings of the 1996 International Symposium on Subsurface Microbiology (ISSM-96). Davos, Switzerland, 15–21 September 1996. *FEMS Microbiol Rev* **20**, 179–638.

Bachofen, R., Ferloni, P. & Flynn, I. (1998). Microorganisms in the subsurface. *Microbiol Res* **153**, 1–22.

Banwart, S., Tullborg, E.-L., Pedersen, K., Gustafsson, E., Laaksoharju, M., Nilsson, A.-C., Wallin, B. & Wikberg, P. (1996). Organic carbon oxidation induced by large-scale shallow water intrusion into a vertical fracture zone at the Äspö Hard Rock Laboratory (Sweden). *J Contam Hydrol* **21**, 115–125.

Butler, R. (1994). Drilling deep holes in the continents. *Geol Today* **10**, 32–35.

Characklis, W. G. & Marshall, K. C. (editors) (1990). *Biofilms*. New York: Wiley.

Des Marais, D. J. (1999). Stable light isotope biogeochemistry of hydrothermal systems. In *Evolution of Hydrothermal Ecosystems on Earth (and Mars?)*, pp. 83–98. Edited by G. R. Bock & J.A. Goode. Chichester: Wiley.

Ekendahl, S. & Pedersen, K. (1994). Carbon transformations by attached bacterial populations in granitic groundwater from deep crystalline bed-rock of the Stripa research mine. *Microbiology* **140**, 1565–1573.

Ekendahl, S., Arlinger, J., Ståhl, F. & Pedersen, K. (1994). Characterization of attached bacterial populations in deep granitic groundwater from the Stripa research mine with 16S rRNA gene sequencing technique and scanning electron microscopy. *Microbiology* **140**, 1575–1583.

Ekendahl, S., O'Neill, A. H., Thomsson, E. & Pedersen, K. (2003). Characterisation of yeasts isolated from deep igneous rock aquifers of the Fennoscandian Shield. *Microb Ecol* **46**, 416–428.

Fredrickson, J. K. & Fletcher, M. (editors) (2001). *Subsurface Microbiology and Biogeochemistry*. New York: Wiley-Liss.

Fuhrman, J. A. & Campbell, L. (1998). Microbial microdiversity. *Nature* **393**, 410.

Ghiorse, W. C. & Wilson, J. T. (1988). Microbial ecology of the terrestrial subsurface. *Adv Appl Microbiol* **33**, 107–172.

Gold, T. (1992). The deep, hot biosphere. *Proc Natl Acad Sci U S A* **89**, 6045–6049.

Haveman, S. A. & Pedersen, K. (2002). Microbially mediated redox processes in natural analogues for radioactive waste. *J Contam Hydrol* **55**, 161–174.

Haveman, S. A., Pedersen, K. & Routsalainen, P. (1999). Distribution and metabolic diversity of microorganisms in deep igneous rock aquifers of Finland. *Geomicrobiol J* **16**, 277–294.

Laaksoharju, M., Skårman, C. & Skårman, E. (1999). Multivariate mixing and mass balance (M3) calculations, a new tool for decoding hydrogeochemical information. *Appl Geochem* **14**, 861–871.

Onstott, T. C., Phelps, T. J., Colwell, F. S., Ringelberg, D., White, D. C. & Boone, D. R. (1998). Observations pertaining to the origin and ecology of microorganisms recovered from the deep subsurface of Taylorsville Basin, Virginia. *Geomicrobiol J* **15**, 353–383.

Pace, N. R. (1997). A molecular view of microbial diversity and the biosphere. *Science* **276**, 734–740.

Pedersen, K. (1993a). The deep subterranean biosphere. *Earth Sci Rev* **34**, 243–260.

Pedersen, K. (1993b). Bacterial processes in nuclear waste disposal. *Microbiol Eur* **1**, 18–23.

Pedersen, K. (2000). Exploration of deep intraterrestrial life: current perspectives. *FEMS Microbiol Lett* **185**, 9–16.

Pedersen, K. (2001). Diversity and activity of microorganisms in deep igneous rock aquifers of the Fennoscandian Shield. In *Subsurface Microbiology and Biogeochemistry*, pp. 97–139. Edited by J. K. Fredrickson & M. Fletcher. New York: Wiley-Liss Inc.

Pedersen, K. & Ekendahl, S. (1992a). Incorporation of CO_2 and introduced organic compounds by bacterial populations in groundwater from the deep crystalline bedrock of the Stripa mine. *J Gen Microbiol* **138**, 369–376.

Pedersen, K. & Ekendahl, S. (1992b). Assimilation of CO_2 and introduced organic compounds by bacterial communities in ground water from Southeastern Sweden deep crystalline bedrock. *Microb Ecol* **23**, 1–14.

Pedersen, K., Arlinger, J., Ekendahl, S. & Hallbeck, L. (1996a). 16S rRNA gene diversity of attached and unattached bacteria in boreholes along the access tunnel to the Äspö hard rock laboratory, Sweden. *FEMS Microbiol Ecol* **19**, 249–262.

Pedersen, K., Arlinger, J., Hallbeck, L. & Pettersson, C. (1996b). Diversity and distribution of subterranean bacteria in groundwater at Oklo in Gabon, Africa, as determined by 16S RNA gene sequencing. *Mol Ecol* **5**, 427–436.

Pedersen, K., Hallbeck, L., Arlinger, J., Erlandson, A.-C. & Jahromi, N. (1997). Investigation of the potential for microbial contamination of deep granitic aquifers during drilling using 16S rRNA gene sequencing and culturing methods. *J Microbiol Methods* **30**, 179–192.

Pedersen, K., Nilsson, E., Arlinger, J., Hallbeck, L. & O'Neill, A. (2004). Distribution, diversity and activity of microorganisms in the hyper-alkaline spring waters of Maqarin in Jordan. *Extremophiles* **8**, 151–164.

Sherwood Lollar, B., Frape, S. K., Fritz, P., Macko, S. A., Welhan, J. A., Blomqvist, R. & Lahermo, P. W. (1993a). Evidence for bacterially generated hydrocarbon gas in

Canadian Shield and Fennoscandian Shield rocks. *Geochim Cosmochim Acta* **57**, 5073–5085.

Sherwood Lollar, B., Frape, S. K., Weise, S. M., Fritz, P., Macko, S. A. & Welhan, J. A. **(1993b)**. Abiogenic methanogenesis in crystalline rocks. *Geochim Cosmochim Acta* **57**, 5087–5097.

Stroes-Gascoyne, S., Pedersen, K., Haveman, S. A. & 8 other authors **(1997)**. Occurrence and identification of microorganisms in compacted clay-based buffer material designed for use in a nuclear fuel waste disposal vault. *Can J Microbiol* **43**, 1133–1146.

Szewzyk, U., Szewzyk, R. & Stenström, T. A. **(1994)**. Thermophilic, anaerobic bacteria isolated from a deep borehole in granite in Sweden. *Proc Natl Acad Sci U S A* **91**, 1810–1813.

Whitman, W. B., Coleman, D. C. & Wiebe, W. J. **(1998)**. Prokaryotes: the unseen majority. *Proc Natl Acad Sci U S A* **95**, 6578–6583.

Iron, nitrogen, phosphorus and zinc cycling and consequences for primary productivity in the oceans

John A. Raven,[1] Karen Brown,[1] Maggie Mackay,[1]
John Beardall,[2] Mario Giordano,[3] Espen Granum,[4]
Richard C. Leegood,[4] Kieryn Kilminster[5] and Diana I. Walker[5]

[1]Plant Research Unit, Division of Environmental and Applied Biology, School of Life Sciences, University of Dundee at SCRI, Scottish Crop Research Institute, Invergowrie, Dundee DD2 5DA, Scotland, UK

[2]School of Biological Sciences, Monash University, Clayton, VIC 3800, Australia

[3]Department of Marine Science, Università Politecnica delle Marche, 60131 Ancona, Italy

[4]Department of Animal and Plant Sciences, University of Sheffield, Sheffield S10 2TN, UK

[5]School of Plant Biology, University of Western Australia, M090 35 Stirling Highway, Crawley, WA 6009, Australia

INTRODUCTION

Primary productivity in the ocean amounts to the net assimilation of CO_2 equivalent to about 50 Pg (petagram, i.e. 10^{15} g) C year^{-1}, while on land this is approximately 60 Pg C year^{-1} (Field et al., 1998). Almost all of this primary productivity involves photosynthesis, and in the ocean it occurs only in the top few hundred metres, even in waters with the smallest light attenuation (Falkowski & Raven, 1997). About 1 Pg C of marine primary productivity involves benthic organisms, i.e. those growing on the substratum (Field et al., 1998), in the very small fraction of the ocean which is close enough to the surface to permit adequate photosynthetically active radiation (PAR) to allow photolithotrophic growth. This depth at which photosynthetic growth is just possible varies in time and space, and defines the bottom of the euphotic zone (Falkowski & Raven, 1997). The remaining ~49 Pg C is assimilated by phytoplankton in the water column (Field et al., 1998). This chapter will concentrate on the planktonic realm, while acknowledging the importance of marine benthic primary producers and their interactions with micro-organisms (e.g. Dudley et al., 2001; Raven et al., 2002; Raven & Taylor, 2003; Cooke et al., 2004; Walker et al., 2004).

The global net primary productivity of the oceans is less than that on land, despite about 70 % of the Earth being covered in ocean and primary productivity over considerable areas of land being limited by water supply. In considering the role of nutrient elements such as iron, nitrogen, phosphorus and zinc in contributing to the restriction of primary productivity in the ocean, we must consider them in the context of other limitations on primary productivity.

SGM symposium 65: Micro-organisms and Earth systems – advances in geomicrobiology.
Editors G. M. Gadd, K. T. Semple & H. M. Lappin-Scott. Cambridge University Press. ISBN 0 521 86222 1 ©SGM 2005

These constraints are frequently divided by oceanographers into 'bottom-up' (physical and chemical) factors and 'top-down' (biotic) factors. This division echoes that of Grime (1974), who separated the constraints on terrestrial primary productivity into 'stress' (factors which constrain the production of primary producer biomass) and 'disturbance' (factors which remove primary producer biomass). The bottom-up (stress) factors include insufficient or excessive PAR, excess UV-B radiation, too high or too low a temperature and insufficient supply of any of the chemical elements essential for growth and completion of the life cycle in a form which is available to the organisms. The top-down (disturbance) factors are the biotic factors of grazing and parasitism (including viral attack) and the physical factor of removal by, in the case of phytoplankton, sinking out of the euphotic zone.

The bottom-up factors determining the growth rate of marine primary producers are intimately related to physical oceanography and to global biogeochemical cycles. Physical oceanography is significant for phytoplankton photosynthesis through the depth of the upper, mixed layer of the ocean in which the phytoplankton organisms are entrained, as modified by movement of organisms relative to the surrounding water as a result of the motility of flagellate organisms and upward and downward movement of non-flagellate organisms whose density is, respectively, lesser or greater than that of the surrounding sea water. When the mixing depth is greater than the so-called 'critical depth', i.e. the depth at which the integrated water-column photosynthesis by phytoplankton equals community respiration, net growth of phytoplankton cannot occur (Falkowski & Raven, 1997). Physical oceanography also impacts on the supply of nutrients to photosynthetic organisms with respect to cycling of nutrients within the ocean. Nutrients are carried to depth by the sinking of live and dead biota and faecal pellets, generically termed 'marine snow'. Below the euphotic zone, most of the particulate organic material is acted on by decomposer organisms which regenerate inorganic nutrients that can then be used by primary producers upon return to the euphotic zone. This return of nutrients to the euphotic zone can occur by seasonal and permanent upwellings and by eddy diffusion across the thermocline at the base of the upper mixed layer (Falkowski & Raven, 1997).

With this background to marine primary productivity and its determinants, we will address the cycles of four nutrient elements, iron, nitrogen, phosphorus and zinc, whose supply has been shown at times to restrict marine primary production in some parts of the ocean. This chapter first addresses the biogeochemical cycles of these elements and the evidence as to their roles in limiting primary productivity, dealing with the elements in the order in which they are thought to be significant in constraining primary productivity. Secondly, we will consider the interactions between these elements in determining primary productivity. Thirdly, we consider some broader aspects of elemental supply and marine primary productivity in a changing world.

BIOGEOCHEMICAL CYCLES OF NITROGEN, IRON, PHOSPHORUS AND ZINC IN RELATION TO MARINE PRIMARY PRODUCTIVITY

Nitrogen

The abundant N_2 in the atmosphere, and dissolved in the ocean, is only directly available to diazotrophic organisms, i.e. those that use the enzyme nitrogenase to reduce N_2 to NH_3, which can then be used to produce the organic N compounds needed by the organism. N_2 fixation by marine organisms involves mainly cyanobacteria, of which the filamentous *Trichodesmium* predominates, although the role of unicellular cyanobacterial diazotrophs has recently been recognized (Capone, 2001). Other inputs of combined N, i.e. nitrogen sources other than N_2, which are available to non-diazotrophic primary producers, occur as atmospheric inputs resulting from lightning converting N_2 to NO_x ($NO + NO_2$) and hence HNO_3, and increasingly as NH_y ($NH_3 + NH_4^+$), NO_x, HNO_3 and organic N from natural and anthropogenic terrestrial sources. Combined N supply also occurs as fluvial inputs of NH_4^+, NO_3^- and organic N from terrestrial ecosystems, again with increases as a result of man's activities (Falkowski & Raven, 1997; Tyrrell, 1999). The losses of combined N from the ocean occur as denitrification, whereby NO_3^- is reduced to N_2O and N_2 in anoxic and hypoxic regions of the ocean, and by the incorporation of particulate organic N into marine sediments (Falkowski & Raven, 1997; Tyrrell, 1999). While it has been suggested that there is a net loss of combined N from the present ocean (Falkowski & Raven, 1997), consideration of the cumulative errors in the estimation of the components of the combined N inputs and outputs for the present ocean suggests that the data are also consistent with a balance of oceanic combined N inputs and outputs (Tyrrell, 1999). Even if there is no net loss of combined N from the ocean at present, there is strong evidence that primary productivity in a large fraction of the world ocean is limited by the supply of combined N (Falkowski, 1997; Falkowski & Raven, 1997; Cullen, 1999; Tyrrell, 1999). This evidence is derived from enrichment experiments in which samples of surface ocean water are incubated for the determination of primary productivity, either unaltered or with the addition of particular nutrients, individually and in combination. The results of such experiments in many parts of the ocean suggest that the nutrient element limiting growth is N, with diazotrophs unable to supply combined N at a sufficiently rapid rate. More generally, the atomic ratio of inorganic combined N to P in the surface ocean is, when averaged over time and space, rather less than the $16:1$ Redfield ratio of the average phytoplankton cell with the elemental composition of $106C:16N:1P$ (Falkowski & Raven, 1997; Falkowski, 2000).

There is a significant difference between the two scenarios for oceanic combined N status (Falkowski, 1997; Tyrrell, 1999) with respect to the possible constraints on biological N_2 fixation in the ocean. We deal first with the possibility that there is not a shortfall in biological fixation, i.e. it is adequate to maintain the oceanic combined N pool by supplying as much combined N as is needed to replace the excess of combined N loss by sedimentation and denitrification over aeolian and riverine inputs (Tyrrell, 1999). If this is the case then it may not be necessary to seek exogenous limitations specific to diazotrophy to account for the magnitude of N_2 fixation other than optimal allocation arguments that the extra machinery required by diazotrophy restricts the maximum specific growth rate of N_2-fixing organisms relative to those using combined N. This would permit the diazotrophs to grow whenever there was a shortfall in combined N supply which constrained the growth of non-diazotrophs, despite the lower maximum specific growth rates, with the non-diazotrophs ousting the slower-growing diazotrophs when combined N is more readily available. However, it is known that there are additional resource costs of diazotrophic growth relative to growth on combined N, such as the increased energy, Fe and Mo (or V) requirement and a less well-defined requirement for additional P (Sañudo-Wilhelmy et al., 2001; Vitousek et al., 2001; Kustka et al., 2003a, b). These constraints on the growth of photosynthetic diazotrophs relative to that of photosynthetic organisms living on combined N could account for any inadequacy of N_2 fixation to maintain the oceanic combined N pool by not allowing the diazotrophs to fix N_2 until there was enough combined N to allow non-diazotrophs to out-compete diazotrophs (Falkowski, 1997). The observation is that Fe is commonly a limiting resource for diazotrophy in the surface ocean, although P and PAR are also significant constraints in some areas (Falkowski, 1997; Sañudo-Wilhelmy et al., 2001; Mills et al., 2004).

The combined N sources for phytoplankton growth in the ocean include nitrate, nitrite, ammonium and organic N (Falkowski & Raven, 1997). As we have noted, nitrate (and nitrite) enters the surface ocean by fluvial and aeolian inputs, as well as in upwellings and eddy diffusion from the deep ocean, where it is produced by nitrification of reduced N sedimented from the euphotic zone. This 'new' (to the euphotic zone) combined N supports what is termed 'new' primary production which, since it can be exported (sedimented) without altering the pre-existing combined N pool in the surface ocean, is also termed 'export production'. Ammonium/ammonia and organic N supply to the surface ocean is predominantly from excretion by higher trophic levels as well as from parasitism of phytoplankton, with only a minor component entering the surface ocean from aeolian and fluvial inputs. The primary productivity supported by the use of this reduced N is termed 'recycled' production, since it depends very largely on combined N recycled from particulate organic matter within the surface ocean. The reason that this 'recycled' production uses reduced combined N directly, rather than after it has been converted to the thermodynamically more stable (in the oxygenated surface ocean) nitrate, seems to be the kinetic constraint on the exergonic conversion of ammonium to nitrate by nitrifying bacteria resulting from light inhibition of growth of the nitrifiers (Falkowski & Raven, 1997). These nitrifiers are chemolithotrophs which can assimilate about 0·19 Pg C from inorganic C per year. The energy used by these organisms can be regarded as the additional energy, ultimately derived from PAR, required for phytoplankton growth on nitrate rather than on reduced N in the euphotic zone. This energy is used to power inorganic C reduction deeper in the ocean (Raven, 1996). The expected lower growth rate of photosynthetic primary producers under light-limiting conditions resulting from the use of nitrate rather than reduced N as the N source is, however, not found routinely (Raven, 1996).

Most marine primary producers examined are able to use all of the combined N forms discussed above, i.e. ammonium/ammonia, one or more species of organic N, nitrite and nitrate (Falkowski & Raven, 1997). However, the very abundant picoplanktonic cyanobacterium *Prochlorococcus* is unable to use nitrate, and some strains are also unable to use nitrite (Bryant, 2003; Dufresne *et al.*, 2003; Palenik *et al.*, 2003; Rocap *et al.*, 2003). This relatively recent loss of metabolic capability during the evolution of *Prochlorococcus* from within the paraphyletic genus *Synechococcus* is clearly not a fatal impediment to *Prochlorococcus*, which is probably the most abundant, in terms of number of individuals, photosynthetic organism on Earth, as well as being amongst the smallest (Bryant, 2003). *Prochlorococcus* is relatively most successful in the oligo-trophic ocean, where it depends mainly on recycled combined N. It also typically occurs deeper in the water column than the metabolically more versatile *Synechococcus*, where the energy-requiring assimilation of oxidized combined N would consume scarce energy in this light-limited environment (Ting *et al.*, 2002; cf. Raven, 1987; MacFarlane & Raven, 1990; Falkowski & Raven, 1997). As we shall see when considering Fe, the absence of nitrate assimilation capacity reduces the Fe requirement of the cells.

A final aspect of the N nutrition of marine primary producers is the possibility of economizing on N by replacing a macromolecule with a high N content (usually a protein) with a macromolecule with a lower N content that performs a similar meta-bolic function (Raven *et al.*, 2004). Exact functional substitutions are uncommon, making the effectiveness of this possible means of economizing on N difficult to evaluate. An example is in photosynthetic light harvesting, where the quantity of apoprotein per molecule of chromophore is at least twice as high for phycobiliproteins as for chlorophyll–protein complexes. However, differences in chromophore absorption spectra and in chromophore-specific absorption coefficients make a like-for-like comparison of the ecophysiological impact of such a substitution difficult to evaluate. Perhaps the best-examined case is that of the almost total replacement of phycobiliproteins by chlorophyll–protein complexes in *Prochlorococcus* relative to the ancestral *Synechococcus* (Ting *et al.*, 2002). Such substitutions can reduce the cell N requirement by over 10 % (Raven *et al.*, 2004). Similar magnitudes of economy in N costs of growth can be achieved by replacing N-containing with N-free low relative molecular mass organic compounds involved in screening UV-B, scavenging and quenching reactive oxygen species, providing osmotic compensation as compatible solutes and restricting the attentions of grazers and parasites using anti-biophage compounds (Raven *et al.*, 2004). Again, there are problems in establishing the effectiveness of the N-free compound relative to the N-containing compound in the organism under natural conditions.

Iron

Fe input to the surface ocean occurs as dissolved and particulate fluxes down rivers, much of which is sedimented in the coastal zone, and as aeolian inputs of dust, either dry or in rain (Berner & Berner, 1996; Chester, 2003; Jickells *et al.*, 2005). There is also some input of Fe to the deep ocean via hydrothermal vents, while the major output of Fe from the ocean is in sinking particles incorporated into marine sediments (Chester, 2003). Although there is some input of Fe into the surface ocean from the deep ocean by upwelling and eddy diffusion, the major inputs of Fe to the ocean surface are from rivers and, more importantly, from the atmosphere. The atmosphere supplies essentially all of the Fe inputs to the non-coastal surface ocean (Chester, 2003).

Fe is a very abundant element on Earth, but the global oxygenation which began over 2 billion years ago has restricted the availability of Fe to organisms by converting it into the insoluble ferric form. While chemical reduction of ferric to ferrous iron can occur in acidic atmospheric droplets, this soluble ferrous iron is converted to the ferric form with a half-time of about 2 min when the droplet equilibrates with sea water at a pH of about 8 (Chester, 2003). The requirement for large quantities of Fe in, for example, nitrogenase and nitrogenase reductase, photosystem I-like photosynthetic reaction centres, NADH dehydrogenase complexes and nitrite reductase (Raven, 1988, 1990; Raven *et al.*, 1999) presumably evolved when Fe was much more readily available, more than 2·5 billion years ago. Squaring the large Fe requirements in these catalysts with the current relative unavailability of Fe is now met by a range of Fe-acquisition mechanisms, as well as a range of means by which the use of Fe is restricted.

Fe acquisition from sea water involves a variety of mechanisms (Völker & Wolf-Gladrow, 1999). Cyanobacteria use siderophores which are secreted by the organism, and some of the molecules are taken up again after they have acquired Fe(III) from colloidal or ligated sources (Chester, 2003). Some eukaryotic marine primary producers can take up Fe(III)-loaded siderophores, although they do not produce the siderophores (Chester, 2003). The genome of the diatom *Thalassiosira pseudonana* gives no indication of the production or uptake of siderophores, but has components of the mechanism which reduces external Fe(III) to Fe(II) followed by uptake of the Fe(II) by a process involving, paradoxically, oxidation to Fe(III) during uptake (Armbrust *et al.*, 2004). This mechanism is widespread in marine primary producers, using Fe(III) from a range of ligands (Falkowski & Raven, 1997). A few marine planktonic primary producers are also phagotrophic, and so are able to take up colloidal or particulate Fe (Raven, 1997).

The constraints on Fe availability to primary producers have evoked a number of ecological and evolutionary responses. One response is the restricted occurrence of diazotrophy and, in some habitats, of the use of nitrate and nitrite as N sources (Falkowski, 1997; Flynn & Hipkin, 1999; Sañudo-Wilhelmy *et al.*, 2001; Kustka *et al.*, 2003a, b; Mills *et al.*, 2004). Another response is the decrease in the Fe requirement for photosystem I on a cell basis in genotypes with low contents of this kind of reaction centre. Examples of low photosystem I contents include the cyanobacterium *Prochloro-coccus* relative to its ancestor *Synechococcus*, and the open-ocean relative to the coastal species of the diatom *Thalassiosira* (Strzepek & Harrison, 2004). In at least some cases, the decreased content of photosystem I constrains acclimation to environments with varying fluxes of PAR (Strzepek & Harrison, 2004). These economies of N use, by changing the exogenous N source or by decreasing the content of photosystem I, can account for several-fold reduction in Fe requirement from the most costly to the least costly resource-acquisition mechanisms (Sunda *et al.*, 1991; Sañudo-Wilhelmy *et al.*, 2001; Kustka *et al.*, 2003a, b; Strzepek & Harrison, 2004; cf. Falkowski *et al.*, 2004). Much smaller economies can be effected by substituting an Fe-containing component which accounts for a relatively small fraction of the total cell N by an Fe-free component which performs essentially the same metabolic function. Examples here are the replacement of the Fe-containing photosynthetic cytochrome c_6 with the Cu-containing plastocyanin and of the Fe-containing ferredoxin with the metal-free flavodoxin (Raven *et al.*, 1999). Here, near-equality of function has been established (Raven *et al.*, 1999). Such replacements can be facultative (in response to Fe deficiency) or obligate, e.g. in the 'red' line of plastid evolution where plastocyanin is absent (Raven *et al.*, 1999; Falkowski *et al.*, 2004). Although the economy in Fe content of the organisms by these substitutions is small, the expression of flavodoxin in natural phytoplankton assemblages can be used as an indicator of actual, or incipient, Fe limitation of growth.

Despite the theoretical likelihood of Fe limitation of primary productivity in at least a part of the surface ocean, problems with the analysis of Fe in sea water, and in carrying out incubations under conditions free of Fe contamination, meant that evidence consistent with Fe limitation in significant areas of the ocean only became available a little over a quarter of a century ago with the insightful work of the late John Martin and his colleagues (Martin & Fitzwater, 1988; Martin *et al.*, 1990, 1991). Martin's conclusions were based on the chemical analysis of sea water and on the sort of incubation experiments described above for N, involving measurements of primary productivity of sea-water samples with and without the addition of a range of nutrients. This work showed that the Fe limitation of primary production occurred in the so-called 'high-nutrient, low-chlorophyll' (HNLC) areas of the ocean, i.e. those parts of the ocean that have relatively high concentrations of the major nutrients nitrate and phosphate in the surface waters, but a low biomass of primary producers as indicated by the chlorophyll content (Chester, 2003). The HNLC areas of the ocean include the north-east sub-Arctic Pacific, the eastern tropical Pacific and the Southern Ocean (Chester, 2003; Jickells *et al.*, 2005). There are also smaller coastal areas of Fe limitation in the eastern Pacific (Hutchins *et al.*, 1998, 2002; Hutchins & Bruland, 1998; Jickells *et al.*, 2005).

The incubation experiments described above did not rule out the possibility of the *in situ* occurrence of deep mixing, leading to light limitation of primary production, as a bottom-up factor, and/or of the impact of grazers or parasites in removing primary producer biomass as a top-down factor. Clarification of the situation came from meso-scale experiments in which Fe (as iron sulfate) was added to several square kilometres of the surface ocean in HNLC areas and the impact of the addition on the chemistry and biology of the Fe-enriched patch was followed for as long as was possible; observations were usually terminated by the subduction or horizontal fragmentation of the patch (Chester, 2003). These experiments have now been carried out more than once in each of the three main HNLC areas of the ocean and, in each case, a decrease in nitrate and phosphate and an increase in primary producer biomass (especially large diatoms) and in primary productivity was found, as well as a delayed increase in the biomass of herbivorous zooplankton (Watson *et al.*, 2000; Boyd *et al.*, 2002, 2004; Boyd, 2002a, b), although light limitation can also contribute to the HNLC state (Boyd, 2002a; van Oijen *et al.*, 2004). However, it has not yet been established whether the increased primary productivity is paralleled by an increased export flux of particulate organic matter, as might be expected from the additional primary productivity resulting from the exogenous input of the productivity-limiting resource (Boyd *et al.*, 2004). It is of interest that the shortage of Fe does not alter the ratio of C to N in phytoplankton cells, so that there seems to be no preferential allocation of what Fe is available to either the C assimilation or the N assimilation pathways. This absence of preferential allocation of Fe could well be a function of homoeostasis of the C to N ratio. Readers are directed to Cooke *et al.* (2004) and Raven *et al.* (2004) for further information.

For diazotrophy in the surface ocean, we have seen in the consideration of N above that Fe is a common limiting resource for N_2 fixation, although there are parts of the ocean in which P or light energy can limit diazotrophy (Falkowski, 1997; Sañudo-Wilhelmy *et al.*, 2001; Mills *et al.*, 2004). As with phytoplankton assimilating combined N, the diazotroph *Trichodesmium* does not exhibit variations in the cellular C : N ratio as a function of Fe availability (see references in Cooke *et al.*, 2004; Raven *et al.*, 2004).

Phosphorus

Almost all of the biology of P occurs at the most oxidized state of phosphate. Phosphate is supplied to the ocean mainly via rivers, although there is some atmospheric input (Berner & Berner, 1996; Chester, 2003). However, atmospheric input of Fe can sequester some of the P in surface water as complexes with ferric iron (Chester, 2003), recalling the decrease in Fe availability with the onset of atmospheric oxygenation more than 2 billion years ago (Bjerrum & Canfield, 2002). P loss from the ocean is in sinking particles which become incorporated into marine sediments (Bjerrum & Canfield, 2002; Chester, 2003). While there has been much emphasis on inorganic phosphate as the dominant source for marine primary producers, there is an increasing appreciation of the role of soluble organic phosphates in the P nutrition of marine primary producers. Generally, organic P sources are used by uptake of inorganic phosphates after the action of extracellular phosphatases. Those marine planktonic primary producers which are also phagotrophic can acquire P by ingestion of particles, with absorption of digestion products across the food vacuole membrane, as well as by transporters in the plasmalemma (Raven, 1997).

There are cogent geochemical arguments that the supply of P, rather than of N, is the long-term determinant of the productivity of Earth, since the supply of P (in the absence of man's activities) is restricted by the rate of weathering on land and is balanced by sedimentation in the ocean, whereas N supply can be maintained by diazotrophy in the face of sedimentation and denitrification (Falkowski & Raven, 1997). However, as we have seen, there are factors in addition to the availability of P that restrict the extent of diazotrophy in the ocean (Falkowski, 1997; Tyrrell, 1999; Sañudo-Wilhelmy *et al.*, 2001; Mills *et al.*, 2004). Nevertheless, there are regions of the ocean in which there is evidence from measurements of nutrient concentrations in the surface ocean, and from nutrient enrichment studies on primary productivity of samples of surface water, consistent with P limitation of primary productivity (Karl *et al.*, 1993; Karl & Tien, 1997; Cullen, 1999; Wu *et al.*, 2000; Benitez-Nelson & Karl, 2002; Ridame & Guieu, 2002; Björkman & Karl, 2003; Heldal *et al.*, 2003; Krom *et al.*, 2004). Particular attention has been paid to P limitation in parts of the Mediterranean Sea (Krom *et al.*, 2004).

The possibilities of economy in the use of P in marine primary producers by sub-stitution of another, more readily available, element for P while maintaining equivalent metabolic function are smaller for P than for N or, especially, Fe. As for reducing the cell quota of a P-containing component, much of the functional (not-stored) P in most marine primary producers is contained in rRNA. In many non-photosynthetic organisms, there is a positive correlation of growth rate and rRNA content even when growth of a given genotype is constrained by environmental factors other than P supply, or when growth stages of a metazoan are compared, or closely related taxa are compared (Sterner & Elser, 2002; Raven *et al.*, 2004, 2005). This correlation is much less clear for photosynthetic organisms, including the marine primary producers (Sterner & Elser, 2002; Elser *et al.*, 2003; Ågren, 2004; Klausmeier *et al.*, 2004a, b; Leonardos & Geider, 2004; Raven *et al.*, 2004, 2005). However, the finite rate of protein synthesis across each rRNA molecule means that there must be a critical rRNA cell quota below which growth rate is constrained by the rRNA content, although the discussion above suggests that the rRNA quota in adequately P-nourished marine primary producers is significantly above this critical level. This apparent excess of rRNA means that a potential P economy measure in marine primary producers has not been widely adopted in these organisms.

The genome is a P-containing cell component that is not generally a large fraction of the non-storage cell P. However, in the very small marine primary producer *Prochloro-coccus* growing under moderate P deficiency, the genome contains over half of the cell P quota (Bertilsson *et al.*, 2003). This is a function of the non-scalable nature of the components of the genome: the size of a gene encoding a given protein is very similar in very small cells and in larger cells. Thus, despite having as few as 1500 protein-coding genes and a high gene density in the genome, the very small size of *Prochlorococcus* means that even its miniaturized genome contains a very significant fraction of the total cell P. The discussion above on N sources for *Prochlorococcus* shows that there are constraints on the nutrition of this organism as a function of gene loss, and some strains have a restricted ability to acclimate to varying light environments.

To a limited extent, the large fraction of cell P in the genome of *Prochlorococcus* can be attributed to the small cell quota of P relative to C and N in well-nourished cells (Bertilsson *et al.*, 2003; Heldal *et al.*, 2003). Although it is tempting to relate these high C:P and N:P ratios to the oligotrophic and potentially P-limited habitat of *Prochlorococcus*, it must be remembered that the low P content in comparison with the Redfield ratio of 106C:16N:1P (by atoms) is not exceptional, granted the variability in the C:P and N:P ratios of marine phytoplankton cultured under nutrient-replete conditions. The Redfield ratio applies to temporal and spatial averages over the world ocean (Falkowski & Raven, 1997; Falkowski, 2000; Geider & La Roche, 2002). A further complication in assessing the P quota of marine phytoplankton is the occurrence, in some organisms, of significant (relative to the intracellular content) surface-bound P (Sañudo-Wilhelmy *et al.*, 2004).

Zinc

Zn enters the ocean by dry and, especially, wet deposition from the atmosphere, as well as in rivers, and leaves it by sedimentation (Chester, 2003). Evidence consistent with Zn limitation in parts of the ocean comes from the work of Crawford *et al.* (2003) and Franck *et al.* (2003) using enrichments with Fe, Zn or Fe + Zn in experiments on primary productivity, or components thereof, using samples of surface ocean water. However, no drawdown of Zn was found during the stimulation of primary productivity in HNLC regions by the addition of Fe, suggesting that Zn was not close to growth-limiting in these surface waters (Frew *et al.*, 2001; Gall *et al.*, 2001). There is evidence consistent with Zn limitation of phytoplankton growth in some freshwater habitats (Ellwood *et al.*, 2001; Sterner *et al.*, 2004).

The enzymes that use Zn in marine primary producers include carbonic anhydrase, alkaline phosphatase and, in seagrasses and in peridinin-containing dinoflagellates, Cu-Zn superoxide dismutase (Raven *et al.*, 1999). Most marine primary producers have only the Fe/Mn superoxide dismutases (Raven *et al.*, 1999). A possible rationale for the lack of Cu-Zn superoxide dismutase in most marine primary producers is the very widespread occurrence of inorganic carbon-concentrating mechanisms (CCMs) in these organisms which, in many cases, involve an accumulation of bicarbonate ions to concentrations in excess of those in sea water. Cu-Zn superoxide dismutases catalyse, in addition to the eponymous dismutation of superoxide to hydrogen peroxide and oxygen, a bicarbonate-mediated peroxidation activity which can damage the enzyme itself or, by reacting with exogenous substrates, protect the enzyme from inactivation (see Elam *et al.*, 2003). Perhaps the absence of Cu-Zn superoxide dismutases in these organisms with CCMs is related to the increased extent of self-oxidation and damage of the enzyme as a function of increasing concentrations of bicarbonate. It is of interest that CCMs in terrestrial vascular plants, i.e. those in plants with C_4 and 'crassulacean acid metabolism' (CAM), and with Cu-Zn superoxide dismutases in their cytosol and stroma, accumulate CO_2 rather than bicarbonate (Giordano *et al.*, 2005).

The quantitatively most significant role of Zn in most marine primary producers is in carbonic anhydrases, although Co or Cd can substitute for Zn *in vitro* and, in some cases, *in vivo* (Giordano *et al.*, 2005). There is also a Cd-specific carbonic anhydrase in the diatom *Thalassiosira weissflogii* (Lane & Morel, 2000b; Giordano *et al.*, 2005). Carbonic anhydrases are involved, inter alia, in CCMs (Giordano *et al.*, 2003, 2005). These CCMs are frequently expressed more strongly at low inorganic C concentrations

(Giordano *et al.*, 2005), and Morel *et al.* (1994) and Lane & Morel (2000a) have shown additional possibilities of Zn limitation at low inorganic C concentrations where carbonic anhydrase is expressed at higher levels. Subsequently, Reinfelder *et al.* (2000, 2004) and Morel *et al.* (2002) have offered evidence consistent with C_4-like photosynthetic metabolism in the planktonic marine diatom *T. weissflogii*, especially at low inorganic C levels, which would decrease the requirement for Zn. The argument here is that, if bicarbonate is the inorganic C form entering the cytosol, no carbonic anhydrase activity is needed in that compartment because bicarbonate is the inorganic C substrate for phosphoenolpyruvate carboxylase, the enzyme which generates the C_4 dicarboxylic acid oxaloacetate (Reinfelder *et al.*, 2000, 2004; Morel *et al.*, 2002). Decarboxylation of oxaloacetate by phosphoenolpyruvate carboxykinase in the chloroplast stroma generates CO_2, the substrate for the core carboxylase of photosynthesis, ribulose bisphosphate carboxylase-oxygenase (Rubisco), which also occurs in the stroma (Reinfelder *et al.*, 2000, 2004; Morel *et al.*, 2002). This scheme for inorganic C assimilation does not require expression of carbonic anhydrase in the cytosol or the stroma, or on the cell surface, since inorganic C enters as bicarbonate, the predominant form of inorganic C in sea water. Although the evidence for C_4 metabolism as an obligatory component of photosynthesis in *T. weissflogii* is incomplete (Johnston *et al.*, 2001; Granum *et al.*, 2005; Giordano *et al.*, 2005), the case is becoming stronger (Reinfelder *et al.*, 2004). Such a mechanism does not necessitate the accumulation of bicarbonate in intracellular compartments, so that the argument suggested above for the absence of Cu-Zn superoxide dismutases from organisms which accumulate bicarbonate is inapplicable, as it is for higher plant C_4 and CAM photosynthesis.

INTERACTIONS AMONG THE AVAILABILITY OF IRON, NITROGEN, PHOSPHORUS AND ZINC AND THAT OF OTHER RESOURCES IN DETERMINING MARINE PRIMARY PRODUCTIVITY AND ITS FATE

Interactions among Fe, N, P and Zn

The discussion above of the interaction of the cycles of the four elements with marine primary producers has perforce included discussion of some of the interactions among these elements, and with some other resources required for growth. An example is the increased Fe requirement for growth when nitrate rather than ammonium is the exogenous N source, with an even greater Fe requirement when diazotrophy supplies ammonia/ammonium from N_2. We also saw that Fe deficiency does not specifically decrease the assimilation of nitrate or N_2 relative to inorganic C, despite the involvement of additional Fe-containing enzymes in the assimilation of these N sources which are not needed for photosynthetic growth with reduced N as the N source. Another example is the role of Fe in atmospheric dust deposited in the ocean on the availability of P from dissolved inorganic phosphate. Finally, Zn is required through its involvement as a cofactor in most carbonic anhydrases in the photosynthetic assimilation of inorganic C in most marine primary producers and through its involvement as a cofactor in extracellular phosphatases in the assimilation of external organic P.

The elementary composition of marine phytoplankton grown under nutrient-sufficient conditions has been investigated by Quigg *et al.* (2003), Ho *et al.* (2003) and Falkowski *et al.* (2004). This work shows that the content of trace elements in the eukaryotes correlates with the evolutionary origin of the plastids, with the 'green' organisms relatively enriched in Fe, Cu and Zn, while the 'red' organisms are relatively enriched in Mn, Co and Cd. These trends in trace metal content cannot all be readily explained by differences in the use of the metals in catalysis, as indicated by the presence or absence of catalysis involving a given metal, and the variations in the quantity of the different metal-containing catalysts, among the taxa (Raven *et al.*, 1999; Quigg *et al.*, 2003; Ho *et al.*, 2003).

Interactions of Fe, N, P and Zn with the assimilation of inorganic C

We have already discussed some interactions of Fe, N, P and Zn with photosynthesis by marine primary producers. Here, we consider some other interactions and comment on the possible impact of increasing surface ocean CO_2 concentrations resulting from anthropogenic CO_2 emissions to the atmosphere on these interactions. We concentrate particularly on the impact on the role of CCMs, since these are the means by which most marine primary producers have their growth rates saturated with inorganic C under the current environmental conditions (Beardall & Raven, 2004; Giordano *et al.*, 2005).

Table 1 shows the effect of the availability of Fe, N, P and Zn on the extent of engagement of CCMs. The evidence as to the extent to which CCMs are involved in inorganic C assimilation generally comes from studies of the concentration of inorganic C required to achieve half of the inorganic-C-saturated rate of photo-synthesis, a lower half-saturation concentration suggesting a greater involvement of CCMs. More rarely, the evidence comes from the measured ratio of the inorganic C concentration in the cells to that in the sea-water medium during steady-state photosynthesis. Some of the data come from work on freshwater rather than marine organisms.

The effect of limitation of photosynthesis by the supply of PAR, e.g. in benthic primary producers growing deep under water or in the shade of other organisms, or planktonic primary producers in a deeply mixed ocean surface layer, is to decrease the engagement

Table 1. Inorganic C affinity and expression of CCMs in cyanobacteria and algae with CCMs as a function of resource limitation for growth

Growth-limiting resource	Effect on inorganic C affinity/CCM expression	Reference(s)
PAR	Decrease	Beardall (1991); Kübler & Raven (1994, 1995); Young & Beardall (2005); Giordano et al. (2005)
Inorganic C	Increase	Giordano et al. (2005)
Inorganic N	Increase (nitrate as N source)	Beardall et al. (1982, 1991); Young & Beardall (2005)
	Decrease (ammonium as N source)	Giordano et al. (2003)
Inorganic P	Increase	Beardall et al. (2005)
	Decrease	Bożena et al. (2000)
Inorganic Fe	Increase	Young & Beardall (2005)
Inorganic Zn	Decrease	Morel et al. (1994); Buitenhuis et al. (2003)

of CCMs in C assimilation. Limitation of photosynthesis by decreased inorganic C supply can occur in marine habitats, e.g. in high intertidal rock pools with high densities of primary producers and infrequent flushing with sea water at neap tides, and in blooms of plankton (Falkowski & Raven, 1997; Raven et al., 2002). Such inorganic C limitation increases the engagement of CCMs (Table 1). Conversely, the increase in sea surface water concentration of CO_2, and to a smaller relative extent in bicarbonate and total inorganic C as well as decrease in pH, predicted to occur by 2100 leads to a decreased CCM engagement and a decreased affinity of photosynthesis for inorganic C (Table 1; Beardall & Raven, 2004). A decreased availability of combined N generally increases the engagement of CCMs, the exception being for an organism grown on ammonium rather than on nitrate, as were all of the other organisms tested (Table 1). There are conflicting data as to the effect of a decreased availability of P on CCM engagement, while the only data available on the effect of limited Fe availability indicate increased CCM engagement (Table 1). Finally, decreased Zn availability decreases the engagement of CCMs (Table 1).

These data (Table 1) are in general agreement with predictions from theoretical considerations on the effect of variations in resource supply. The effects of light availability can be rationalized in terms of the relatively greater impact of leakage from the accumulated inorganic C pool as the energy supply for (re-)accumulation is decreased, and there is an increased 'affinity' for PAR in driving photosynthesis (higher rate at a given low photon flux density) paralleling the decreased CCM expression for organisms acclimated to low light, and the reverse for organisms acclimated to low inorganic C and high light (Kübler & Raven, 1994, 1995). The theoretical rationale for the commonly observed increase in CCM expression when N availability is restricted relates to the N costs of the additional protein components related to the CCM relative to the N costs of the alternative of Rubisco oxygenase and photorespiratory metabolism (Beardall *et al.*, 1982). Similarly, the increased CCM engagement under low Fe conditions can be related to Fe requirements for thylakoid components involved in CCMs and in Rubisco oxygenase activity and photorespiration (Raven & Johnston, 1991). The decreased CCM engagement in organisms not thought to use a C_4-like metabolism when Zn availability is restricted can be related to the Zn in the increased levels of carbonic anhydrases attendant on increased CCM activity.

While these rationalizations are useful in interpreting the data, further mechanistic studies are needed, especially related to the interaction of the various resource supply factors that determine CCM engagement. Also required is work in which the possibility of long-term genetic adaptation, as well as of short-term acclimation, is addressed for increased CO_2 supply (Collins & Bell, 2004) in relation to other resource supply factors.

Interaction of Fe, N, P and Zn supply with the fate of marine primary producers

So far we have concentrated on the impact of the availability of resources on the rate of primary production, i.e. a consideration of bottom-up or stress factors determining the accumulation of biomass. We now consider the implications of resource availability for the fate of the primary producer biomass. This includes the biotic interactions of grazing and parasitism, i.e. the classic top-down factors, as well as the abiotic factor of sinking in the case of phytoplankton; all of these would be subsumed under Grime's concept of disturbance.

Dealing first with the effect of nutrient limitation on grazing, restricted availability of a nutrient can alter the chemical composition of primary producers and hence their nutritional value to herbivores (grazers). Most of the work on the impact of the chemical composition of aquatic primary producers on their nutritional value has involved freshwaters (Sterner & Elser, 2002). While there are influences on the ingestion rate of primary producers by grazers as a function of the resource supply conditions during the growth of the photosynthetic organisms, it is not clear whether these changes routinely alter the impact of grazers on primary producer populations (Sterner & Elser, 2002). The situation is made more complex by the general effect of nutrient deficiency not just on the ratio of major nutritional components such as proteins, carbohydrates and lipids, and the content of essential trace elements, but by additional responses of primary producers. The further responses to nutrient deficiency include the production of additional quantities of organic chemicals which deter or poison grazers or render the ingested material indigestible. These responses by the primary producers are a matter of some controversy. One debate concerns the extent to which the production of 'defence compounds' is selected for because of their effect in restricting removal of biomass when the rate of biomass production is already constrained by limited nutrient availability (Sterner & Elser, 2002). What some workers perceive as an alternative explanation is that the additional production of at least the N-free 'defence compounds' is related to the excess of photosynthate over the capacity of the organism to use the photosynthate in growth when nutrient deficiency constrains photosynthate use in growth more than it does photosynthate production (Sterner & Elser, 2002). This contrast may be more apparent than real, with 'excess' organic C in nutrient-limited primary producers used to make compounds which limit the loss to grazers of a population whose capacity for replacement is resource-limited. The arguments become especially complex when very small organisms, such as phytoplankton, are considered (Wolfe *et al.*, 1997; Ianora *et al.*, 2004), as the effectiveness of defence compounds in restricting grazing in terms of natural selection is a function of the ability of the grazers to distinguish among prey, and then to restrict the ingestion of the better-defended cells. Resolution of these problems involves not just consideration of the sensory and manipulative abilities of the grazers in relation to their learned or innate behaviour in detecting and avoiding better-defended cells, but also the nature of the phytoplankton populations (e.g. clonal or outbreeding) and the unit of natural selection in these organisms (Thornton, 2002; Raven & Waite, 2004). The implications of these 'defence compounds' in avoiding parasitism, including viral infection, are even less clear than their effects on grazers.

Additional effects of nutrient deficiency on the palatability and nutritional value of marine primary producers include the impact of nutrient deficiency on the extent of mineralization. Tables 2 and 3 show the effects of resource deficiency on the ratio of mineral material to organic matter in two groups of quantitatively very important marine primary producers, the silicified diatoms (planktonic and benthic) and the calcified coccolithophores (planktonic, with benthic stages in the life cycle of a few coastal species). The data in Table 2 indicate that restricted availability of Fe, N, P and Zn, as well as of PAR and of inorganic C, increases the content of silica relative to that of organic matter. Indeed, every factor which increases the length of the vegetative life cycle of diatoms increases the time of the G2 phase of the life cycle. The G2 phase is when silicification occurs, so that slowing cell growth increases the relative amount of silica per cell (Martin-Jézéquel *et al.*, 2000). The situation is rather less clear for calcification in coccolithophores (Table 3). Here, restricted light availability decreases calcification relative to organic C production, as does restricted bicarbonate

Table 2. Silicification by diatoms as a function of resource limitation for growth

Growth limitation by supply of the named resource	Effect on particulate silica/particulate organic C	Reference(s)
PAR	Increase	Martin-Jézéquel et al. (2000); Claquin et al. (2002)
Inorganic C	Increase*	Milligan et al. (2004)
Inorganic N	Increase	Martin-Jézéquel et al. (2000); Claquin et al. (2002)
Inorganic P	Increase	Martin-Jézéquel et al. (2000); Claquin et al. (2002)
Inorganic Fe	Increase	Hutchins & Bruland (1998); Takeda (1998); De La Rocha et al. (2000); Franck et al. (2003); Leynaert et al. (2004)
Inorganic Zn	Increase	De La Rocha et al. (2000); Franck et al. (2003)

*Increased cell Si quota at low CO_2 is a result of decreased loss of Si by efflux and dissolution rather than increased influx of $Si(OH)_4$.

Table 3. Calcification by coccolithophores as a function of resource limitation for growth

Growth limitation by supply of the named resource	Effect on particulate inorganic C/particulate organic C	Reference(s)
PAR	Decrease	Paasche (1964, 1999, 2001)
Inorganic C	Increase with decreased CO_2	Riebesell et al. (2000)
	Decrease with decreased bicarbonate	Sekino & Shiraiwa (1994); Shiraiwa (2003)
Inorganic N	Increase	Paasche (1998, 2001)
Inorganic P	Increase	Paasche & Bruback (1994); Paasche (1998, 2001)
Inorganic Fe	No effect	Schulz et al. (2004); cf. Crawford et al. (2003)
Inorganic Zn	Increase	Schulz et al. (2004); cf. Crawford et al. (2003)

supply, while restricted CO_2 supply increases calcification. Restricted supply of N, P and Zn increase calcification, while there is no significant effect of restricted supply of Fe.

Does the increased mineralization as a result of nutrient deficiency restrict mortality due to grazing and parasitism via mechanical effects or, in the case of calcification, via challenges to the maintenance of low pH in any acid parts of the grazer's digestive system? There is little direct evidence even on how effective the 'nutrient-sufficient' quantities of silica and of calcite in diatoms and coccolithophores are in decreasing losses due to grazing and parasitism relative to notionally comparable organisms lacking silicification and calcification (Raven & Waite, 2004).

Increased mineralization must increase the density of the organisms, since silica and calcite are each two to three times as dense as the cell protoplast, thus increasing the sinking rate of planktonic organisms. Raven & Waite (2004) consider the role of the increased rate of sinking resulting from the additional 'nutrient-deficient' quantities of silica and calcite in diatoms and coccolithophores in the context of movement down the water column to where there might be higher concentrations of nutrients. However, the vertical water movements in the upper mixed layer typical of habitats favoured by planktonic diatoms mean that such effects are essentially limited to an increased chance of falling out of the upper, mixed layer into the more nutrient-rich thermocline, with the possibility of the less-dense cells resulting from (light-limited but nutrient-sufficient) growth in the thermocline becoming reincorporated into the upper mixed layer (see Rodriguez *et al.*, 2001; Ptacnik *et al.*, 2003). The variations in density which permit very large-celled planktonic diatoms such as *Ethmodiscus* and some species of *Rhizosolenia* to make vertical migrations taking days within one cell division cycle in oligotrophic waters, gaining photosynthate near the surface and nutrients at depth, must depend on changes in protoplast density, since silicification-related changes in density are not manifest within a cell cycle covering a range of light and nutrient-supply conditions (Raven & Waite, 2004).

Aside from the effects of mineralization and its variation with resource supply in the life of diatoms and coccolithophores, the increased ballast per unit organic matter in nutrient-limited cells increases the potential for sinking of dead cells. In the case of diatoms, this effect is exacerbated by the absence of the mechanisms regulating (typically by lowering) the density of the protoplast in living cells (Raven & Waite, 2004). This effect could be multiplicative with the effect of nutrient deficiency in enhancing production of extracellular polysaccharides by algal cells (Fogg & Westlake, 1955; Hellebust, 1974; Wetz & Wheeler, 2003; but see Engel, 2002; Engel *et al.*, 2004). These exopolysaccharides are important in the flocculation of particles in sea water (Engel *et al.*, 2004), which increases their sinking rate for a given density by decreasing the surface area per unit volume (Stokes' law). While terrigenous mineral, e.g. clay, particles are also important as mineral ballast in the sinking of particulate organic matter in marine snow, the double effect of nutrient deficiency, by increasing the amount of biomineral ballast and the amount of floc-forming extracellular polysaccharides, could enhance not only the sinking of particulate organic matter in the 'biological pump', but also the removal of Fe, N, P and Zn from already nutrient-deficient surface waters. A further feedback on nutrient removal as a result of low nutrient availability is the capacity of the extracellular polysaccharides in the flocs to bind Fe and Zn from sea water (Engel *et al.*, 2004). While Passow (2004) suggests that the organic C fluxes may drive mineral particle fluxes rather than vice versa, it seems inescapable that mineral particles increase the density, and sinking rate, of organic particles (see Klaas & Anchor, 2002; Boyd *et al.*, 2004; Jickells *et al.*, 2005).

CONCLUSIONS

Fe, N and P have important roles in constraining marine primary productivity; the role of Zn in limiting the growth of primary producers is less clearly established. There are very substantial ecological, physiological and biochemical interactions among these nutrients, and also interactions with the supply of other resources. Decreased availability of these nutrient elements also potentially increases the sedimentation of particulate material, including these four nutrient elements; further work is needed to establish the significance of the potential feedbacks.

ACKNOWLEDGEMENTS

The work of K. B. and E. G. on inorganic C assimilation by diatoms is funded by a grant to R. C. L. and J. A. R. from the Natural Environment Research Council UK, and that of M. M. on mechanisms of inorganic C assimilation in coccolithophores is funded by the Biotechnology and Biological Sciences Research Council UK. J. B.'s work on algal photosynthesis is funded by the Australian Research Council. Discussion with Professor Geoff Codd was very helpful.

REFERENCES

Ågren, G. I. (2004). The C : N : P stoichiometry of autotrophs – theory and observations. *Ecol Lett* **7**, 185–191.

Armbrust, E. V., Berges, J. A., Bowler, C. & 42 other authors (2004). The genome of the diatom *Thalassiosira pseudonana:* ecology, evolution, and metabolism. *Science* **306**, 79–86.

Beardall, J. (1991). Effects of photon flux density on the CO_2-concentrating mechanism in the cyanobacterium *Anabaena variabilis. J Plankton Res* **13** (Suppl.), 133–146.

Beardall, J. & Raven, J. A. (2004). The potential effects of global climate change on microalgal photosynthesis, growth and ecology. *Phycologia* **43**, 26–40.

Beardall, J., Griffiths, H. & Raven, J. A. (1982). Carbon isotope discrimination and the CO_2 accumulating mechanism in *Chlorella emersonii. J Exp Bot* **33**, 729–737.

Beardall, J., Roberts, S. & Millhouse, J. (1991). Effects of nitrogen limitation on inorganic carbon uptake and specific activity of ribulose-1,5-P_2 carboxylase in green micro-algae. *Can J Bot* **69**, 1146–1150.

Beardall, J., Roberts, S. & Raven, J. A. (2005). Regulation of inorganic carbon acquisition by phosphorus limitation in the green algae, *Chlorella emersonii. Can J Bot* (in press).

Benitez-Nelson, C. R. & Karl, D. M. (2002). Phosphorus cycling in the North Pacific subtropical gyre using cosmogenic ^{32}P and ^{33}P. *Limnol Oceanogr* **47**, 762–770.

Berner, E. K. & Berner, R. A. (1996). *Global Environment. Water, Air, and Geochemical Cycles.* Upper Saddle River, NJ & London: Prentice Hall.

Bertilsson, S., Berglund, O., Karl, D. M. & Chisholm, S. W. (2003). Elemental composition of marine *Prochlorococcus* and *Synechococcus*: implications for ecological stoichiometry of the sea. *Limnol Oceanogr* **48**, 1721–1731.

Bjerrum, C. J. & Canfield, D. E. (2002). Ocean productivity before about 1·9 Gyr ago limited by phosphorus absorption onto iron oxides. *Nature* **417**, 159–162.

Björkman, K. M. & Karl, D. M. (2003). Bioavailability of dissolved organic phosphorus in the euphotic zone at station ALOHA, North Pacific subtropical gyre. *Limnol Oceanogr* **48**, 1049–1057.

Boyd, P. W. (2002a). Environmental factors controlling phytoplankton processes in the Southern Ocean. *J Phycol* **38**, 844–861.

Boyd, P. W. (2002b). The role of iron in the biogeochemistry of the Southern Ocean and equatorial Pacific: a comparison of *in situ* iron enrichments. *Deep Sea Res II Top Stud Oceanogr* **49** 1803–1821.

Boyd, P. W., Jackson, G. A. & Waite, A. M. (2002). Are mesoscale perturbation experiments in polar waters prone to physical artefacts? Evidence from algal aggregation modelling studies. *Geophys Res Lett* **29**, art. no. 1541.

Boyd, P. W., Law, C. S., Wong, C. S. & 35 other authors (2004). The decline and fate of an iron-induced subarctic phytoplankton bloom. *Nature* **428**, 549–553.

Bożena, K.-S., Zielinski, P. & Maleszewski, S. (2000). Involvement of glycolate metabolism in acclimation of *Chlorella vulgaris* cultures to low phosphate supply. *Plant Physiol Biochem* **38**, 727–734.

Bryant, D. A. (2003). The beauty in small things revealed. *Proc Natl Acad Sci U S A* **100**, 9647–9649.

Buitenhuis, E. I., Timmermans, K. R. & van de Baar, H. J. W. (2003). Zinc-bicarbonate colimitation of *Emiliania huxleyii*. *Limnol Oceanogr* **48**, 1575–1582.

Capone, D. G. (2001). Marine nitrogen fixation: what's the fuss? *Curr Opin Microbiol* **4**, 341–348.

Chester, R. (2003). *Marine Geochemistry*, 2nd edn. Oxford: Blackwell Science.

Claquin, P., Martin-Jézéquel, V., Kromkamp, J. C., Veldhuis, M. J. W. & Kraay, G. W. (2002). Uncoupling of silicon compared with carbon and nitrogen metabolisms and the role of the cell cycle in continuous cultures of *Thalassiosira pseudonana* (Bacillariophyceae) under light, nitrogen, and phosphorus control. *J Phycol* **38**, 922–930.

Collins, S. & Bell, G. (2004). Phenotypic consequences of 1,000 generations of selection at elevated CO_2 in a green alga. *Nature* **431**, 566–569.

Cooke, R. R. M., Hurd, C. L., Lord, J. M., Peake, B. M., Raven, J. A. & Rees, T. A. V. (2004). Iron and zinc content of *Hormosira banksii* in New Zealand. *N Z J Mar Freshw Res* **38**, 73–85.

Crawford, D. W., Lipsen, M. S., Purdie, D. A. & 9 other authors (2003). Influence of zinc and iron enrichments on phytoplankton growth in the northeastern subarctic Pacific. *Limnol Oceanogr* **48**, 1583–1600.

Cullen, J. J. (1999). Iron, nitrogen and phosphorus in the ocean. *Nature* **402**, 372.

De La Rocha, C. L., Hutchins, D. A., Brzezinski, M. A. & Zhang, Y. H. (2000). Effects of iron and zinc deficiency on elemental composition and silica production by diatoms. *Mar Ecol Prog Ser* **195**, 71–79.

Dudley, B. J., Gahnström, A. M. E. & Walker, D. I. (2001). The role of benthic vegetation as a sink for elevated inputs of ammonium and nitrate in a mesotrophic estuary. *Mar Ecol Prog Ser* **219**, 99–107.

Dufresne, A., Salanoubat, M., Partensky, F. & 18 other authors (2003). Genome sequence of the cyanobacterium *Prochlorococcus marinus* SS120, a nearly minimal oxyphototrophic genome. *Proc Natl Acad Sci U S A* **100**, 10020–10025.

Elam, J. S., Malek, K., Rodrigez, J. A., Doucette, P. A., Taylor, A. B., Hayward, L. J., Cabelli, D. E., Valentine, J. S. & Hart, P. J. (2003). An alternative mechanism of bicarbonate-mediated peroxidation by copper-zinc superoxide dismutase: rates enhanced via proposed enzyme-associated peroxycarbonate intermediate. *J Biol Chem* **278**, 21032–21039.

Ellwood, M. J., Hunter, K. A. & Kim, J. P. (2001). Zinc speciation in Lakes Manapouri and Hayes, New Zealand. *Mar Freshw Res* **52**, 217–222.

Elser, J. J., Acharya, K., Kyle, M. & 9 other authors (2003). Growth rate–stoichiometry couplings in diverse biota. *Ecol Lett* **6**, 936–943.

Engel, A. (2002). Direct relationship between CO_2 uptake and transparent exopolymer particles production in natural phytoplankton. *J Plankton Res* **24**, 49–53.

Engel, A., Thoms, S., Riebesell, U., Rochelle-Newall, E. & Zondervan, I. (2004). Polysaccharide aggregation as a potential sink of marine dissolved organic carbon. *Nature* **428**, 929–932.

Falkowski, P. G. (1997). Evolution of the nitrogen cycle and its influence on the biological sequestration of CO_2 in the ocean. *Nature* **387**, 272–275.

Falkowski, P. G. (2000). Rationalizing element ratios in unicellular algae. *J Phycol* **36**, 3–6.

Falkowski, P. G. & Raven, J. A. (1997). *Aquatic Photosynthesis*. Oxford: Blackwell Science.

Falkowski, P. G., Katz, M. E., Knoll, A. H., Quigg, A., Raven, J. A., Schofield, O. & Taylor, F. J. R. (2004). The evolution of modern eukaryotic phytoplankton. *Science* **305**, 354–360.

Field, C. B., Behrenfeld, M. J., Randerson, J. T. & Falkowski, P. G. (1998). Primary production of the biosphere: integrating terrestrial and oceanic components. *Science* **281**, 237–240.

Flynn, K. J. & Hipkin, C. R. (1999). Interactions between iron, light, ammonium, and nitrate: insights from the construction of a dynamic model of algal physiology. *J Phycol* **35**, 1171–1190.

Fogg, G. E. & Westlake, D. F. (1955). The importance of extracellular products of algae in freshwater. *Verhein Int Verein Limnol* **12**, 219–232.

Franck, V. M., Bruland, K. W., Hutchins, D. A. & Brzezinski, M. A. (2003). Iron and zinc effects on silicic acid and nitrate uptake kinetics in three high-nutrient, low-chlorophyll (HNLC) regions. *Mar Ecol Prog Ser* **252**, 15–33.

Frew, R., Bowie, A., Croot, P. & Pickmere, S. (2001). Macronutrient and trace-metal geochemistry of an *in situ* iron-induced Southern Ocean bloom. *Deep Sea Res II Top Stud Oceanogr* **48**, 2467–2481.

Gall, M. P., Strzepek, R., Maldonado, M. & Boyd, P. W. (2001). Phytoplankton processes. Part 2. Rates of primary production and factors controlling algal growth during the Southern Ocean Iron RElease Experiment (SOIREE). *Deep Sea Res II Top Stud Oceanogr* **48**, 2571–2590.

Geider, R. J. & La Roche, J. (2002). Redfield revisited: variability of $C:N:P$ in marine microalgae and its biochemical basis. *Eur J Phycol* **37**, 1–17.

Giordano, M., Norici, A., Forssen, M., Eriksson, M. & Raven, J. A. (2003). An anaplerotic role for mitochondrial carbonic anhydrase in *Chlamydomonas reinhardtii*. *Plant Physiol* **132**, 2126–2134.

Giordano, M., Beardall, J. & Raven, J. A. (2005). CO_2 concentrating mechanisms in algae: mechanisms, environmental modulation, and evolution. *Annu Rev Plant Biol* **56**, 99–131.

Granum, E., Raven, J. A. & Leegood, R. C. (2005). How do diatoms fix ten billion tonnes of inorganic carbon per year? *Can J Bot* (in press).

Grime, J. P. (1974). Vegetation classification by reference to strategies. *Nature* **250**, 26–31.

Heldal, M., Scanlan, D. J., Norland, S., Thingstad, F. & Mann, N. H. (2003). Elemental composition of single cells of various strains of marine *Prochlorococcus* and *Synechococcus* using X-ray microanalysis. *Limnol Oceanogr* **48**, 1732–1743.

Hellebust, J. A. (1974). Extracellular products. In *Algal Physiology and Biochemistry*, pp. 838–863. Edited by W. D. P. Stewart. Oxford: Blackwell Scientific.

Ho, T.-Y., Quigg, A., Finkel, Z. V., Milligan, A. J., Wyman, K., Falkowski, P. G. & Morel, F. M. M. (2003). The elemental composition of some marine phytoplankton. *J Phycol* **39**, 1145–1159.

Hutchins, D. A. & Bruland, K. W. (1998). Iron-limited growth and Si : N uptake ratios in a coastal upwelling regime. *Nature* **393**, 561–564.

Hutchins, D. A., DiTullio, G. R., Zhang, Y. & Bruland, K. W. (1998). An iron limitation mosaic in the California upwelling regime. *Limnol Oceanogr* **43**, 1037–1054.

Hutchins, D. A., Hare, C. E., Weaver, R. S. & 12 other authors (2002). Phytoplankton iron limitation in the Humboldt Current and Peru Upwelling. *Limnol Oceanogr* **47**, 997–1011.

Ianora, A., Miralto, A., Poulet, S. A. & 9 other authors (2004). Aldehyde suppression of copepod recruitment in blooms of a ubiquitous planktonic diatom. *Nature* **429**, 403–407.

Jickells, T. D., An, Z. S., Andersen, K. K. & 16 other authors (2005). Global iron connections between desert dust, ocean biogeochemistry, and climate. *Science* **308**, 67–71.

Johnston, A. M., Raven, J. A., Beardall, J. & Leegood, R. C. (2001). Photosynthesis in a marine diatom. *Nature* **412**, 40–41.

Karl, D. M. & Tien, G. (1997). Temporal variability in dissolved phosphorus concentration in the subtropical North Pacific Ocean. *Mar Chem* **56**, 77–96.

Karl, D. M., Tien, G., Dore, J. & Winn, C. D. (1993). Total dissolved nitrogen and phosphorus concentrations at US-JGOFS Station ALOHA: Redfield reconciliation. *Mar Chem* **41**, 203–208.

Klaas, C. & Anchor, D. E. (2002). Association of sinking organic matter with various types of mineral ballast in the deep sea: implications for the rain ratio. *Global Biogeochem Cycles* **16**, art. no. 116.

Klausmeier, C. A., Litchman, E., Dufresne, T. & Levin, S. A. (2004a). Optimal nitrogen-to phosphorus stoichiometry of phytoplankton. *Nature* **429**, 171–174.

Klausmeier, C. A., Litchman, E. & Levin, S. A. (2004b). Phytoplankton growth and stoichiometry under multiple nutrient limitation. *Limnol Oceanogr* **49**, 1463–1470.

Krom, M. D., Herut, B. & Mantoura, R. F. C. (2004). Nutrient budgets for the Eastern Mediterranean: implications for phosphorus limitation. *Limnol Oceanogr* **49**, 1582–1592.

Kübler, J. E. & Raven, J. A. (1994). Consequences of light-limitation for carbon acquisition in three rhodophytes. *Mar Ecol Prog Ser* **110**, 203–208.

Kübler, J. E. & Raven, J. A. (1995). Interaction between inorganic carbon acquisition and light supply in *Palmaria palmata* (Rhodophyta). *J Phycol* **31**, 369–375.

Kustka, A., Sañudo-Wilhelmy, S., Carpenter, E. J., Capone, D. G. & Raven, J. A. (2003a). A revised estimate of the iron use efficiency of nitrogen fixation, with special reference to the marine cyanobacterium *Trichodesmium* spp. (Cyanophyta). *J Phycol* **39**, 12–25.

Kustka, A., Sañudo-Wilhelmy, S. A., Carpenter, E. J., Capone, D., Burns, J. & Sunda, W. G. (2003b). Iron requirements for dinitrogen- and ammonium-supported growth in cultures of *Trichodesmium* (IMS 101): comparison with nitrogen fixation rates and iron : carbon ratios of field populations. *Limnol Oceanogr* **48**, 1869–1884.

Lane, T. W. & Morel, F. M. M. (2000a). Regulation of carbonic anhydrase expression by zinc, cobalt, and carbon dioxide in the marine diatom *Thalassiosira weissflogii*. *Plant Physiol* **123**, 345–352.

Lane, T. W. & Morel, F. M. M. (2000b). A biological function for cadmium in the marine diatoms. *Proc Natl Acad Sci U S A* **97**, 4627–4631.

Leonardos, N. & Geider, R. J. (2004). Response of elemental and biochemical composition of *Chaetoceros muelleri* to growth under varying light and nitrate : phosphate supply ratios and their influence on critical N : P. *Limnol Oceanogr* **49**, 2105–2114.

Leynaert, A., Bucciarelli, E., Claquin, P., Dugdale, R. C., Martin-Jézéquel, V., Pondaven, P. & Ragueneau, O. (2004). Effect of iron deficiency on diatom size and silicic acid uptake kinetics. *Limnol Oceanogr* **49**, 1134–1143.

MacFarlane, J. J. & Raven, J. A. (1990). C, N and P nutrition in *Lemanea mammilosa* Kutz. (Batrachospermales, Rhodophyta) in the Dighty Burn, Angus, Scotland. *Plant Cell Environ* **13**, 1–13.

Martin, J. H. & Fitzwater, S. E. (1988). Iron deficiency limits phytoplankton growth in the north-east Pacific subarctic. *Nature* **331**, 341–343.

Martin, J. H., Fitzwater, S. E. & Gordon, R. M. (1990). Iron deficiency limits phytoplankton growth in Antarctic waters. *Global Biogeochem Cycles* **4**, 5–13.

Martin, J. H., Gordon, R. E. & Fitzwater, S. E. (1991). The case for iron. *Limnol Oceanogr* **36**, 1793–1802.

Martin-Jézéquel, V., Hildebrand, M. & Brzezinski, M. A. (2000). Silicon metabolism in diatoms: implications for growth. *J Phycol* **36**, 821–840.

Milligan, A. J., Varela, D. E., Brzezinski, M. A. & Morel, F. M. M. (2004). Dynamics of silicon metabolism and silicon isotopic discrimination in a marine diatom as a function of pCO_2. *Limnol Oceanogr* **49**, 322–329.

Mills, M. M., Ridame, C., Davey, M., La Roche, J. & Geider, R. J. (2004). Iron and phosphorus co-limit nitrogen fixation in the eastern tropical North Atlantic. *Nature* **429**, 292–294.

Morel, F. M. M., Reinfelder, J. R., Roberts, S. B., Chamberlain, C. P., Lee, J. G. & Yee, D. (1994). Zinc and carbon co-limitation of marine phytoplankton. *Nature* **369**, 740–742.

Morel, F. M. M., Cox, E. H., Kraepiel, A. M. L., Lane, T. W., Milligan, A. J., Schaperdoth, I., Reinfelder, J. R. & Tortell, P. D. (2002). Acquisition of inorganic carbon by the marine diatom *Thalassiosira weissflogii*. *Funct Plant Biol* **29**, 301–308.

Paasche, E. (1964). A tracer study of the inorganic carbon uptake during coccolith formation and photosynthesis in the coccolithophorid *Coccolithus huxleyi*. *Physiol Plant Suppl* III, 12–82.

Paasche, E. (1998). Roles of nitrogen and phosphorus in coccolith formation in *Emiliania huxleyi* (Prymnesiophyceae). *Eur J Phycol* **33**, 33–42.

Paasche, E. (1999). Reduced coccolith calcite production under light-limited growth: a comparative study of three clones of *Emiliania huxleyi* (Prymnesiophyceae). *Phycologia* **38**, 508–516.

Paasche, E. (2001). A review of the coccolithophorid *Emilianina huxleyi* (Prymnesiophyceae), with particular reference to growth, coccolith formation, and calcification–photosynthesis interactions. *Phycologia* **40**, 503–529.

Paasche, E. & Bruback, S. (1994). Enhanced calcification in the coccolithophorid *Emiliania huxleyi* (Haptophyceae) under phosphorus limitation. *Phycologia* **33**, 324–330.

Palenik, B., Brahamsha, B., Larimer, F. W. & 12 other authors (2003). The genome of a motile marine *Synechococcus*. *Nature* **424**, 1037–1042.

Passow, U. (2004). Switching perspectives: do mineral fluxes determine particulate organic carbon fluxes or vice versa? *Geochem Geophys Geosyst* **5**, art. no. Q04002.

Ptacnik, R., Diehl, S. & Berger, S. (2003). Performance of sinking and nonsinking phytoplankton taxa in a gradient of mixing depths. *Limnol Oceanogr* **48**, 1903–1912.

Quigg, A., Finkel, Z. V., Irwin, A. J., Rosenthal, Y., Ho, T.-Y., Reinfelder, J. R., Schofield, O., Morel, F. M. M. & Falkowski, P. G. (2003). The evolutionary inheritance of elementary stoichiometry in marine phytoplankton. *Nature* **425**, 291–294.

Raven, J. A. (1987). The role of vacuoles. *New Phytol* **106**, 358–422.

Raven, J. A. (1988). The iron and molybdenum use efficiencies of plant growth with different energy, carbon and nitrogen sources. *New Phytol* **109**, 279–287.

Raven, J. A. (1990). Predictions of Mn and Fe use efficiencies of phototrophic growth as a function of light availability for growth and C assimilation pathway. *New Phytol* **116**, 1–18.

Raven, J. A. (1996). The role of autotrophs in global CO_2 cycling. In *Microbial Growth on C_1 Compounds*, pp. 105–144. Edited by M. E. Lidstrom & F. R. Tabita. Dordrecht, Netherlands: Kluwer Scientific.

Raven, J. A. (1997). Phagotrophy in phototrophs. *Limnol Oceanogr* **42**, 198–205.

Raven, J. A. & Johnston, A. M. (1991). Mechanisms of inorganic-carbon acquisition in marine phytoplankton and their implications for the use of other resources. *Limnol Oceanogr* **36**, 1701–1714.

Raven, J. A. & Taylor, R. (2003). Macroalgal growth in nutrient-enriched estuaries: a biogeochemical and evolutionary perspective. *Water Air Soil Pollut Focus* **3** (1), 7–26.

Raven, J. A. & Waite, A. M. (2004). The evolution of silicification of diatoms: inescapable sinking and sinking as escape? *New Phytol* **162**, 45–61.

Raven, J. A., Evans, M. C. W. & Korb, R. (1999). The role of trace metals in photosynthetic electron transport in O_2-evolving organisms. *Photosynth Res* **60**, 111–150.

Raven, J. A., Johnston, A. M., Kübler, J. E. & 9 other authors (2002). Mechanistic interpretation of carbon isotope discrimination by marine macroalgae and sea-grasses. *Funct Plant Biol* **29**, 355–378.

Raven, J. A., Handley, L. L. & Andrews, M. (2004). Global aspects of C/N interactions determining plant-environment interactions. *J Exp Bot* **55**, 11–25.

Raven, J. A., Andrews, M. & Quigg, A. S. (2005). The evolution of oligotrophy: implications for the breeding of crop plants for low input agricultural systems. *Ann Appl Biol* **146**, 261–280.

Reinfelder, J. R., Kraepiel, A. M. L. & Morel, F. M. M. (2000). Unicellular C_4 photosynthesis in a marine diatom. *Nature* **407**, 996–999.

Reinfelder, J. R., Milligan, A. J. & Morel, F. M. M. (2004). The role of the C_4 pathway in carbon accumulation and fixation in a marine diatom. *Plant Physiol* **135**, 2106–2111.

Ridame, C. & Guieu, C. (2002). Saharan input of phosphate to the oligotrophic water of the open western Mediterranean Sea. *Limnol Oceanogr* **47**, 856–869.

Riebesell, U., Zondervan, I., Rist, B., Tortell, P. D., Zeebe, R. E. & Morel, F. M. M. (2000). Reduced calcification of marine phytoplankton in response to increased atmospheric CO_2. *Nature* **407**, 364–367.

Rocap, G., Larimer, F. W., Lamerdin, J. & 21 other authors (2003). Genome divergence in two *Prochlorococcus* ecotypes reflects oceanic niche differentiation. *Nature* **424**, 1042–1047.

Rodriguez, J., Tintore, J., Allen, J. T. & 7 other authors (2001). Mesoscale vertical motion and the size of structure of phytoplankton in the ocean. *Nature* **410**, 360–363.

Sañudo-Wilhelmy, S. A., Kustka, A. B., Gobler, C. J. & 7 other authors (2001). Phosphorus limitation of nitrogen fixation by *Trichodesmium* in the central Atlantic Ocean. *Nature* **411**, 66–69.

Sañudo-Wilhelmy, S. A., Tovar-Sanchez, A., Fu, F.-X., Capone, D. G., Carpenter, E. J. & Hutchins, D. A. (2004). The impact of surface-absorbed phosphorus on phyto-plankton Redfield stoichiometry. *Nature* **432**, 897–901.

Schulz, K. G., Zondervan, I., Gerringa, L. J. A., Timmermans, K. R., Veldhuis, M. J. W. & Riebesell, U. (2004). Effect of trace metal availability on coccolithophorid calcification. *Nature* **430**, 673–676.

Sekino, K. & Shiraiwa, Y. (1994). Accumulation and utilization of dissolved inorganic carbon by a marine unicellular coccolithophorid, *Emiliania huxleyi*. *Plant Cell Physiol* **35**, 353–361.

Shiraiwa, Y. (2003). Physiological regulation of carbon fixation in the photosynthesis and calcification of coccolithophorids. *Comp Biochem Physiol B Biochem Mol Biol* **136**, 775–783.

Sterner, R. W. & Elser, J. J. (2002). *Ecological Stoichiometry*. Princeton, NJ: Princeton University Press.

Sterner, R. W., Smutka, T. M., McKay, R. M. L., Qin, X., Brown, E. T. & Sherrell, R. M. (2004). Phosphorus and trace metal limitation of algae and bacteria in Lake Superior. *Limnol Oceanogr* **49**, 495–507.

Strzepek, R. F. & Harrison, P. J. (2004). Photosynthetic architecture differs in coastal and oceanic diatoms. *Nature* **431**, 689–692.

Sunda, W. G., Swift, D. G. & Huntsman, S. A. (1991). Low iron requirement for growth in oceanic phytoplankton. *Nature* **351**, 55–57.

Takeda, S. (1998). Influence of iron availability on nutrient consumption ratio of diatoms in oceanic waters. *Nature* **393**, 774–777.

Thornton, D. C. O. (2002). Individuals, clones or groups? Phytoplankton behaviour and units of selection. *Ethol Ecol Evol* **14**, 165–173.

Ting, C. S., Rocap, G., King, J. & Chisholm, S. W. (2002). Cyanobacterial photosynthesis in the oceans: the origin and significance of divergent light-harvesting strategies. *Trends Microbiol* **10**, 134–142.

Tyrrell, T. (1999). The relative influences of nitrogen and phosphorus on oceanic primary production. *Nature* **400**, 525–531.

van Oijen, T., van Leeuwe, M. A., Granum, E., Weissing, F. J., Bellerby, R. G. J., Gieskes, W. W. C. & de Baar, H. J. W. (2004). Light rather than iron controls photosynthate production and allocation in Southern Ocean phytoplankton populations during the austral autumn. *J Plankton Res* **26**, 885–900.

Vitousek, P. M., Cassman, K., Cleveland, C. & 8 other authors (2001). Towards an eco-logical understanding of biological nitrogen fixation. *Biogeochemistry* **57/58**, 1–45.

Völker, C. & Wolf-Gladrow, D. A. (1999). Physical limits on iron uptake mediated by siderophores or surface reductases. *Mar Chem* **65**, 227–244.

Walker, D. I., Campey, M. L. & Kendrick, G. A. (2004). Nutrient dynamics in two seagrass species, *Posidonia coriacea* and *Zostera tasmanica* on Success Bank, Western Australia. *Estuar Coastal Shelf Sci* **60**, 251–260.

Watson, A. J., Bakker, D. C. E., Ridgwell, A. J., Boyd, P. W. & Law, C. S. (2000). Effect of iron supply on Southern Ocean CO_2 uptake and implications for glacial atmospheric CO_2. *Nature* **407**, 730–733.

Wetz, M. S. & Wheeler P. A. (2003). Production and partitioning of organic matter during simulated phytoplankton blooms. *Limnol Oceanogr* **48**, 1808–1817.

Wolfe, G. V., Steinke, M. & Kirst, G. O. (1997). Grazing-activated chemical defence in a unicellular marine alga. *Nature* **387**, 894–897.

Wu, J., Sunda, W., Boyle, E. A. & Karl, D. M. (2000). Phosphate depletion in the western North Atlantic Ocean. *Science* **289**, 759–762.

Young, E. B. & Beardall, J. (2005). Modulation of photosynthesis and inorganic carbon acquisition in a marine microalga by nitrogen, iron and light availability. *Can J Bot* (in press).

Mechanisms and environmental impact of microbial metal reduction

Jonathan R. Lloyd

The Williamson Research Centre for Molecular Environmental Studies and the School of Earth, Atmospheric and Environmental Sciences, University of Manchester, Manchester M13 9PL, UK

INTRODUCTION

Although it has been known for over a century that micro-organisms have the potential to reduce metals, more recent observations showing that a diversity of specialist bacteria and archaea can use such activities to conserve energy for growth under anaerobic conditions have opened up new and fascinating areas of research with potentially exciting practical applications (Lloyd, 2003). Micro-organisms have also evolved metal-resistance processes that often incorporate changes in the oxidation state of toxic metals. Several such resistance mechanisms, which do not support anaerobic growth, have been studied in detail by using the tools of molecular biology. Three obvious examples include resistance to Hg(II), As(V) and Ag(II) (Bruins *et al.*, 2000). The molecular bases of respiratory metal-reduction processes have not, however, been studied in such fine detail, although rapid advances are expected in this area with the imminent availability of complete genome sequences for key metal-reducing bacteria, in combination with genomic, proteomic and metabolomic tools. This research is being driven forward both by the need to understand the fundamental basis of a range of biogeochemical cycles, and also by the possibility of harnessing such activities for a range of biotechnological applications. These include the bio-remediation of metal-contaminated land and water (Lloyd & Lovley, 2001), the oxidation of xenobiotics under anaerobic conditions (Lovley & Anderson, 2000), metal recovery in combination with the formation of novel biocatalysts (Yong *et al.*, 2002a) and even the generation of electricity from sediments (Bond *et al.*, 2002). The aim of this review is to give an overview of the range of metals (and metalloids) reduced by micro-organisms, the mechanisms involved and the environmental impact of

SGM symposium 65: Micro-organisms and Earth systems – advances in geomicrobiology.
Editors G. M. Gadd, K. T. Semple & H. M. Lappin-Scott. Cambridge University Press. ISBN 0 521 86222 1 ©SGM 2005

such transformations. Where appropriate, possible applications for these processes will also be discussed.

REDUCTION OF Fe(III) AND Mn(IV)

A wide range of archaea and bacteria are able to conserve energy though the reduction of Fe(III) (ferric iron) to Fe(II) (ferrous iron). Many of these organisms are also able to grow through the reduction of Mn(IV) to Mn(II). The environmental relevance of Fe(III) and Mn(IV) reduction has been well-documented (Thamdrup, 2000), whilst geochemical and microbiological evidence suggests that the reduction of Fe(III) may have been an early form of respiration on Earth (Vargas *et al.*, 1998). Some workers have even proposed Fe(III) reduction as a candidate process for the basis of life on other planets (Nealson & Cox, 2002). On modern Earth, Fe(III) can be the dominant electron acceptor for microbial respiration in many subsurface environments (Lovley & Chapelle, 1995). As such, Fe(III)-reducing communities can be responsible for the majority of organic matter oxidized in such environments. Recent studies have shown that a range of important xenobiotics that contaminate aquifers can also be degraded under anaerobic conditions by Fe(III)-reducing micro-organisms (Lovley, 1997; Anderson *et al.*, 1998; Lovley & Anderson, 2000). Transformations of inorganic contaminants are also possible, and Fe(III)-reducing micro-organisms can also have an impact on the fate of other high-valency contaminant metals through direct enzymic reduction or via indirect reduction catalysed by biogenic Fe(II).

Focusing on the enzymic transformations of Fe(III) and Mn(IV), these organisms can also influence the mineralogy of sediments through the reductive dissolution of insoluble Fe(III) and Mn(IV) oxides (Fig. 1a). These processes can result in the release of potentially toxic levels of reduced Fe(II) and Mn(II), and also trace metals that were bound by the host Fe(III) or Mn(IV) minerals. Depending on the chemistry of the water, a range of reduced minerals can also be formed, including, for Fe(III) reduction, magnetite (Fe_3O_4), siderite ($FeCO_3$) and vivianite [$Fe_3(PO_4)_2 \cdot 8H_2O$] (see Fig. 1b–d), resulting in a change in structure of the sediments. There is a considerable level of interest in these end products of Fe(III) reduction, as they are nanoscale and have properties that may make them useful for a range of biotechnological processes. For example, large quantities of regular-shaped, nanosized (5–10 nm) crystals of the ferromagnetic mineral magnetite (Fe_3O_4) are formed when Fe(II)-reducing bacteria respire by using Fe(III) oxides in laboratory culture (Fig. 2). By controlling and manipulating this process of biomineral production, a low-cost, low-energy, environmentally friendly method of manufacture of nanoparticles could be developed to replace existing practices. Of particular appeal is the potential to convert natural and waste Fe(III) oxides (e.g. from mining/water industries), which are bulky and difficult to handle, to a high-value product that is easy to process. Indeed, ferrite spinels such as

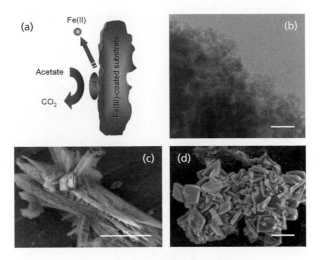

Fig. 1. Mechanism of Fe(III) reduction (a) and the biogenic Fe(II)-bearing minerals magnetite (b), vivianite (c) and siderite (d). Bars, 50 nm (b); 10 µm (c); 20 µm (d). Figures were generously provided by M. Wilkins, V. Coker, F. Islam and L. Adams (University of Manchester, UK).

magnetite have low coercivity (i.e. low power needed to magnetize or demagnetize them), high permeability, high magnetic saturation and low conductivity. These properties make them ideal for use in ultrahigh-density data-storage media, frequency-selective circuits, radio-receiver antennae, microwave waveguides and other high-frequency devices. Indeed, recent work from our laboratory has also demonstrated that 'designer' magnets can be made by Fe(III)-reducing bacteria by incorporating other transition metals into the spinel structure in place of Fe, in some cases enhancing the magnetic properties of the biomineral (Coker *et al.*, 2004).

Diversity of Fe(III)-reducing organisms

The first organisms were shown to conserve energy for growth through the reduction of Fe(III) [and Mn(IV)] in the 1980s. These model organisms were *Shewanella oneidensis* (formerly *Alteromonas putrefaciens* and then *Shewanella putrefaciens*) and *Geobacter metallireducens* (formerly strain GS-15) (Lovley *et al.*, 1987, 1989a; Myers & Nealson, 1988). Earlier studies had focused on organisms that grow predominantly via fermentation of sugars such as glucose, with metals utilized as minor electron acceptors (Roberts, 1947); typically, < 5 % of the reducing equivalents is used for metal reduction (Lovley, 1991).

Over the last 20 years, numerous organisms have been isolated that can grow by using Fe(III) as an electron acceptor; more than 90 are listed in a recent review (Lovley *et al.*, 2004). In many freshwater subsurface environments, the most abundant seem to be

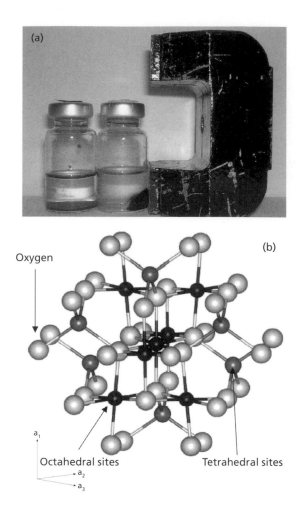

Fig. 2. Production (a) and structure (b) of magnetite produced by *Geobacter sulfurreducens*. Magnetite exhibits a classical spinel crystalline structure, with the oxygen ions forming a compact, face-centred cubic assembly and the iron cations occupying octahedral and tetrahedral sites. Figures were generously provided by V. Coker and R. Pattrick (University of Manchester, UK).

relatives of *Geobacter metallireducens* and fall within the family *Geobacteraceae*, in the delta subdivision of the *Proteobacteria* (e.g. Rooney-Varga *et al.*, 1999; Snoeyenbos-West *et al.*, 2000; Röling *et al.*, 2001; Stein *et al.*, 2001; Holmes *et al.*, 2002; Islam *et al.*, 2004). This group comprises the genera *Geobacter*, *Desulfuromonas*, *Desulfuromusa* and *Pelobacter* (Lovley *et al.*, 2004). These organisms, with the exception of *Pelobacter* species, are able to oxidize a wide range of organic compounds completely, including acetate, when respiring by using Fe(III); *Pelobacter* species are more restricted in the range of electron donors utilized, although they can couple hydrogen oxidation to metal reduction. Some members of the family *Geobacteraceae* are also able to use aromatic compounds, including toluene, phenol and benzoate, as electron donors for metal reduction. This is in contrast to *Shewanella oneidensis* and close relatives in the gamma subdivision of the *Proteobacteria* (a range of *Shewanella*, *Ferrimonas* and *Aeromonas* species) that are generally able to use only a restricted range of small organic acids and hydrogen as electron donors for Fe(III) and Mn(IV) reduction. The full range of other prokaryotes able to reduce Fe(III) is extensive and increasing steadily, and crosses a wide range of environments (including extremes of pH and temperature) and, indeed, taxonomic groupings. Although *Shewanella* and *Geobacter* species have been the most intensively studied model Fe(III)-reducing bacteria, the reader is referred to an excellent overview of the wide range of currently identified Fe(III)-reducing bacteria (Lovley *et al.*, 2004). Perhaps one of the most interesting new isolates that can grow through the reduction of Fe(III) is archaeal 'strain 121', which has pushed the upper temperature limit for life to 121 °C (Kashefi & Lovley, 2003).

Mechanisms of Fe(III) and Mn(IV) reduction: electron transfer to insoluble minerals

The mechanisms of Fe(III) reduction and, to a lesser degree, Mn(IV) reduction have been studied in most detail in *Shewanella oneidensis* and *Geobacter sulfurreducens*. Indeed, research on these organisms has been given added impetus through the availability of their genome sequences (available at http://www.tigr.org) and suitable genetic systems for the generation of deletion mutants for both of these organisms (Myers & Myers, 2000; Coppi *et al.*, 2001). Although the terminal reductase has yet to be identified unequivocally in either organism, the involvement of *c*-type cytochromes has been implicated in electron transport to Fe(III) and Mn(IV) by several studies (Myers & Myers, 1993, 1997; Gaspard *et al.*, 1998; Magnuson *et al.*, 2000; Beliaev *et al.*, 2001; Lloyd *et al.*, 2003). In some examples, activities have also been localized to the outer membrane or surface of the cell, consistent with a role in direct transfer of electrons to Fe(III) and Mn(IV) oxides that are highly insoluble at circumneutral pH (Myers & Myers, 1992, 2001; Gaspard *et al.*, 1998; DiChristina *et al.*, 2002; Lloyd *et al.*, 2002). In addition to the proposed direct transfer of electrons to Fe(III) and Mn(IV) minerals, soluble 'electron shuttles' are also able to transfer electrons between metal-reducing prokaryotes and the mineral surface. This mechanism alleviates the requirement for direct contact between the micro-organism and mineral. For example, humics and other extracellular quinones are utilized as electron acceptors by Fe(III)-reducing bacteria (Lovley *et al.*, 1996) and the reduced hydroquinone moieties are able to transfer electrons abiotically to Fe(III) minerals. The oxidized humic is then available for reduction by the micro-organism, leading to further rounds of electron shuttling to the insoluble mineral (Nevin & Lovley, 2002). Very low concentrations of an electron shuttle, e.g. 100 nM of the humic analogue anthraquinone-2,6-disulfonate (AQDS), can rapidly accelerate the reduction of Fe(III) oxides (Lloyd *et al.*, 1999a) and possibly other insoluble metal oxides, such as Mn(IV). The environmental significance of such processes, however, remains to be confirmed. The secretion of soluble electron shuttles by actively respiring Fe(III) and Mn(IV) reducers has also been proposed for both *Shewanella oneidensis* and *Geobacter sulfurreducens*, and remains hotly debated in *Geobacter* species. Early studies suggested the release of a small, soluble, *c*-type cytochrome by *Geobacter sulfurreducens* (Seeliger *et al.*, 1998), but more recent studies have suggested that this protein is not an effective electron shuttle (Lloyd *et al.*, 1999a). Studies have also suggested that a small, quinone-containing, extracellular electron shuttle is released by *Shewanella oneidensis* and may also promote electron transfer to Fe(III) and Mn(IV) minerals (Newman & Kolter, 2000). Finally, an important new discovery was made recently when it was shown that *Geobacter metallireducens* synthesized pili and flagella when grown on insoluble Fe(III) or Mn(IV) minerals, but not soluble forms of the metals (Childers *et al.*, 2002). These results suggest that *Geobacter* species sense when soluble electron acceptors are depleted and synthesize the appropriate appendages that allow movement to Fe(III) and Mn(IV) minerals and subsequent attachment. Pili may also play a direct role in electron transfer to the extracellular electron acceptor (Reguera *et al.*, 2005).

REDUCTION OF OTHER TRANSITION METALS

In addition to Fe(III) and Mn(IV), dissimilatory metal-reducing prokaryotes are able to respire by using a wide range of transition metals, including high-valency ions of vanadium, chromium, molybdenum, cobalt, palladium, silver, gold and mercury. In many cases, reduction leads to a dramatic change in solubility, can potentially lead to the precipitation of metal-containing ores and may also offer routes to the bioremediation of metal-contaminated water.

Vanadium reduction

Early studies showed V(V) reduction by '*Micrococcus lactilyticus*', *Desulfovibrio desulfuricans* and *Clostridium pasteurianum* (Woolfolk & Whiteley, 1962), followed by the observation that the ability to reduce V(V) was widespread amongst soil bacteria and fungi (Bautista & Alexander, 1972). More recent work has focused on two pseudomonads: '*Pseudomonas vanadiumreductans*' and '*Pseudomonas isachenkovii*', isolated from a waste stream from a ferrovanadium factory and sea water, respectively (Yurkova & Lyalikova, 1991). Anaerobic cells were able to utilize a wide range of electron donors, including hydrogen, sugars and amino acids. V(V) was reduced to blue-coloured V(IV) and possibly further to V(III), the latter indicated by the formation of a black precipitate and by its reaction with the reagent Tairon (Yurkova & Lyalikova, 1991). *Geobacter metallireducens* also reduces V(V) and this form of metabolism has been suggested as a mechanism for remediating vanadium-contaminated water (Ortiz-Bernad *et al.*, 2004a).

Chromium reduction

The widespread use of chromium in the metal industries and subsequent contamination problems have led to a lot of interest in this metal. Although trace quantities are required for some metabolic activities, e.g. glucose and lipid metabolism, chromium is considered toxic and is designated a priority pollutant in many countries. Two oxidation states dominate: Cr(VI) is the most toxic and mobile form encountered commonly, with Cr(III) being less soluble and less toxic. Indeed, Cr(III) is considered 1000 times less mutagenic than Cr(VI) (Wang, 2000). Current treatment involves reduction of Cr(VI) to Cr(III) by using chemical reductants at low pH, followed by adjustment to near-neutral pH and subsequent precipitation of Cr(III). Recent studies, however, have shown that micro-organisms can also reduce Cr(VI) efficiently at circumneutral pH, and could be used to treat Cr(VI)-contaminated water. In most cases, Cr(VI) reduction does not support anaerobic growth, although a few publications have suggested that conservation of energy is possible through this form of anaerobic metabolism (e.g. Tebo & Obraztsova, 1998).

A wide range of facultative anaerobes are able to reduce Cr(VI) to Cr(III), including *Escherichia coli*, pseudomonads, *Shewanella oneidensis* and *Aeromonas* species [see Wang (2000) for a more exhaustive list]. Anaerobic conditions are generally required to induce maximum activity against Cr(VI), but some enzyme systems operate under aerobic conditions, e.g. the soluble NAD(P)H-dependent reductases of '*Pseudomonas ambigua*' G-1 (Suzuki *et al.*, 1992) and *Pseudomonas putida* (Park *et al.*, 2000). Obligate anaerobes are also able to reduce Cr(VI) enzymically and anaerobic growth coupled to Cr(VI) reduction has been reported for a sulfate-reducing bacterium (Tebo & Obraztsova, 1998). The reduction of Cr(VI) by sulfate-reducing bacteria is particularly well-studied (e.g. Lloyd *et al.*, 2001) and has been shown to be catalysed by cytochrome c_3 (Lovley & Phillips, 1994). Other studies have also implicated the involvement of cytochromes in Cr(VI) reduction by bacteria: cytochrome *c* in *Enterobacter cloacae* (Wang *et al.*, 1989) and cytochromes *b* and *d* in *Escherichia coli* (Shen & Wang, 1993). Environmental factors that affect Cr(VI) reduction include competing electron acceptors, pH, temperature, redox potential and the presence of other metals (Wang, 2000). A recent study has also demonstrated that the presence of complexing agents can promote Cr(VI) reduction, possibly through protection of the metal reductase by chelation of Cr(III) or intermediates formed (Mabbett *et al.*, 2002). The type of electron donor supplied can also have an effect on the rate and extent of Cr(VI) reduction. Optimal electron donors, in keeping with other dissimilatory metal-reduction processes described in this review, are low-molecular-mass carbohydrates, amino acids and fatty acids. Finally, indirect mechanisms that also promote Cr(VI) reduction in contaminated sediments are catalysed by biogenic sulfide (Smillie *et al.*, 1981; Fude *et al.*, 1994) and Fe(II) (Fendorf & Li, 1996). Experiments using contaminated sediments from

Norman, OK, USA, have, however, confirmed that indirect mechanisms may not always be the critical control on chromium solubility, with direct enzymic Cr(VI) reduction by a consortium of methanogens being implicated (Marsh *et al.*, 2000).

Molybdenum reduction

Although microbial Mo(VI) reduction could play a role in the molybdenum cycle, for example leading to the concentration of insoluble molybdenum in anaerobic marine sediments and reduction spots in rocks (Lovley, 1993), comparatively few studies have addressed this process. Early work suggested that '*Pseudomonas guillermondii*' and a *Micrococcus* species could reduce Mo(VI) to molybdenum blue (Bautista & Alexander, 1972). More recently, similar activities have been identified in cultures of a molybdenum-resistant *Enterobacter* species (Ghani *et al.*, 1993). The organism was grown under anaerobic conditions in glucose-containing medium supplemented with 200 mM Mo(VI). Reduction of Mo(VI) was accompanied by a change in colour, as Mo(V) formed and complexed with phosphate in the medium to form methylene blue (Ghani *et al.*, 1993). The use of metabolic inhibitors suggested that the electrons for Mo(VI) reduction were derived from the glycolytic pathway, whilst NADH functioned as an electron donor in broken cells *in vitro* (Ghani *et al.*, 1993). The ability to reduce Mo(VI) has also been identified in pre-grown cells of the sulfate-reducing bacterium *Desulfovibrio desulfuricans*, both through direct enzymic mechanisms and indirectly via sulfide under sulfate-reducing conditions (Tucker *et al.*, 1997). Cells of *Desulfovibrio desulfuricans* immobilized in a bioreactor have also been used to remove Mo(VI) from solution at high efficiency (Tucker *et al.*, 1998). The organism was unable to grow using Mo(VI) as an electron acceptor (Tucker *et al.*, 1997) and it is unlikely that actively growing cultures of sulfate-reducing bacteria would play a direct role in reducing high concentrations of Mo(VI) in the environment, given the toxicity of molybdate to these organisms (Oremland & Capone, 1988).

Cobalt reduction

The reduction of Co(III) has received recent attention because radioactive ^{60}Co can be a problematic contaminant at sites where radioactive waste has been stored. Co(III) is especially mobile when complexed with EDTA and several studies have focused on the ability of Fe(III)-reducing bacteria to retard the mobility of the metal through reduction to Co(II) (Caccavo *et al.*, 1994; Gorby *et al.*, 1998). The Co(II) formed does not associate strongly with EDTA [it is over 25 orders of magnitude less thermodynamically stable than Co(III)EDTA] and absorbs to soils, offering the potential for *in situ* immobilization of the metal in contaminated soils. The precise mechanisms of dissimilatory Co(III) reduction remain to be investigated.

Palladium reduction

The reduction of soluble Pd(II) to insoluble Pd(0) has also attracted interest, as this enzymic process may be used to recover palladium from industrial catalysts (Lloyd *et al.*, 1998a) and may also be used to synthesize nanoscale bioinorganic catalysts of considerable commercial potential (Yong *et al.*, 2002a). Interest in this area is driven by the widespread use of platinum-group metals (PGMs), including palladium, in automotive catalytic converters required to reduce gaseous emissions, and problems associated with their recycling. With approximately 5 g PGM per catalytic converter, the consumption of PGMs was altogether 70 312·5 kg in 1994, with only 11 250 kg recovered (Lloyd *et al.*, 1998a). The lifetime of a catalytic converter is only approximately 80 000 km (50 000 miles), although many fail sooner, and future shortages and higher prices may be predicted. Chemical and electrochemical treatments are made difficult by complex solution chemistry.

Initial experiments aimed at reducing and recovering palladium were based on the use of *Desulfovibrio desulfuricans* because it is active against a wide range of metals, including Fe(III), Mn(IV), U(VI), Cr(VI) and Tc(VII), via hydrogenase or cytochrome c_3 (Lloyd *et al.*, 1998a). Cells were able to reduce 0·5 mM Pd(II) [as Pd(NH$_3$)$_4$Cl] with a range of electron donors, including pyruvate, formate and H$_2$. Although the enzyme responsible for Pd(II) reduction has not been identified, the involvement of a periplasmic hydrogenase is implicated by the use of hydrogen as electron donor and inhibition by treatment with 0·5 mM Cu^{2+} (Lloyd *et al.*, 1998a). Transmission electron microscopic studies, in combination with energy-dispersive X-ray microanalysis, confirmed precipitation in the periplasm, with X-ray diffraction studies confirming reduction to Pd(0). More recent studies have focused on the recovery of Pd(II) at a range of pH values and also from acid (aqua regia) leachates from spent automotive catalysts (Yong *et al.*, 2002b). Inhibition by chloride ions was reported and this may therefore necessitate leaching from catalysts to minimize the formation of PdCl$_4^{2-}$ prior to bioreduction and recovery. Delivery of reducing power to an immobilized biocatalyst has also been studied in a novel electrobioreactor (Yong *et al.*, 2002b) containing a biofilm of *Desulfovibrio desulfuricans* immobilized on a Pd–Ag membrane that transported atomic hydrogen to the cells, minimizing loss of gaseous hydrogen. Pd(0) recovered in the electrobioreactor proved a better catalyst that its chemical counterpart, as determined by hydrogen liberation from hypophosphite (Yong *et al.*, 2002a) and reduction of several target organics (L. E. Macaskie, personal communication).

Gold and silver reduction

It has been argued that Fe(III)-reducing bacteria may play a role in the deposition of gold ores, as these organisms are present in high- and moderate-temperature sedimentary environments where gold deposits have been recovered (Kashefi *et al.*,

2001). Indeed, several dissimilatory Fe(III)-reducing bacteria and archaea, including the hyperthermophilic archaea *Pyrobaculum islandicum* and *Pyrococcus furiosus*, the hyperthermophilic bacterium *Thermotoga maritima* and the mesophilic bacteria *Shewanella algae* and *Geovibrio ferrireducens*, were shown to reduce Au(III) (as gold chloride) to insoluble Au(0) (Kashefi *et al.*, 2001). The ability to reduce Au(III) seems to be species-specific, and closely related organisms with similar activities against a range of other metals have differing activities against Au(III) (Kashefi *et al.*, 2001). For example, unlike *Pyrobaculum islandicum*, a close relative, *Pyrobaculum aerophilum*, is unable to reduce Au(III). Also, there is an obligate requirement for hydrogen as an electron donor in organisms that can reduce Au(III), suggesting the involvement of a hydrogenase. Given the direct reduction of other metals by hydrogenase (Lloyd *et al.*, 1997), it is tempting to hypothesize that hydrogenases may play a direct role in Au(III) reduction.

Microbial reduction of Ag(I) has also been studied, but in little detail. Early reports noted that the reduction of Ag(I) may account for resistance to silver in some micro-organisms (Belly & Kydd, 1982), but more recent studies have uncovered alternative strategies for resistance to Ag(I) in organisms isolated from hospital burns wards, where silver may be used as a biocide (Gupta *et al.*, 1999), with a recent review of this new area presented by Silver (2003). Several studies, including that by Fu *et al.* (2000), have shown biosorption of Ag(I) to the surface of cells (in this case, a *Lactobacillus* species), followed by reduction to Ag(0). Here, the mechanism for Ag(I) reduction remains unknown.

Mercury reduction

A well-studied metal-resistance system is encoded by genes of the *mer* or mercury-resistance operon, which relies upon the reduction of Hg(II) (mercuric ions). Here, Hg(II) is transported into the cell via the MerT transporter protein and detoxified by reduction to relatively non-toxic, volatile elemental mercury by an intracellular mercuric reductase (MerA) (Hobman & Brown, 1997). The biotechnological potential of this process has been described recently, focusing on the use of mercury-resistant bacteria and the proteins that they encode (Lloyd *et al.*, 2004). Applications include the bioremediation of mercury-contaminated water and the development of Hg(II)-detecting biosensors. Finally, in addition to the MerA-mediated mechanism of mercury reduction, other enzymes are also able to reduce Hg(II). A novel, Fe^{2+}-dependent mechanism for mercury reduction has been characterized in the membrane fraction of *Thiobacillus ferrooxidans*, which may involve cytochrome *c* oxidase (Iwahori *et al.*, 2000), and *c*-type cytochromes of *Geobacter metallireducens* also reduce Hg(II) (Lovley *et al.*, 1993). Results from our laboratory have also shown that whole cells of *Geobacter sulfurreducens* are able to reduce Hg(II) without the involvement of *mer* mercury-resistance genes (N. Law & J. R. Lloyd, unpublished data).

REDUCTION OF METALLOIDS

Arsenic reduction: a role in mass poisoning worldwide?

Contamination of groundwaters, abstracted for drinking and irrigation, by sediment-derived arsenic threatens the health of tens of millions of people worldwide, most notably in Bangladesh and West Bengal (Chakraborti *et al.*, 2002; Smedley & Kinniburgh, 2002). Despite the calamitous human-health impacts arising from the extensive use of arsenic-enriched groundwaters in these regions, the mechanisms of arsenic release from sediments remain poorly characterized and have been a topic of intense debate (Nickson *et al.*, 1998; Chowdhury *et al.*, 1999; Oremland & Stolz, 2003). However, recent results have suggested that metal-reducing bacteria may be a major cause of this humanitarian catastrophe.

The immediate source of arsenic in these groundwaters in Bengal is widely considered to be the host sediments, which are transported by the rivers Ganges, Brahmaputra and Meghna and derived from the weathering of the Himalayas within the $1.5 \times 10^6 \, km^2$ catchment area of these three great river systems. Depending upon the sediment type, arsenic typically occurs at concentrations of 2–100 parts per million and is found in, and adsorbed onto, a variety of mineralogical hosts, including hydrated ferric oxides, phyllosilicates and sulfide minerals (Smedley & Kinniburgh, 2002). The mechanism of arsenic release from contaminated sediments remains controversial, but microbially mediated release of arsenic from hydrated ferric oxides is gaining the consensus as the dominant mechanism for the mobilization of arsenic into groundwater systems of the Ganges delta. For example, we recently provided direct evidence for the role of indigenous metal-reducing bacteria in the formation of toxic, mobile As(III) in sediments from the Ganges delta (Islam *et al.*, 2004). In this study, sediment samples were collected at a depth of 13 m from a site in West Bengal known to have relatively high concentrations of arsenic in the groundwater. Arsenic mobilization was optimal under anaerobic conditions, and addition of acetate to anaerobic sediments, as a proxy for organic matter and a potential electron donor for metal reduction, resulted in stimulation of microbial reduction of Fe(III), followed by As(V) reduction and release of mobile As(III). Microbial communities responsible for metal reduction and arsenic mobilization in the stimulated anaerobic sediments were analysed by using molecular (PCR-based) and cultivation-dependent techniques. Both approaches confirmed an increase in numbers of Fe(III)-reducing bacteria and suggested a vital role for metal-reducing bacteria in mediating arsenic release. Indeed, the PCR studies showed that the microbial communities in these sediments were dominated by *Geobacter* species. We have also recently obtained similar results using Cambodian sediments liberating arsenic into groundwater (H. A. L. Rowland, R. L. Pederick, D. A. Polya, R. D. Pancost, B. E. van Dongen, A. G. Gault, J. M. Charnock, D. J. Vaughan & J. R. Lloyd, unpublished data). However, we have also noted recently that *Geobacter* species (e.g. *Geobacter sulfurreducens*) cannot reduce As(V) enzymically and may actually promote the retention of arsenic in sediments, through the formation of Fe(II) minerals that also sorb the metalloid (F. S. Islam, R. L. Pederick, A. G. Gault, L. K. Adams, D. A. Polya, J. M. Charnock & J. R. Lloyd, unpublished data). These results also show that the reduction of Fe(III) alone is not sufficient to mobilize sorbed arsenic, suggesting that specialized As(V)-respiring bacteria play a very important role in the reductive mobilization of arsenic as mobile, toxic As(III).

Several organisms capable of growing through dissimilatory reduction of As(V) have now been isolated, although none from south-east Asian aquifers. *Chrysiogenes arsenatis* is a strict anaerobe that was isolated from wastewater from a gold mine (Macy *et al.*, 1996). This organism contains a periplasmic arsenate reductase consisting of two subunits (masses 87 and 29 kDa) that contain molybdenum, iron, sulfur and zinc co-factors (Krafft & Macy, 1998). *Sulfurospirillum arsenophilum* (previously strain MIT 13^T) is a Gram-negative, vibrioid, microaerobic, sulfur-reducing bacterium isolated from arsenic-contaminated watershed sediments in eastern Massachusetts, USA. It is a very close relative of *Sulfurospirillum barnesii*, which can also reduce As(V), and has a broad activity against metals, including Fe(III) and Se(VI) (see sections on these metals/metalloids). A recent review has mentioned the preliminary purification and characterization of an arsenate reductase from this organism, which constituted a trimeric complex of 120 kDa, consisting of subunits of 65, 31 and 22 kDa (Oremland & Stolz, 2000). A Gram-positive, sulfate-reducing bacterium, *Desulfotomaculum auripigmentum*, has also been described, which reduces As(V) followed by sulfate, resulting in the formation of orpiment (As$_2$S$_3$) (Newman *et al.*, 1997).

Reduction of As(V) to As(III) by the ArsC reductase also forms the basis of a well-studied microbial arsenic-resistance mechanism, preceding efflux of As(III) from the cell (Mukhopadhyay *et al.*, 2002). It should be noted, however, that ArsC-mediated As(V) reduction does not support microbial growth and there is currently little evidence linking this mechanism of As(V) reduction to the biogeochemical cycling of arsenic. Finally, on this note, it is worth mentioning that a novel organism has recently been isolated that is able to use As(III) as an electron donor for aerobic growth (Santini *et al.*, 2000), thus closing the biological arsenic cycle.

Finally, As(V) is actually a minor (but important) electron acceptor for anaerobic respiration in aquifers. However, in some environments, such as Mono Lake, CA, USA, As(V) can be the dominant electron acceptor for carbon oxidation (Oremland *et al.*, 2000) and here, the role of microbial metabolism in controlling arsenic speciation is far clearer. This fascinating environment and the organisms that it supports have been reviewed recently (Oremland *et al.*, 2004).

Reduction of Se(VI) and Se(IV) and other group VIB elements

In contrast to As(V) reduction, the biotransformation of Se(VI) and Se(IV) to relatively unreactive Se(0) results in its removal from water. Several studies have demonstrated that these transformations can be catalysed by microbes (Oremland & Stolz, 2000). For example, the ability to reduce Se(VI) is widespread in sediments, with biological reduction demonstrated unequivocally in 10 out of 11 sediment types (Steinberg & Oremland, 1990). Also, Se(VI) is not reduced chemically under physiological conditions of pH and temperature, and Se(VI) reduction is inhibited by autoclaving of sediments.

Organisms that are known to reduce Se(VI) enzymically include *Wolinella succinogenes* (Tomei *et al.*, 1992), *Desulfovibrio desulfuricans* (Tomei *et al.*, 1995), *Pseudomonas stutzeri* (Lortie *et al.*, 1992), *Enterobacter cloacae* (Losi & Frankenberger, 1997) and *Escherichia coli* (Avazéri *et al.*, 1997). In these examples, Se(VI) reduction does not support growth and seems to be incidental to the physiology of the organism. In at least one organism (*Escherichia coli*), the involvement of broad-specificity nitrate reductases is implicated by biochemical studies (Avazéri *et al.*, 1997). In addition to these rather non-specific reactions, specialist organisms are known to conserve energy through Se(VI) reduction, including *Thauera selenatis* (Macy & Lawson, 1993), *Sulfuro-spirillum barnesii* [originally '*Geospirillum barnesii*' strain SES-3 (Stolz *et al.*, 1997)] and two bacilli (*Bacillus arseniciselenatis* and *Bacillus selenitireducens*), both isolated from Mono Lake, CA, USA (Switzer Blum *et al.*, 1998). Of these four model organisms, the mechanism of Se(VI) reduction is best understood in *Thauera selenatis* (Schröder *et al.*, 1997). A periplasmic complex of approximately 180 kDa (with subunits of masses 96, 40 and 23 kDa) has been characterized and shown to contain molybdenum, iron and acid-labile sulfur. Specificity for Se(VI) is high, with a K_m of 16 µM. The enzyme was unable to reduce nitrate, nitrite, chlorate, chlorite or sulfate. Biochemical studies are not so advanced in *Sulfurospirillum barnesii*, although the enzyme activity contrasts with that characterized in *Thauera selenatis,* as it is localized in the membrane fraction and may have a wider substrate specificity (Stolz *et al.*, 1997).

Tellurite (TeO_3^{2-}) reduction has also been studied in several organisms, mainly in the context of resistance to this toxic oxyanion. Indeed, the antibacterial properties of Te(IV) have been known for more than 70 years: in the pre-antibiotic era, Te(IV) was used to treat a range of bacterial infections and Te(IV) remains an ingredient of several selective media (e.g. for verocytotoxigenic *Escherichia coli* O157). Basal levels of resistance to toxic Te(IV) have been attributed to the activity of a membrane-bound nitrate reductase in *Escherichia coli* (Avazéri *et al.*, 1997). An additional Te(IV) reductase was detected in the soluble fraction of anaerobically grown cells. Growth by using Te(IV) as an electron acceptor was also reported in an engineered strain over-expressing nitrate reductase, but was not thought to be physiologically relevant in wild-type cells of *Escherichia coli* (Avazéri *et al.*, 1997). *Rhodobacter sphaeroides* has also been reported to reduce Te(IV) (as well as oxyanions of selenium), with an absolute requirement for a functional photosynthetic electron-transfer chain under photo-synthetic (anaerobic) growth conditions, or functional cytochromes bc_1 and c_2 under aerobic growth conditions (Moore & Kaplan, 1992). Again, metal reduction was discussed in the context of resistance to the metalloids. Finally, plasmids are known to encode several distinct resistance determinants for Te(IV) and, again, Te(IV) reduction is implicated as the resistance mechanism, as elemental tellurium is deposited within tellurite-resistant bacteria (Taylor, 1999). However, other mechanisms of resistance, involving cysteine-metabolizing enzymes and methyl transferases, may be important (Taylor, 1999). Finally, Te(IV) [and Se(VI)/Se(IV)] reduction and precipitation by sulfate-reducing bacteria has also been reported, in the order Te(IV) > Se(VI) > Se(IV), which is in contrast to that predicted by the redox potentials alone (Lloyd *et al.*, 2001). To date, there have been no reports of microbial growth coupled to the reduction of Te(IV) by non-genetically engineered micro-organisms.

REDUCTION OF ACTINIDES AND FISSION PRODUCTS AND THE BIOREMEDIATION OF RADIOACTIVE WASTE

The release of radionuclides from nuclear sites and their subsequent mobility in the environment is a subject of intense public concern and has prompted much recent research on the environmental fate of key radionuclides (Lloyd & Renshaw, 2005a). The major burden of anthropogenic environmental radioactivity is from the controlled discharge of process effluents produced by industrial activities allied to the generation of nuclear power, although significant quantities of natural and artificial radionuclides were also released as a consequence of nuclear weapons testing in the 1950s and 1960s via accidental release, e.g. from Chernobyl in 1986, and from the ongoing storage of nuclear materials amassed over the last 60 years of nuclear activities. Indeed, the scale of our nuclear legacy is enormous, including 120 Department of Energy sites in the USA alone, and other facilities in Europe and the former USSR (Lloyd & Renshaw, 2005a). In several cases, storage has been compromised, leading to contamination of trillions of litres of groundwater and millions of cubic metres of contaminated soil and debris. The costs of cleaning up these sites are estimated to be in excess of a trillion US dollars in the USA alone, and 50 billion pounds sterling in the UK. Given these high costs and the technical limitations of current chemical-based approaches, there has been an unprecedented interest in the interactions of micro-organisms with key radionuclides, in the hope of developing cost-effective bioremediation approaches for decontamination of sediments and waters affected by nuclear waste (Lloyd *et al.*, 2004).

Because many radionuclides of concern are both redox-active and less soluble when reduced, bioreduction offers much promise for controlling the solubility and mobility of target radionuclides in contaminated sediments, e.g. the reduction of U(VI) (the uranyl ion; UO_2^{2+}) to U(IV) (uraninite; UO_2) (Lovley *et al.*, 1991; Lovley & Phillips, 1992b) or the reduction of the fission product Tc(VII) (the pertechnetate ion; TcO_4^-) to Tc(IV) (TcO_2) (Lloyd *et al.*, 2000b). Several studies have also addressed the colonization of radioactive environments (see Lloyd & Renshaw, 2005b) and it would seem that the radioactive burden of several nuclear-waste types is not necessarily inhibitory to all microbial life. For example, a recent study using pure cultures of bacteria proposed for application in bioremediation programmes has addressed the toxicity of actinides, metals and chelators. The model organisms tested include *Deinococcus radiodurans*, *Pseudomonas putida* and *Shewanella putrefaciens* CN32 (Ruggiero *et al.*, 2005). Actinides, including chelated Pu(IV), U(VI) and Np(V), inhibited growth at millimolar concentrations, suggesting that actinide toxicity is primarily chemical (not radio-logical) and that radiation resistance (e.g. in *Deinococcus* species) does not necessarily ensure radionuclide tolerance. The author proposes that actinide toxicity will not impede bioremediation using naturally occurring bacteria, although the toxicity of these key radionuclides remains to be determined in sedimentary environments under field conditions.

Uranium reduction

The first demonstration of dissimilatory U(VI) reduction was by Lovley *et al.* (1991), who reported that the Fe(III)-reducing bacteria *Geobacter metallireducens* (previously designated strain GS-15^T) and *Shewanella oneidensis* (formerly *Alteromonas putre-faciens* and then *Shewanella putrefaciens*) can conserve energy for anaerobic growth via the reduction of U(VI). It should be noted, however, that the ability to reduce U(VI) enzymically is not restricted to Fe(III)-reducing bacteria. Other organisms, including a *Clostridium* species (Francis, 1994) and the sulfate-reducing bacteria *Desulfovibrio desulfuricans* (Lovley & Phillips, 1992a) and *Desulfovibrio vulgaris* (Lovley & Phillips, 1994), also reduce uranium, but are unable to conserve energy for growth via this transformation. To date, the enzyme system responsible for U(VI) reduction has been best studied in *Desulfovibrio vulgaris*. Purified tetrahaem cytochrome c_3 was shown to function as a U(VI) reductase *in vitro*, in combination with hydrogenase, its physio-logical electron donor (Lovley & Phillips, 1994). *In vivo* studies using a cytochrome c_3

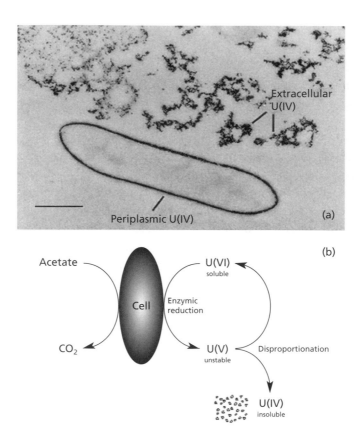

Fig. 3. Reduction of U(VI) to insoluble U(IV) (electron-dense deposits) in thin sections of *Geobacter sulfurreducens*, viewed by using transmission electron microscopy (a). U(VI) reduction by this organism is via unstable U(V), which disproportionates to give insoluble U(IV) [and U(VI) for further reduction] (b). Transmission electron microscopy by S. Glasauer (University of Guelph, Canada). Bar, 0·5 μm.

mutant of the close relative *Desulfovibrio desulfuricans* strain G20 confirmed a role for cytochrome c_3 in hydrogen-dependent U(VI) reduction, but suggested additional pathways from organic electron donors to U(VI) that bypassed the cytochrome (Payne *et al.*, 2002). Similar cytochrome-mediated mechanisms have been proposed in *Geobacter* species, whilst U(VI) reduction in a *Shewanella putrefaciens* strain shares components of the nitrite-reducing pathway in this organism (Wade & DiChristina, 2000). The mechanism of U(VI) reduction by *Geobacter sulfurreducens* has also been studied recently in detail by using X-ray absorbance spectroscopy, which showed the formation of an unstable U(V) intermediate (Renshaw *et al.*, 2005). This organism was unable to reduce the stable analogue Np(V), suggesting that the further reduction of U(V) is by disproportionation to U(IV) [with U(VI) generated available for further reduction] (Fig. 3). This surprising level of specificity of hexavalent actinides illustrates a need for detailed investigations on the impact of micro-organisms on complex, actinide-containing wastes.

Field studies on uranium bioreduction *in situ*

There has been a considerable level of interest in harnessing the metabolism of U(VI)-reducing bacteria for the bioremediation of uranium-contaminated aquifers. For example, recent studies focused on biostimulation of U(VI)-reducing bacteria at a 'Uranium Mill Tailings Remedial Action' (UMTRA) site in Rifle, CO, USA, through the injection of an electron donor (acetate) into the subsurface (Anderson *et al.*, 2003). The decrease in soluble U(VI) was coincident with an increase in Fe(II) in the groundwater and a significant enrichment of *Geobacter* species. However, after 39 days, the composition of the microbial community began to change as sulfate-reducing organisms dominated, and soluble U(VI) increased with a decrease in Fe(II), acetate and sulfate and an accumulation of sulfite. Thus, the precise constituents of the microbial communities present in the sediments clearly exert control on U(VI) speciation and require careful optimization. The geochemistry of the groundwaters should also not be overlooked. For example, Ca^{2+} cations at millimolar concentrations cause a significant decrease in the rate and extent of bacterial U(VI) reduction by a range of metal-reducing bacteria, suggesting that U(VI) is a less effective electron acceptor when present as the $Ca_2UO_2(CO_3)_3$ complex (Brooks *et al.*, 2003). High nitrate concentrations can also inhibit U(VI) reduction by acting as a competing electron acceptor, and may even promote reoxidation of reduced U(IV) (Istok *et al.*, 2004). Mineralogical constraints can also be important in controlling the end points for uranium bioremediation. For example, although much work has focused on the reduction of soluble U(VI), the fate of sorbed U(VI), which can be appreciable in sediments, is also potentially important. For example, although soluble U(VI) was reduced in a slurry prepared from sediments from Rifle, CO, USA, sorbed U(VI) was not reduced and was deemed 'not bioavailable' for microbial reduction (Ortiz-Bernad *et al.*, 2004b).

Reduction of other actinides (plutonium and neptunium)

Although ^{238}U remains the priority pollutant in most medium- and low-level radioactive wastes, other actinides, including ^{230}Th, ^{237}Np, ^{241}Pu and ^{241}Am, can also be present (Macaskie, 1991; Lloyd & Macaskie, 2000). Th(IV) and Am(III) are stable across most E_h values encountered in radionuclide-contaminated waters, but the potentials for Pu(V)/Pu(IV) and Np(V)/Np(IV), in common with that of U(VI)/U(IV), are more electropositive than the standard redox potential of ferrihydrite/Fe^{2+} (approx. 0 V; Thamdrup, 2000). Thus, Fe(III)-reducing bacteria have the metabolic potential to reduce these radionuclides enzymically or via Fe(II) produced from the reduction of Fe(III) oxides. This is significant because the tetravalent actinides are amenable to bioremediation, due to their high ligand-complexing abilities (Lloyd & Macaskie,

2000), and are also immobilized in sediments containing active biomass (Peretrukhin *et al.*, 1996). Thus, although it is possible for Fe(III)-reducing bacteria to reduce and precipitate actinides directly, e.g. the reduction of soluble U(VI) to insoluble U(IV) (see above), some transformations do not result in formation of an insoluble mineral phase, but in the formation of a cation more amenable to bioprecipitation. This is illustrated when considering highly soluble Np(V) (NpO_2^+), which was reduced to soluble Np(IV) by *Shewanella putrefaciens*, with the Np(IV) removed as an insoluble phosphate biomineral by a phosphate-liberating *Citrobacter* species (Lloyd *et al.*, 2000a). This is in sharp contrast to the case in *Geobacter sulfurreducens*, which is unable to reduce Np(V) (see above). Also, some studies have suggested that the reduction of Pu(IV) to Pu(III) can be achieved by Fe(III)-reducing bacteria, although the Pu(III) was reported to reoxidize spontaneously (Rusin *et al.*, 1994). Although this may lead to solublization of sediment-bound Pu(IV), it will yield a trivalent actinide that is also amenable to bioremediation by using a range of microbially produced ligands (Lloyd & Macaskie, 2000). The biochemical basis of these transformations remains uncharacterized.

Technetium reduction

The fission product technetium is another long-lived radionuclide that is present in nuclear waste and has attracted considerable recent interest. This is due to a combination of its mobility as the soluble pertechnetate ion [Tc(VII); TcO_4^-], bioavailability as an analogue of sulfate and long half-life ($2 \cdot 13 \times 10^5$ years) (Wildung *et al.*, 1979). Like Np(V), Tc(VII) has weak ligand-complexing capabilities and is difficult to remove from solution by using conventional 'chemical' approaches. Several reduced forms of the radionuclide are insoluble, however, and metal-reducing micro-organisms can reduce Tc(VII) and precipitate the radionuclide as a low-valency oxide [Tc(IV); TcO_2].

In an early study on Tc(VII) bioreduction, a novel phosphorimaging technique was used to show reduction of the radionuclide by *Shewanella putrefaciens* and *Geobacter metallireducens*, with similar activities subsequently detected in laboratory cultures of *Rhodobacter sphaeroides*, *Paracoccus denitrificans*, some pseudomonads (Lloyd *et al.*, 2002), *Escherichia coli* (Lloyd *et al.*, 1997) and a range of sulfate-reducing bacteria (Lloyd *et al.*, 1998b, 1999b, 2001). Other workers have used this technique to show that *Thiobacillus ferrooxidans* and *Thiobacillus thiooxidans* (Lyalikova & Khizhnyak, 1996) and the hyperthermophile *Pyrobaculum islandicum* (Kashefi & Lovley, 2000) are also able to reduce Tc(VII). It should be stressed that Tc(VII) reduction has not been shown to support growth in any of these studies, and seems to be a fortuitous biochemical side reaction in the organisms studied to date. Recent work has also shown that Tc(VII) can be reduced through indirect microbial processes via, for example, biogenic sulfide (Lloyd *et al.*, 1998b), Fe(II) (Lloyd *et al.*, 2000b) or U(IV) (Lloyd *et al.*, 2002). Tc(VII) reduction and precipitation by biogenic Fe(II) are particularly efficient and may offer a potentially useful mechanism for the remediation of technetium-contaminated sediments containing active Fe(III)-reducing bacteria (Lloyd *et al.*, 2000b).

This latter point has been confirmed by several recent studies using a range of sediment materials. In one study, sediments from the Humber Estuary, UK, were left to age and exhibited a clear progression of terminal electron-accepting processes (Burke *et al.*, 2005). The reduction and precipitation of Tc(VII) were associated with the formation of biogenic Fe(II) and were catalysed by pure cultures of Fe(III)-reducing prokaryotes inoculated in sterilized microcosms. Technetium solubility has also been studied by using core samples from a shallow, sandy aquifer located on the US Atlantic Coastal Plain (Wildung *et al.*, 2004). The dominant electron donor in the sediments was Fe(II) (0·5 M HCl-extractable), with Tc(IV) hydrous oxide being the major solid-phase reduction product. The authors noted presumptive evidence for direct enzymic reduction in only a few key sand samples. The potential for biogenic Fe(II)-mediated reduction of Tc(VII) has also been assessed in detail in other studies, e.g. by using sediments from the US Department of Energy's Hanford and Oak Ridge sites (Fredrickson *et al.*, 2004).

The biochemical basis of Tc(VII) reduction has been best studied in *Escherichia coli*. Initial studies demonstrated that anaerobic, but not aerobic, cultures of *Escherichia coli* reduced Tc(VII), with the reduced radionuclide precipitated within the cell (Lloyd *et al.*, 1997). Results obtained from studies conducted with wild-type cells and 34 defined mutants defective in the synthesis of regulatory or electron-transfer proteins were used to construct a model for Tc(VII) reduction by *Escherichia coli*. The central tenet of this model is that the hydrogenase 3 component of formate hydrogenlyase catalyses the transfer of electrons from dihydrogen to Tc(VII) (Fig. 4). According to this model, the formate dehydrogenase component (FdhH) is required only if formate, or a precursor, is supplied as an electron donor for Tc(VII) reduction in place of hydrogen. This model has been validated by the observations that a mutant unable to synthesize hydrogenase 3 was unable to reduce Tc(VII) when either hydrogen or formate was supplied as an electron donor (Lloyd *et al.*, 1997). Hydrogenase-mediated Tc(VII) reduction has also been noted in sulfate-reducing bacteria (Lloyd *et al.*, 1999b; De Luca *et al.*, 2001) and *Geobacter sulfurreducens* (J. C. Renshaw & J. R. Lloyd, unpublished observations).

DEGRADATION OF XENOBIOTICS BY METAL-REDUCING BACTERIA

Some metal-reducing bacteria, most notably *Geobacter* species, have the ability to couple Fe(III) reduction to the complete oxidation of aromatic contaminants (Lovley *et al.*, 1989b; Lovley & Lonergan, 1990; Coates *et al.*, 2001). It is possible to stimulate

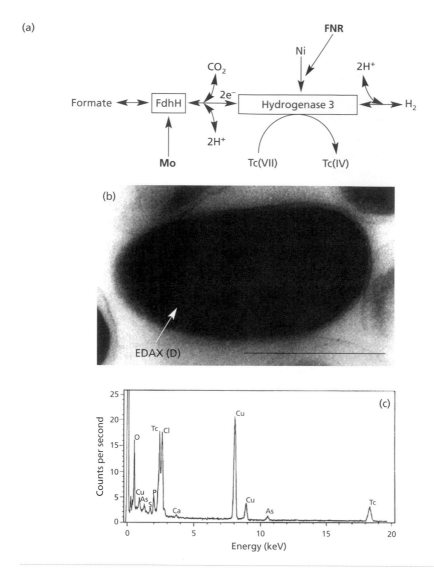

Fig. 4. Mechanism of Tc(VII) reduction by *E. coli*. Hydrogenase 3 of the formate hydrogenlyase complex is able to catalyse the reduction of soluble Tc(VII) to insoluble Tc(IV) (a). Hydrogen or formate are suitable electron donors; with the latter a formate dehydrogenase (FdhH) is also required for reduction of Tc(VII). The insoluble Tc(IV) is precipitated within the cell, visible as an electron dense deposit in TEM images of thin sections of the cells (b) and confirmed by EDS analysis (c). Bar, 1 μm.

these activities by increasing the bioavailability of Fe(III) oxides in subsurface sediments by, for example, using Fe(III) chelators that solublize Fe(III) (Lovley *et al.*, 1994) or humic acids that can act as electron shuttles between the Fe(III)-reducing species and Fe(III) oxides (Lovley *et al.*, 1996). Both approaches eliminate the need for the organism to contact the insoluble Fe(III) oxide directly to reduce it.

(a)

Xenobiotic hydrazone and azo bonds are part of the chromophore

Xenobiotic aromatic sulfonic acid groups make the dye highly soluble

(b)

Fig. 5. Reduction and decoloration of the azo dye remazol black B (a) by anaerobically grown cells of *Shewanella* sp. J18143 (b). Figures were generously provided by C. Pearce (University of Manchester, UK).

In addition to coupling the oxidation of aromatics to metal reduction, Fe(III)-reducing bacteria can also reduce redox-active organic xenobiotics in lieu of Fe(III). A good example is the reduction of recalcitrant, coloured azo dyes, the disposal of which poses a considerable problem to textile-dyeing industries worldwide. Due to the relatively low levels of dye–fibre fixation in current reactive dyeing processes, as much as 50 % of the dye that is present in the original dyebath is lost to the wastewater. Physical and/or chemical processes are available, but are costly and can generate problematic concentrated sludges for disposal. An alternative procedure utilizes anaerobically grown cultures of a *Shewanella* species (designated strain J18143), isolated from soil contaminated with textile dyes, to reduce and decolorize textile wastewaters (Nelson *et al.*, 2000). This highly efficient biocatalyst has been incorporated into the recently developed BIOCOL commercial process for the treatment of azo dyes (Conlon & Khraisheh, 2002). In this process, the bacterial cells are immobilized on an activated carbon support that adsorbs the target dye molecules and the potentially toxic amine breakdown products for further biodegradation. Recent studies from our laboratory have studied the underlying physiology of this organism, which is able to reduce a wide range of azo compounds (e.g. remazol black B; see Fig. 5) and is compatible with realistic process conditions, including moderate temperatures and alkaline pH (C. Pearce, J. Guthrie & J. R. Lloyd, unpublished data).

CONCLUSIONS

Although the environmental relevance of microbial metal-reduction processes has only recently become apparent, rapid advances in the understanding of these important biotransformations have been made. However, we still have much to learn about the precise mechanisms involved and the full impact of such reactions on a range of biogeochemical cycles. Given the availability of genomic sequences for key metal-reducing micro-organisms, new post-genomic approaches and the possibility of combining these tools with advanced techniques from other branches of science and technology (e.g. isotopic, spectroscopic and computational tools), rapid advances in these areas are predicted.

ACKNOWLEDGEMENTS

The author thanks the UK Natural Environment Research Council, Biotechnology and Biological Sciences Research Council and Engineering and Physical Sciences Research Council and the Natural and Accelerated Bioremediation Research (NABIR) programme of the US Department of Energy for financial support.

REFERENCES

Anderson, R. T., Rooney-Varga, J. N., Gaw, C. V. & Lovley, D. R. (1998). Anaerobic benzene oxidation in the Fe(III) reduction zone of petroleum-contaminated aquifers. *Environ Sci Technol* **32**, 1222–1229.

Anderson, R. T., Vrionis, H. A., Ortiz-Bernad, I. & 10 other authors (2003). Stimulating the in situ activity of *Geobacter* species to remove uranium from the groundwater of a uranium-contaminated aquifer. *Appl Environ Microbiol* **69**, 5884–5891.

Avazéri, C., Turner, R. J., Pommier, J., Weiner, J. H., Giordano, G. & Verméglio, A. (1997). Tellurite reductase activity of nitrate reductase is responsible for the basal resistance of *Escherichia coli* to tellurite. *Microbiology* **143**, 1181–1189.

Bautista, E. M. & Alexander, M. (1972). Reduction of inorganic compounds by soil microorganisms. *Soil Sci Soc Am Proc* **36**, 918–920.

Beliaev, A. S., Saffarini, D. A., McLaughlin, J. L. & Hunnicutt, D. (2001). MtrC, an outer membrane decahaem c cytochrome required for metal reduction in *Shewanella putrefaciens* MR-1. *Mol Microbiol* **39**, 722–730.

Belly, R. T. & Kydd, G. C. (1982). Silver resistance in microorganisms. *Dev Ind Microbiol* **23**, 567–577.

Bond, D. R., Holmes, D. E., Tender, L. M. & Lovley, D. R. (2002). Electrode-reducing microorganisms that harvest energy from marine sediments. *Science* **295**, 483–485.

Brooks, S. C., Fredrickson, J. K., Carroll, S. L., Kennedy, D. W., Zachara, J. M., Plymale, A. E., Kelly, S. D., Kemner, K. M. & Fendorf, S. (2003). Inhibition of bacterial U(VI) reduction by calcium. *Environ Sci Technol* **37**, 1850–1858.

Bruins, M. R., Kapil, S. & Oehme, F. W. (2000). Microbial resistance to metals in the environment. *Ecotoxicol Environ Saf* **45**, 198–207.

Burke, I. T., Boothman, C., Lloyd, J. R., Mortimer, R. J. G., Livens, F. R. & Morris, K. (2005). Effects of progressive anoxia on the solubility of technetium in sediments. *Environ Sci Technol* **39**, 4109–4116.

Caccavo, F., Jr, Lonergan, D. J., Lovley, D. R., Davis, M., Stolz, J. F. & McInerney, M. J. (1994). *Geobacter sulfurreducens* sp. nov., a hydrogen- and acetate-oxidizing dissimilatory metal-reducing microorganism. *Appl Environ Microbiol* **60**, 3752–3759.

Chakraborti, D., Rahman, M. M., Paul, K., Chowdhury, U. K., Sengupta, M. K., Lodh, D., Chanda, C. R., Saha, K. C. & Mukherjee, S. C. (2002). Arsenic calamity in the Indian subcontinent: what lessons have been learned? *Talanta* **58**, 3–22.

Childers, S. E., Ciufo, S. & Lovley, D. R. (2002). *Geobacter metallireducens* accesses insoluble Fe(III) oxide by chemotaxis. *Nature* **416**, 767–769.

Chowdhury, T. R., Basu, G. K., Mandal, B. K. & 11 other authors (1999). Arsenic poisoning in the Ganges delta. *Nature* **401**, 545–546.

Coates, J. D., Bhupathiraju, V. K., Achenbach, L. A., McInerney, M. J. & Lovley, D. R. (2001). *Geobacter hydrogenophilus*, *Geobacter chapellei* and *Geobacter grbiciae*, three new, strictly anaerobic, dissimilatory Fe(III)-reducers. *Int J Syst Evol Microbiol* **51**, 581–588.

Coker, V., Pattrick, R. A. D., van der laan, G. & Lloyd, J. R. (2004). *Use of Bacteria to Produce Spinel Nanoparticles*. UK Patent 0424636.9.

Conlon, M. & Khraisheh, M. (2002). *Bioadsorption Process for the Removal of Colour from Textile Effluent*. Patent WO0242228.

Coppi, M. V., Leang, C., Sandler, S. J. & Lovley, D. R. (2001). Development of a genetic system for *Geobacter sulfurreducens*. *Appl Environ Microbiol* **67**, 3180–3187.

De Luca, G., de Philip, P., Dermoun, Z., Rousset, M. & Verméglio, A. (2001). Reduction of technetium(VII) by *Desulfovibrio fructosovorans* is mediated by the nickel-iron hydrogenase. *Appl Environ Microbiol* **67**, 4583–4587.

DiChristina, T. J., Moore, C. M. & Haller, C. A. (2002). Dissimilatory Fe(III) and Mn(IV) reduction by *Shewanella putrefaciens* requires *ferE*, a homolog of the *pulE* (*gspE*) type II protein secretion gene. *J Bacteriol* **184**, 142–151.

Fendorf, S. E. & Li, G. (1996). Kinetics of chromate reduction by ferrous iron. *Environ Sci Technol* **30**, 1614–1617.

Francis, A. J. (1994). Microbial transformations of radioactive wastes and environmental restoration through bioremediation. *J Alloys Comp* **213/214**, 226–231.

Fredrickson, J. K., Zachara, J. M., Kennedy, D. W., Kukkadapu, R. K., McKinley, J. P., Heald, S. M., Liu, C. & Plymale, A. E. (2004). Reduction of TcO_4^- by sediment-associated biogenic Fe(II). *Geochim Cosmochim Acta* **68**, 3171–3187.

Fu, J. K., Liu, Y. Y., Gu, P. Y., Tang, D. L., Lin, Z. Y., Yao, B. X. & Weng, S. Z. (2000). Spectroscopic characterization on the biosorption and bioreduction of Ag^+ by *Lactobacillus* A09. *Acta Phys-Chim Sin* **16**, 779–782.

Fude, L., Harris, B., Urrutia, M. M. & Beveridge, T. J. (1994). Reduction of Cr(VI) by a consortium of sulfate-reducing bacteria (SRB III). *Appl Environ Microbiol* **60**, 1525–1531.

Gaspard, S., Vazquez, F. & Holliger, C. (1998). Localization and solubilization of the iron(III) reductase of *Geobacter sulfurreducens*. *Appl Environ Microbiol* **64**, 3188–3194.

Ghani, B., Takai, M., Hisham, N. Z., Kishimoto, N., Ismail, A. K. M., Tano, T. & Sugio, T. (1993). Isolation and characterization of a Mo^{6+}-reducing bacterium. *Appl Environ Microbiol* **59**, 1176–1180.

Gorby, Y. A., Caccavo, F., Jr & Bolton, H., Jr (1998). Microbial reduction of cobalt[III]EDTA⁻ in the presence and absence of manganese(IV) oxide. *Environ Sci Technol* **32**, 244–250.

Gupta, A., Matsui, K., Lo, J.-F. & Silver, S. (1999). Molecular basis for resistance to silver cations in *Salmonella*. *Nat Med* **5**, 183–188.

Hobman, J. L. & Brown, N. L. (1997). Bacterial mercury-resistance genes. *Met Ions Biol Syst* **34**, 527–568.

Holmes, D. E., Finneran, K. T., O'Neil, R. A. & Lovley, D. R. (2002). Enrichment of members of the family *Geobacteraceae* associated with stimulation of dissimilatory metal reduction in uranium-contaminated aquifer sediments. *Appl Environ Microbiol* **68**, 2300–2306.

Islam, F. S., Gault, A. G., Boothman, C., Polya, D. A., Charnock, J. M., Chatterjee, D. & Lloyd, J. R. (2004). Role of metal-reducing bacteria in arsenic release from Bengal delta sediments. *Nature* **430**, 68–71.

Istok, J. D., Senko, J. M., Krumholz, L. R., Watson, D., Bogle, M. A., Peacock, A., Chang, Y.-J. & White, D. C. (2004). In situ bioreduction of technetium and uranium in a nitrate-contaminated aquifer. *Environ Sci Technol* **38**, 468–475.

Iwahori, K., Takeuchi, F., Kamimura, K. & Sugio, T. (2000). Ferrous iron-dependent volatilization of mercury by the plasma membrane of *Thiobacillus ferrooxidans*. *Appl Environ Microbiol* **66**, 3823–3827.

Kashefi, K. & Lovley, D. R. (2000). Reduction of Fe(III), Mn(IV), and toxic metals at 100 °C by *Pyrobaculum islandicum*. *Appl Environ Microbiol* **66**, 1050–1056.

Kashefi, K. & Lovley, D. R. (2003). Extending the upper temperature limit for life. *Science* **301**, 934.

Kashefi, K., Tor, J. M., Nevin, K. P. & Lovley, D. R. (2001). Reductive precipitation of gold by dissimilatory Fe(III)-reducing *Bacteria* and *Archaea*. *Appl Environ Microbiol* **67**, 3275–3279.

Krafft, T. & Macy, J. M. (1998). Purification and characterization of the respiratory arsenate reductase of *Chrysiogenes arsenatis*. *Eur J Biochem* **255**, 647–653.

Lloyd, J. R. (2003). Microbial reduction of metals and radionuclides. *FEMS Microbiol Rev* **27**, 411–425.

Lloyd, J. R. & Lovley, D. R. (2001). Microbial detoxification of metals and radionuclides. *Curr Opin Biotechnol* **12**, 248–253.

Lloyd, J. R. & Macaskie, L. E. (2000). Bioremediation of radionuclide-containing wastewaters. In *Environmental Microbe–Metal Interactions*, pp. 277–327. Edited by D. R. Lovley. Washington, DC: American Society for Microbiology.

Lloyd, J. R. & Renshaw, J. C. (2005a). Microbial transformations of radionuclides: fundamental mechanisms and biogeochemical implications. *Met Ions Biol Syst* **44**, 205–240.

Lloyd, J. R. & Renshaw, J. C. (2005b). Bioremediation of radioactive waste: radionuclide–microbe interactions in laboratory and field-scale studies. *Curr Opin Biotechnol* **16**, 254–260.

Lloyd, J. R., Cole, J. A. & Macaskie, L. E. (1997). Reduction and removal of heptavalent technetium from solution by *Escherichia coli*. *J Bacteriol* **179**, 2014–2021.

Lloyd, J. R., Yong, P. & Macaskie, L. E. (1998a). Enzymatic recovery of elemental palladium by using sulfate-reducing bacteria. *Appl Environ Microbiol* **64**, 4607–4609.

Lloyd, J. R., Nolting, H.-F., Solé, V. A., Bosecker, K. & Macaskie, L. E. (1998b). Technetium reduction and precipitation by sulfate-reducing bacteria. *Geomicrobiol J* **15**, 45–58.

Lloyd, J. R., Blunt-Harris, E. L. & Lovley, D. R. (1999a). The periplasmic 9·6-kilodalton c-type cytochrome of *Geobacter sulfurreducens* is not an electron shuttle to Fe(III). *J Bacteriol* **181**, 7647–7649.

Lloyd, J. R., Ridley, J., Khizniak, T., Lyalikova, N. N. & Macaskie, L. E. (1999b). Reduction of technetium by *Desulfovibrio desulfuricans*: biocatalyst characterization and use in a flowthrough bioreactor. *Appl Environ Microbiol* **65**, 2691–2696.

Lloyd, J. R., Yong, P. & Macaskie, L. E. (2000a). Biological reduction and removal of Np(V) by two microorganisms. *Environ Sci Technol* **34**, 1297–1301.

Lloyd, J. R., Sole, V. A., Van Praagh, C. V. G. & Lovley, D. R. (2000b). Direct and Fe(II)-mediated reduction of technetium by Fe(III)-reducing bacteria. *Appl Environ Microbiol* **66**, 3743–3749.

Lloyd, J. R., Mabbett, A. N., Williams, D. R. & Macaskie, L. E. (2001). Metal reduction by sulphate-reducing bacteria: physiological diversity and metal specificity. *Hydrometallurgy* **59**, 327–337.

Lloyd, J. R., Chesnes, J., Glasauer, S., Bunker, D. J., Livens, F. R. & Lovley, D. R. (2002). Reduction of actinides and fission products by Fe(III)-reducing bacteria. *Geomicrobiol J* **19**, 103–120.

Lloyd, J. R., Leang, C., Hodges Myerson, A. L., Coppi, M. V., Ciufo, S., Methe, B., Sandler, S. J. & Lovley, D. R. (2003). Biochemical and genetic characterization of PpcA, a periplasmic c-type cytochrome in *Geobacter sulfurreducens*. *Biochem J* **369**, 153–161.

Lloyd, J. R., Lovley, D. R. & Macaskie, L. E. (2004). Biotechnological application of metal-reducing microorganisms. *Adv Appl Microbiol* **53**, 85–128.

Lortie, L., Gould, W. D., Rajan, S., McCready, R. G. L. & Cheng, K.-J. (1992). Reduction of selenate and selenite to elemental selenium by a *Pseudomonas stutzeri* isolate. *Appl Environ Microbiol* **58**, 4042–4044.

Losi, M. E. & Frankenberger, W. T., Jr (1997). Reduction of selenium oxyanions by *Enterobacter cloacae* SLD1a-1: isolation and growth of the bacterium and its expulsion of selenium particles. *Appl Environ Microbiol* **63**, 3079–3084.

Lovley, D. R. (1991). Dissimilatory Fe(III) and Mn(IV) reduction. *Microbiol Rev* **55**, 259–287.

Lovley, D. R. (1993). Dissimilatory metal reduction. *Annu Rev Microbiol* **47**, 263–290.

Lovley, D. R. (1997). Potential for anaerobic bioremediation of BTEX in petroleum-contaminated aquifers. *J Ind Microbiol Biotechnol* **18**, 75–81.

Lovley, D. R. & Anderson, R. T. (2000). Influence of dissimilatory metal reduction on fate of organic and metal contaminants in the subsurface. *Hydrogeol J* **8**, 77–88.

Lovley, D. R. & Chapelle, F. H. (1995). Deep subsurface microbial processes. *Rev Geophys* **33**, 365–382.

Lovley, D. R. & Lonergan, D. J. (1990). Anaerobic oxidation of toluene, phenol, and p-cresol by the dissimilatory iron-reducing organism, GS-15. *Appl Environ Microbiol* **56**, 1858–1864.

Lovley, D. R. & Phillips, E. J. P. (1992a). Reduction of uranium by *Desulfovibrio desulfuricans*. *Appl Environ Microbiol* **58**, 850–856.

Lovley, D. R. & Phillips, E. J. P. (1992b). Bioremediation of uranium contamination with enzymatic uranium reduction. *Environ Sci Technol* **26**, 2228–2234.

Lovley, D. R. & Phillips, E. J. P. (1994). Reduction of chromate by *Desulfovibrio vulgaris* and its c_3 cytochrome. *Appl Environ Microbiol* **60**, 726–728.

Lovley, D. R., Stolz, J. F., Nord, G. L. & Phillips, E. J. P. (1987). Anaerobic production of magnetite by a dissimilatory iron-reducing microorganism. *Nature* **330**, 252–254.

Lovley, D. R., Phillips, E. J. P. & Lonergan, D. J. (1989a). Hydrogen and formate oxidation coupled to dissimilatory reduction of iron or manganese by *Alteromonas putrefaciens*. *Appl Environ Microbiol* **55**, 700–706.

Lovley, D. R., Baedecker, M. J., Lonergan, D. J., Cozzarelli, I. M., Phillips, E. J. P. & Siegel, D. I. (1989b). Oxidation of aromatic contaminants coupled to microbial iron reduction. *Nature* **339**, 297–300.

Lovley, D. R., Phillips, E. J. P., Gorby, Y. A. & Landa, E. R. (1991). Microbial reduction of uranium. *Nature* **350**, 413–416.

Lovley, D. R., Giovannoni, S. J., White, D. C., Champine, J. E., Phillips, E. J. P., Gorby, Y. A. & Goodwin, S. (1993). *Geobacter metallireducens* gen. nov. sp. nov., a microorganism capable of coupling the complete oxidation of organic compounds to the reduction of iron and other metals. *Arch Microbiol* **159**, 336–344.

Lovley, D. R., Woodward, J. C. & Chapelle, F. H. (1994). Stimulated anoxic biodegradation of aromatic hydrocarbons using Fe(III) ligands. *Nature* **370**, 128–131.

Lovley, D. R., Coates, J. D., Blunt-Harris, E. L., Phillips, E. J. P. & Woodward, J. C. (1996). Humic substances as electron acceptors for microbial respiration. *Nature* **382**, 445–448.

Lovley, D. R., Holmes, D. E. & Nevin, K. P. (2004). Dissimilatory Fe(III) and Mn(IV) Reduction. *Adv Microb Physiol* **49**, 219–286.

Lyalikova, N. N. & Khizhnyak, T. V. (1996). Reduction of heptavalent technetium by acidophilic bacteria of the genus *Thiobacillus*. *Mikrobiologiia* **65**, 468–473 (in Russian).

Mabbett, A. N., Lloyd, J. R. & Macaskie, L. E. (2002). Effect of complexing agents on reduction of Cr(VI) by *Desulfovibrio vulgaris* ATCC 29579. *Biotechnol Bioeng* **79**, 389–397.

Macaskie, L. E. (1991). The application of biotechnology to the treatment of wastes produced from the nuclear fuel cycle: biodegradation and bioaccumulation as a means of treating radionuclide-containing streams. *Crit Rev Biotechnol* **11**, 41–112.

Macy, J. M. & Lawson, S. (1993). Cell yield (Y_M) of *Thauera selenatis* grown anaerobically with acetate plus selenate or nitrate. *Arch Microbiol* **160**, 295–298.

Macy, J. M., Nunan, K., Hagen, K. D., Dixon, D. R., Harbour, P. J., Cahill, M. & Sly, L. I. (1996). *Chrysiogenes arsenatis* gen. nov., sp. nov., a new arsenate-respiring bacterium isolated from gold mine wastewater. *Int J Syst Bacteriol* **46**, 1153–1157.

Magnuson, T. S., Hodges-Myerson, A. L. & Lovley, D. R. (2000). Characterization of a membrane-bound NADH-dependent Fe^{3+} reductase from the dissimilatory Fe^{3+}-reducing bacterium *Geobacter sulfurreducens*. *FEMS Microbiol Lett* **185**, 205–211.

Marsh, T. L., Leon, N. M. & McInerney, M. J. (2000). Physiochemical factors affecting chromate reduction by aquifer materials. *Geomicrobiol J* **17**, 291–303.

Moore, M. D. & Kaplan, S. (1992). Identification of intrinsic high-level resistance to rare-earth oxides and oxyanions in members of the class *Proteobacteria*: characterization of tellurite, selenite, and rhodium sesquioxide reduction in *Rhodobacter sphaeroides*. *J Bacteriol* **174**, 1505–1514.

Mukhopadhyay, R., Rosen, B. P., Phung, L. T. & Silver, S. (2002). Microbial arsenic: from geocycles to genes and enzymes. *FEMS Microbiol Rev* **26**, 311–325.

Myers, C. R. & Myers, J. M. (1992). Localization of cytochromes to the outer membrane of anaerobically grown *Shewanella putrefaciens* MR-1. *J Bacteriol* **174**, 3429–3438.

Myers, C. R. & Myers, J. M. (1993). Ferric reductase is associated with the membranes of anaerobically grown *Shewanella putrefaciens* MR-1. *FEMS Microbiol Lett* **108**, 15–22.

Myers, C. R. & Myers, J. M. (1997). Cloning and sequence of *cymA*, a gene encoding a tetraheme cytochrome *c* required for reduction of iron(III), fumarate, and nitrate by *Shewanella putrefaciens* MR-1. *J Bacteriol* **179**, 1143–1152.

Myers, C. R. & Nealson, K. H. (1988). Bacterial manganese reduction and growth with manganese oxide as the sole electron acceptor. *Science* **240**, 1319–1321.

Myers, J. M. & Myers, C. R. (2000). Role of the tetraheme cytochrome CymA in anaerobic electron transport in cells of *Shewanella putrefaciens* MR-1 with normal levels of menaquinone. *J Bacteriol* **182**, 67–75.

Myers, J. M. & Myers, C. R. (2001). Role of outer membrane cytochromes OmcA and OmcB of *Shewanella putrefaciens* MR-1 in reduction of manganese dioxide. *Appl Environ Microbiol* **67**, 260–269.

Nealson, K. H. & Cox, B. L. (2002). Microbial metal-ion reduction and Mars: extraterrestrial expectations? *Curr Opin Microbiol* **5**, 296–300.

Nelson, G., Wilmott, N. & Guthrie, J. (2000). *Degradative Bacteria*. UK patent GB2316684B.

Nevin, K. P. & Lovley, D. R. (2002). Mechanisms for Fe(III) oxide reduction in sedimentary environments. *Geomicrobiol J* **19**, 141–159.

Newman, D. K. & Kolter, R. (2000). A role for excreted quinones in extracellular electron transfer. *Nature* **405**, 94–97.

Newman, D. K., Kennedy, E. K., Coates, J. D., Ahmann, D., Ellis, D. J., Lovley, D. R. & Morel, F. M. M. (1997). Dissimilatory arsenate and sulfate reduction in *Desulfotomaculum auripigmentum* sp. nov. *Arch Microbiol* **168**, 380–388.

Nickson, R., McArthur, J., Burgess, W., Ahmed, K. M., Ravenscroft, P. & Rahman, M. (1998). Arsenic poisoning of Bangladesh groundwater. *Nature* **395**, 338.

Oremland, R. S. & Capone, D. G. (1988). Use of "specific" inhibitors in biogeochemistry and microbial ecology. *Adv Microb Ecol* **10**, 285–383.

Oremland, R. S. & Stolz, J. F. (2000). Dissimilatory reduction of selenate and arsenate in nature. In *Environmental Microbe–Metal Interactions*, pp. 199–224. Edited by D. R. Lovley. Washington, DC: American Society for Microbiology.

Oremland, R. S. & Stolz, J. F. (2003). The ecology of arsenic. *Science* **300**, 939–944.

Oremland, R. S., Dowdle, P. R., Hoeft, S. & 7 other authors (2000). Bacterial dissimilatory reduction of arsenate and sulfate in meromictic Mono Lake, California. *Geochim Cosmochim Acta* **64**, 3073–3084.

Oremland, R. S., Stolz, J. F. & Hollibaugh, J. T. (2004). The microbial arsenic cycle in Mono Lake, California. *FEMS Microbiol Ecol* **48**, 15–27.

Ortiz-Bernad, I., Anderson, R. T., Vrionis, H. A. & Lovley, D. R. (2004a). Vanadium respiration by *Geobacter metallireducens*: novel strategy for in situ removal of vanadium from groundwater. *Appl Environ Microbiol* **70**, 3091–3095.

Ortiz-Bernad, I., Anderson, R. T., Vrionis, H. A. & Lovley, D. R. (2004b). Resistance of solid-phase U(VI) to microbial reduction during in situ bioremediation of uranium-contaminated groundwater. *Appl Environ Microbiol* **70**, 7558–7560.

Park, C. H., Keyhan, M., Wielinga, B., Fendorf, S. & Matin, A. (2000). Purification to homogeneity and characterization of a novel *Pseudomonas putida* chromate reductase. *Appl Environ Microbiol* **66**, 1788–1795.

Payne, R. B., Gentry, D. M., Rapp-Giles, B. J., Casalot, L. & Wall, J. D. (2002). Uranium reduction by *Desulfovibrio desulfuricans* strain G20 and a cytochrome c_3 mutant. *Appl Environ Microbiol* **68**, 3129–3132.

Peretrukhin, V. F., Khizhnyak, N. N., Lyalikova, N. N. & German, K. E. (1996). Biosorption of technetium-99 and some actinides by bottom sediments of Lake Belsso Kosino of the Moscow region. *Radiochemistry* **38**, 440–443.

Reguera, G., McCarthy, K. D., Mehta, T., Nicoll, J. S., Tuominen, M. T. & Lovley, D. R. (2005). Extracellular electron transfer via microbial nanowires. *Nature* **435**, 1098–1101.

Renshaw, J. C., Butchins, L. J. C., Livens, F. R., May, I., Charnock, J. M. & Lloyd, J. R. (2005). Bioreduction of uranium: environmental implications of a pentavalent intermediate. *Environ Sci Technol* (in press).

Roberts, J. L. (1947). Reduction of ferric hydroxide by strains of *Bacillus polymyxa*. *Soil Sci* **63**, 135–140.

Röling, W. F. M., van Breukelen, B. M., Braster, M., Lin, B. & van Verseveld, H. W. (2001). Relationships between microbial community structure and hydrochemistry in a landfill leachate-polluted aquifer. *Appl Environ Microbiol* **67**, 4619–4629.

Rooney-Varga, J. N., Anderson, R. T., Fraga, J. L., Ringelberg, D. & Lovley, D. R. (1999). Microbial communities associated with anaerobic benzene degradation in a petroleum-contaminated aquifer. *Appl Environ Microbiol* **65**, 3056–3063.

Ruggiero, C. E., Boukhalfa, H., Forsythe, J. H., Lack, J. G., Hersman, L. E. & Neu, M. P. (2005). Actinide and metal toxicity to prospective bioremediation bacteria. *Environ Microbiol* **7**, 88–97.

Rusin, P. A., Quintana, L., Brainard, J. R., Strietelmeier, B. A., Tait, C. D., Ekberg, S. A., Palmer, P. D., Newton, T. W. & Clark, D. L. (1994). Solubilization of plutonium hydrous oxide by iron-reducing bacteria. *Environ Sci Technol* **28**, 1686–1690.

Santini, J. M., Sly, L. I., Schnagl, R. D. & Macy, J. M. (2000). A new chemolithoautotrophic arsenite-oxidizing bacterium isolated from a gold mine: phylogenetic, physiological, and preliminary biochemical studies. *Appl Environ Microbiol* **66**, 92–97.

Schröder, I., Rech, S., Krafft, T. & Macy, J. M. (1997). Purification and characterization of the selenate reductase from *Thauera selenatis*. *J Biol Chem* **272**, 23765–23768.

Seeliger, S., Cord-Ruwisch, R. & Schink, B. (1998). A periplasmic and extracellular c-type cytochrome of *Geobacter sulfurreducens* acts as a ferric iron reductase and as an electron carrier to other acceptors or to partner bacteria. *J Bacteriol* **180**, 3686–3691.

Shen, H. & Wang, Y.-T. (1993). Characterization of enzymatic reduction of hexavalent chromium by *Escherichia coli* ATCC 33456. *Appl Environ Microbiol* **59**, 3771–3777.

Silver, S. (2003). Silver resistance. *FEMS Microbiol Rev* **27**, 341–353.

Smedley, P. L. & Kinniburgh, D. G. (2002). A review of the source, behaviour and distribution of arsenic in natural waters. *Appl Geochem* **17**, 517–568.

Smillie, R. H., Hunter, K. & Loutit, M. (1981). Reduction of chromium(VI) by bacterially produced hydrogen sulphide in a marine environment. *Water Res* **15**, 1351–1354.

Snoeyenbos-West, O. L., Nevin, K. P., Anderson, R. T. & Lovley, D. R. (2000). Enrichment of *Geobacter* species in response to stimulation of Fe(III) reduction in sandy aquifer sediments. *Microb Ecol* **39**, 153–167.

Stein, L. Y., La Duc, M. T., Grundl, T. J. & Nealson, K. H. (2001). Bacterial and archaeal populations associated with freshwater ferromanganous micronodules and sediments. *Environ Microbiol* **3**, 10–18.

Steinberg, N. A. & Oremland, R. S. (1990). Dissimilatory selenate reduction potentials in a diversity of sediment types. *Appl Environ Microbiol* **56**, 3550–3557.

Stolz, J. F., Gugliuzza, T., Switzer Blum, J., Oremland, R. S. & Murillo, F. M. (1997). Differential cytochrome content and reductase activity in *Geospirillum barnesii* strain SeS3. *Arch Microbiol* **167**, 1–5.

Suzuki, T., Miyata, N., Horitsu, H., Kawai, K., Takamizawa, K., Tai, Y. & Okazaki, M. (1992). NAD(P)H-dependent chromium (VI) reductase of *Pseudomonas ambigua* G-1: a Cr(V) intermediate is formed during the reduction of Cr(VI) to Cr(III). *J Bacteriol* **174**, 5340–5345.

Switzer Blum, J., Burns Bindi, A., Buzzelli, J., Stolz, J. F. & Oremland, R. S. (1998). *Bacillus arsenicoselenatis*, sp. nov., and *Bacillus selenitireducens* sp. nov.: two halo-alkaliphiles from Mono Lake, California, that respire oxyanions of selenium and arsenic. *Arch Microbiol* **171**, 19–30.

Taylor, D. E. (1999). Bacterial tellurite resistance. *Trends Microbiol* **7**, 111–115.

Tebo, B. M. & Obraztsova, A. Ya. (1998). Sulfate-reducing bacterium grows with Cr(VI), U(VI), Mn(IV), and Fe(III) as electron acceptors. *FEMS Microbiol Lett* **162**, 193–198.

Thamdrup, B. (2000). Bacterial manganese and iron reduction in aquatic sediments. *Adv Microb Ecol* **16**, 41–84.

Tomei, F. A., Barton, L. L., Lemanski, C. L. & Zocco, T. G. (1992). Reduction of selenate and selenite to elemental selenium by *Wolinella succinogenes*. *Can J Microbiol* **38**, 1328–1333.

Tomei, F. A., Barton, L. L., Lemanski, C. L., Zocco, T. G., Fink, N. H. & Sillerud, L. O. (1995). Transformation of selenate and selenite to elemental selenium by *Desulfovibrio desulfuricans*. *J Ind Microbiol* **14**, 329–336.

Tucker, M. D., Barton, L. L. & Thomson, B. M. (1997). Reduction and immobilization of molybdenum by *Desulfovibrio desulfuricans*. *J Environ Qual* **26**, 1146–1152.

Tucker, M. D., Barton, L. L. & Thomson, B. M. (1998). Reduction of Cr, Mo, Se and U by *Desulfovibrio desulfuricans* immobilized in polyacrylamide gels. *J Ind Microbiol Biotechnol* **20**, 13–19.

Vargas, M., Kashefi, K., Blunt-Harris, E. L. & Lovley, D. R. (1998). Microbiological evidence for Fe(III) reduction on early Earth. *Nature* **395**, 65–67.

Wade, R., Jr & DiChristina, T. J. (2000). Isolation of U(VI) reduction-deficient mutants of *Shewanella putrefaciens*. *FEMS Microbiol Lett* **184**, 143–148.

Wang, Y.-T. (2000). Microbial reduction of chromate. In *Environmental Microbe–Metal Interactions*, pp. 225–235. Edited by D. R. Lovley. Washington, DC: American Society for Microbiology.

Wang, P.-C., Mori, T., Komori, K., Sasatsu, M., Toda, K. & Ohtake, H. (1989). Isolation and characterization of an *Enterobacter cloacae* strain that reduces hexavalent chromium under anaerobic conditions. *Appl Environ Microbiol* **55**, 1665–1669.

Wildung, R. E., McFadden, K. M. & Garland, T. R. (1979). Technetium sources and behaviour in the environment. *J Environ Qual* **8**, 156–161.

Wildung, R. E., Li, S. W., Murray, C. J., Krupka, K. M., Xie, Y., Hess, N. J. & Roden, E. E. (2004). Technetium reduction in sediments of a shallow aquifer exhibiting dissimilatory iron reduction potential. *FEMS Microbiol Ecol* **49**, 151–162.

Woolfolk, C. A. & Whiteley, H. R. (1962). Reduction of inorganic compounds with molecular hydrogen by *Micrococcus lactilyticus*. I. Stoichiometry with compounds of arsenic, selenium, tellurium, transition and other elements. *J Bacteriol* **84**, 647–658.

Yong, P., Farr, J. P. G., Harris, I. R. & Macaskie, L. E. (2002a). Palladium recovery by immobilized cells of *Desulfovibrio desulfuricans* using hydrogen as the electron donor in a novel electrobioreactor. *Biotechnol Lett* **24**, 205–212.

Yong, P., Rowson, N. A., Farr, J. P. G., Harris, I. R. & Macaskie, L. E. (2002b). Bioaccumulation of palladium by *Desulfovibrio desulfuricans*. *J Chem Technol Biotechnol* **77**, 593–601.

Yurkova, N. A. & Lyalikova, N. N. (1991). New vanadate-reducing facultative chemolithotrophic bacteria. *Mikrobiologiia* **59**, 672–677 (in Russian).

New insights into the physiology and regulation of the anaerobic oxidation of methane

Martin Krüger[1,2] and Tina Treude[2]

[1]Federal Institute for Geosciences and Resources (BGR), Stilleweg 2, D-30655 Hannover, Germany

[2]Max-Planck-Institute for Marine Microbiology, Celsiusstrasse 1, D-28359 Bremen, Germany

INTRODUCTION

Methane is an important link within the global carbon cycle and has become a major focus for scientific investigations over the last decades, especially since the discovery of large deposits of methane hydrates in continental margins. The majority of recent methane production is biogenic, i.e. produced either by thermogenic transformation of organic material or by methanogenesis as the final step in fermentation of organic matter carried out by methanogenic archaea in anoxic habitats (Reeburgh, 1996). There are also abiotic sources of methane, e.g. at mid-oceanic ridges, where serpentinization takes place. In marine environments, the bulk of the methane is produced in shelf and upper continental-margin sediments, which receive large amounts of organic matter from deposition (Reeburgh, 1996). As methane builds up, it migrates upwards and may reach the sediment surface. Here, its ebullition and oxidation can lead to the formation of complex geostructures, such as pockmarks or carbonate chimneys and platforms, as well as large-scale topographies, such as mud volcanoes and carbonate mounds (Ivanov et al., 1991; Milkov, 2000). In most of the deeper continental margin and the abyssal plain sediments, methane production is low, as only 1–5 % of the surface primary production reaches the bathyal and abyssal seabed, due to degradation processes in the water column (Gage & Tyler, 1996).

Despite the high rates of methane production in shallow marine regions, the contribution of the oceans, with around 3–5 % to the global methane emission into the atmosphere, is extremely low compared with major methane sources, such as wetlands, rice fields or ruminants (IPCC, 1994; Reeburgh, 1996). The reason for this is the

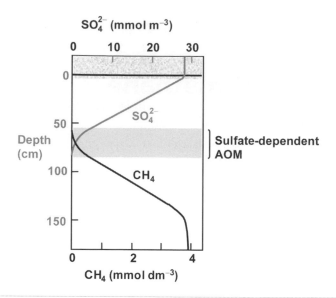

Fig. 1. Scheme depicting typical profiles of sulfate and methane concentrations in porewater of an anoxic marine sediment, indicating a distinct zone of AOM activity.

presence of sulfate in marine systems, which is an electron acceptor used for microbial sulfate reduction (Jørgensen & Fenchel, 1974; Jørgensen, 1982). As long as sulfate is present in the sediment, methanogenesis is shifted to deeper sediment layers due to substrate competition between sulfate-reducing bacteria and methanogens (Zehnder, 1988). When migrating towards the sediment surface, methane is consumed by two microbial pathways: anaerobic oxidation of methane (AOM) and aerobic oxidation of methane. Reeburgh (1996) proposed that, by these processes, up to 80 % of the methane produced in the marine sediments is oxidized prior to its release into the hydrosphere.

Unlike freshwater and terrestrial habitats, aerobic oxidation of methane is less important in marine ecosystems, as oxygen availability in the sediments is low compared to sulfate, the electron acceptor for AOM (D'Hondt *et al.*, 2002). However, information about rates and micro-organisms involved in aerobic oxidation of methane in the ocean is scarce. So far, only a few studies have been published (e.g. Lidstrom, 1988; Holmes *et al.*, 1996; Valentine *et al.*, 2001; Krüger *et al.*, 2005), of which the majority deal mainly with pelagic processes.

In marine sediments, the bulk of upward-migrating methane is consumed during AOM using sulfate instead of oxygen as electron acceptor (equation 1; Zehnder & Brock, 1980; Hoehler *et al.*, 1994; Reeburgh, 1996; Valentine & Reeburgh, 2000; Hinrichs &

Table 1. Characteristics of habitats investigated for AOM (see also Fig. 1).

Characteristic	Hydrate Ridge	Eckernförde Bay	Chilean continental margin
Type of seep	Gas seep/gas hydrates	Gassy coastal sediment	Diffusive system
Methane transport	Advective	Diffusive/advective	Diffusive
Sulfate penetration depth (cm)	10	30	150–350
Sediment depth at which methane is completely consumed (cm)	Methane reaches hydrosphere	0–5	110–360
Thickness of AOM zone (cm)	0–10	20–25	4–40
Measured areal AOM rate (mmol m^{-2} day^{-1})	56–100	0·4–1·5*	0·2–5·7
Calculated methane flux (mmol m^{-2} day^{-1})	56–200†	0·6–1·3‡	0·07–0·13
Retention of methane in sediment (%)	50–100	100 (except gas) bubbles	100

*Rates determined by Treude et al. (2005b).

†Calculated from rate measurements (Treude et al., 2003) and methane effluxes (Torres et al., 2002).

‡Estimated from the methane profiles of Abegg & Anderson (1997).

Boetius, 2002). The high availability of sulfate enables methane consumption far below the sediment–water interface (Fig. 1; Table 1).

$$CH_4 + SO_4^{2-} \rightarrow HCO_3^- + H_2O + HS^-$$
(equation 1)

It has been proposed that the only major escape route for methane into the hydrosphere or atmosphere is ebullition of free gas or gas hydrates floating from the seabed. Wherever sulfate is available, AOM communities control the release of dissolved methane from the sediment (Treude, 2003). Without this mechanism, the contribution of the world's oceans to global methane emission would approximately equal the amount of methane emanating from ruminants, one of the biggest sources of today's methane emission into the atmosphere (Reeburgh, 1996).

After a historical introduction, this review focuses on recent findings concerning the micro-organisms involved in AOM and its environmental regulation and new information concerning the mechanism of this still enigmatic process.

HISTORY OF AOM

The first published investigation of AOM was by Martens & Berner (1974), who described conspicuous methane and sulfate profiles in organic-rich sediments (Fig. 1), in which methane did not accumulate before sulfate depletion from porewater. From the decrease of methane concentrations in the sulfate-reducing zone, they concluded that methane must be consumed with sulfate instead of oxygen as terminal electron acceptor.

Since then, more biogeochemical evidence has been published confirming that the process of methane consumption in marine sediments takes place at the base of the sulfate zone, linked directly or indirectly to the activity of sulfate-reducing bacteria (Reeburgh, 1980; Iversen & Jørgensen, 1985; Hoehler et al., 1994; Hansen et al., 1998; Niewöhner et al., 1998; Borowski et al., 2000). These findings were based on depth profiles of methane and sulfate/sulfide (Fig. 1), $^{13}C : ^{12}C$ ratios in carbon dioxide and methane in sediment profiles, and labelling studies with sediment samples (e. g. Iversen & Blackburn, 1981; Alperin & Reeburgh, 1985; Iversen & Jørgensen, 1985; Hoehler et al., 1994). However, Zehnder & Brock (1979, 1980) were the first to demonstrate methane oxidation under anoxic conditions by methanogenic archaea and hypothesized a coupled two-step mechanism of AOM. They postulated that methane is first activated by methanogenic archaea working in reverse, leading to the formation of intermediates, e.g. acetate or methanol. In a second step, the intermediates are oxidized to CO_2 under concurrent sulfate reduction by other non-methanogenic members of the microbial community.

Since these pioneering studies, knowledge of AOM has increased substantially, involving biogeochemical, microbiological and molecular methods. Radiotracer measurements enabled the first direct quantification of AOM and concurrent sulfate reduction in anoxic marine sediments (Fig. 2; Reeburgh, 1976; Iversen & Blackburn, 1981; Devol, 1983). Iversen & Blackburn (1981) measured a 1:1 ratio of AOM and sulfate reduction in the sulfate–methane transition zone of Danish sediments, demonstrating the close coupling between these processes.

Hoehler et al. (1994) confirmed by thermodynamic modelling that a consortium of methanogenic archaea and sulfate-reducing bacteria could gain energy from AOM (Fig. 3). In situ and inhibitor studies (Hoehler et al., 1994), as well as laboratory experiments with growing methanogens converting [^{14}C]methane to $^{14}CO_2$ during methanogenesis (Harder, 1997; Zehnder & Brock, 1979, 1980), further stimulated the discussion about AOM being a reverse process of methanogenesis.

MICRO-ORGANISMS INVOLVED IN AOM

It has only been during the last 5–10 years that the identification of the micro-organisms responsible for AOM has been possible. It was advanced by investigations of lipid biomarkers in sediments from methane seeps. In the search for these organisms,

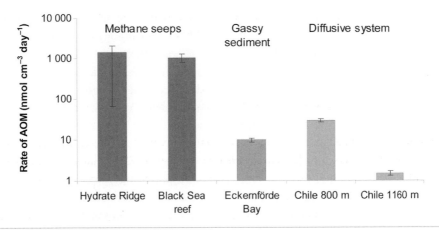

Fig. 2. Rates of AOM at methane hot spots in selected habitats differing in methane availability and flux rates (mean ± SEM, $n = 3$–18). Data are from Treude *et al.* (2003) (Hydrate Ridge), Michaelis *et al.* (2002) (Black Sea), Treude *et al.* (2005b) (Eckernförde Bay) and Treude *et al.* (2005a) (Chile).

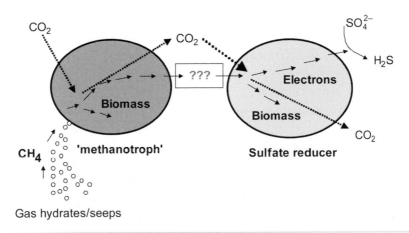

Fig. 3. Interactions between methanotrophic archaea and sulfate-reducing bacteria according to the hypothesis of Hoehler *et al.* (1994). Scheme by K. Nauhaus.

scientists found methanogen-associated lipids, named crocetane, archaeol and hydroxy-archaeol, in active methane seeps, revealing extremely light $\delta^{13}C$ values down to −110 ‰, giving evidence for an involvement of archaea in methane consumption (Elvert *et al.*, 1999; Hinrichs *et al.*, 1999; Pancost *et al.*, 2000; Thiel *et al.*, 2001; Schouten *et al.*, 2003). These archaeal lipids were also found in association with isotopically light bacterial lipids, commonly found in sulfate-reducing bacteria (Hinrichs *et al.*, 2000; Hinrichs & Boetius, 2002; Elvert *et al.*, 2003). Similar to archaeal lipids, this relative enrichment in ^{12}C indicated the incorporation of methane-derived carbon into bacterial cells.

Boetius *et al.* (2000) presented the first microscopic pictures of an AOM consortium visualized by fluorescence *in situ* hybridization (FISH), showing aggregates of archaeal cells surrounded by a shell of sulfate-reducing bacteria. The aggregates grow to a size of about 6–10 μm before they break apart into subaggregates, implicating the need to keep short distances between cells during substrate exchange. These consortia were discovered in surface sediment overlying methane hydrates at Hydrate Ridge, where they represented > 90 % of the microbial biomass.

After revealing the conspicuous morphology of the AOM consortium, further methods were used to obtain direct evidence for methanotrophy of the AOM consortium. A combination of FISH and secondary-ion mass spectrometry allowed the measurement of $\delta^{13}C$ profiles of the biomass of single aggregates (Orphan *et al.*, 2001a). A high depletion in ^{13}C, with values down to −96 and −62‰, was detected in archaeal and bacterial cells, respectively. These results confirmed the assimilation of isotopically light methane by the consortium.

Molecular studies showed that the anaerobic methanotrophs (ANME) were affiliated most closely with methanogenic archaea of the order *Methanosarcinales* (Hinrichs *et al.*, 1999; Orphan *et al.*, 2001b) and have frequently been associated with sulfate-reducing bacteria of the genera *Desulfosarcina* and *Desulfococcus* (Boetius *et al.*, 2000; Michaelis *et al.*, 2002; Knittel *et al.*, 2003). Today, three major groups of methane-oxidizing archaea have been identified: ANME-1, ANME-2 and ANME-3 (Hinrichs *et al.*, 1999; Boetius *et al.*, 2000; Orphan *et al.*, 2001b; Knittel *et al.*, 2005). ANME-2 and ANME-3 belong to the order *Methanosarcinales*. The ANME-2 group has recently been divided into three subgroups, ANME-2a to -2c (Knittel *et al.*, 2005), which seem to exhibit differences in environmental preferences and the structure of their aggregation with sulfate reducers. ANME-1 is distinct from, but related to, methanogenic archaea of the orders *Methanomicrobiales* and *Methanosarcinales*.

The bacterial diversity at methane seeps is high, especially within the δ-proteobacteria (Knittel *et al.*, 2003), including members of the genera *Desulfosarcina*, *Desulfo-rhopalus*, *Desulfocapsa* and *Desulfobulbus*. Comprehensive overviews on the diversity and phylogeny of archaea and sulfate reducers involved in AOM and associated with methane seeps have been published recently by Knittel *et al.* (2003, 2005), respectively.

HOT SPOTS FOR THE STUDY OF AOM

In general, AOM can be expected wherever methane and sulfate coexist in anoxic environments. One main factor determining the magnitude of AOM is the methane supply, because methane-turnover rates were found to increase with methane concentration and methane flux (Nauhaus *et al.*, 2002; Treude, 2003). Hot spots for AOM have been found in diverse habitats, characterized by a wide range of environmental characteristics. Hinrichs & Boetius (2002), Treude (2003) and Krüger *et al.* (2005) have reviewed AOM rates in marine sediments of different water depths, as well as methane seeps. These first compilations of AOM field measurements and modelling have suggested a direct coupling between methane supply and methane consumption in the habitat. At methane seeps of ancient reservoirs or gas-hydrate locations, AOM rates were found to be 10–100 times higher than those in non-seep regions. However, the data of environmental AOM rates are still fragmentary. Below, two methane-rich marine environments, which have so far played a major role during the investigation of AOM, are briefly introduced: they are located at Hydrate Ridge, off the coast of Oregon, USA (Boetius & Suess, 2004 and references therein) and on the north-western shelf of the Black Sea (Michaelis *et al.*, 2002).

Hydrate Ridge

At Hydrate Ridge, gas-hydrate deposits are located a few centimetres below the sediment surface in a water depth of 600–800 m, corresponding to the hydrate-stability zone (Suess *et al.*, 1999). These layers lead to very high methane fluxes (up to 200 mmol $m^{-2} day^{-1}$; Table 1), which fuel a diverse, seep-associated community (Sahling *et al.*, 2002; Treude *et al.*, 2003), including zones covered by thick mats of sulfide-oxidizing members of the genus *Beggiatoa* or inhabited by different dwelling clams, such as *Calyptogena* and *Acharax* species. In these surface sediments, aggregates of methanotrophic ANME-2 and sulfate-reducing *Desulfococcus*/*Desulfosarcina* cells dominate the microbial biomass (Boetius *et al.*, 2000). Their maximum abundance is located within the upper 10 cm below the sea floor, where methane and sulfate meet (Treude *et al.*, 2003). In this zone, some of the highest densities of ANME cells and methane-turnover rates known from marine environments have been found (Boetius *et al.*, 2000; Boetius & Suess, 2004). Elevated HCO_3^- concentrations caused by this high AOM activity result in an increase in alkalinity and support carbonate precipitation, forming large carbonate landscapes at Hydrate Ridge.

Black Sea

In the north-western Black Sea, hundreds of active gas seeps occur along the shelf edge west of the Crimea peninsula, at water depths between 35 and 800 m (Ivanov *et al.*, 1991). Within the anoxic zone, massive carbonate accumulations up to 4 m high and 1 m in diameter have been found associated with these seeps (Pimenov *et al.*, 1997; Thiel *et al.*, 2001; Lein *et al.*, 2002; Michaelis *et al.*, 2002; Blumenberg *et al.*, 2004). These build-ups are covered by up to 10 cm thick microbial methanotrophic mats. From holes in these structures, streams of gas bubbles emanate into the water column. Strong ^{13}C depletions indicate an incorporation of methane carbon into carbonates, bulk microbial biomass and specific lipids. The main matrix of the microbial mats consists of densely aggregated cells of methanotrophic ANME-1 and sulfate-reducing *Desulfococcus*/*Desulfosarcina*, but many other bacteria of unknown diversity and function co-occur in the mats (Michaelis *et al.*, 2002; Blumenberg *et al.*, 2004; Knittel *et al.*, 2005). The physiology of the methanotrophic mats has been studied in greater detail (Pimenov *et al.*, 1997; Michaelis *et al.*, 2002; Treude, 2003; Nauhaus *et al.*, 2005), as described below.

ENVIRONMENTAL REGULATION OF AOM

The most important questions regarding the functioning of AOM in the ocean concern its regulation and the growth and environmental adaptation of the communities mediating AOM. According to thermodynamic calculations, whether free energy is available from AOM depends on the environmental settings (Zehnder & Brock, 1980; Iversen & Blackburn, 1981). So far, only limited experimental data are available to investigate the effect of variable environmental factors on the efficiency of AOM (Valentine & Reeburgh, 2000; Nauhaus *et al.*, 2002, 2005). For example, the balance between the concentrations of sulfate and sulfide might be an important factor regulating microbial methane consumption (Treude, 2003; Treude *et al.*, 2003).

So far, no methanotrophic archaea are available for cultivation, probably because of their extremely slow growth rate (Girguis *et al.*, 2003). Hence, the investigation of their physiological capabilities and adaptations is only possible by *in vitro* studies with naturally enriched samples from the environment. At many sites studied extensively for AOM, including Hydrate Ridge, the Gulf of Mexico and the Black Sea, ANME-1 and ANME-2 have been found to co-occur. However, the dominance of either group varies; for example, in Hydrate Ridge samples, ANME-2 populations dominated the community (Boetius *et al.*, 2000) whereas, in the Black Sea mats, ANME-1 far outnumbered ANME-2 (Knittel *et al.*, 2005). These differences in the community composition might be due to differences in environmental parameters, such as temperature or the availability of methane and sulfate.

The stoichiometry of AOM, which has been estimated from porewater studies and rate measurements in the field (Iversen & Blackburn, 1981), has been confirmed by *in vitro* studies with environmental samples. Simultaneous measurements of methane oxidation and sulfide production in samples from Hydrate Ridge and the Black Sea have revealed a molar ratio of 1 : 1 between the two processes (Nauhaus *et al.*, 2002, 2005). Interestingly, the comparison of methane-driven sulfate reduction per cell revealed that ANME-2 communities (Hydrate Ridge) were up to 20 times more active than the ANME-1 communities in the microbial mats from the Black Sea. However, it is not known whether all methanotrophic cells within a cell aggregate or within a mat are equally active, as cells with no contact to the bacterial partner might be inactive if not situated within an optimal substrate-concentration range required for sufficient energy conservation (Sørensen *et al.*, 2001). Therefore, it remains an interesting question for future research, ideally with pure cultures, whether this difference in cell-specific activity between ANME-1 and -2 can be attributed to specific substrate kinetics or enzymic mechanisms of the ANME groups.

It can be assumed that the efficiency of AOM in mitigating methane emissions is influenced by environmental parameters, such as pH, temperature and methane and sulfate fluxes (Joye *et al.*, 2004). Methane availability *in situ* depends on the methane flux from subsurface reservoirs, as well as methane solubility, which is in turn influenced by hydrostatic pressure and temperature (Yamamoto *et al.*, 1976). Consequently, it is essential to investigate the response of the AOM organisms to changes in these parameters, to be able to estimate the effects of environmental or climatic changes on AOM efficiency. An increase of methane partial pressure from 0·1 to 1·1 MPa resulted in a fivefold increase of AOM rates in ANME-2-dominated samples (Nauhaus *et al.*, 2002) and a twofold increase in ANME-1-dominated samples (Nauhaus *et al.*, 2005). A similar stimulation was also detected in samples from shallow water depths (Krüger *et al.*, 2005), which generally do not encounter such high concentrations of methane. It is remarkable that it was also possible to induce AOM in formerly inactive sediments by increasing the methane availability (Girguis *et al.*, 2003; Krüger *et al.*, 2005).

The free gas ebullition observed at seeps such as the microbial reefs in the Black Sea and at Hydrate Ridge indicates methane saturation, with theoretical values of 2·3 MPa (40 mM) and 8 MPa (140 mM) for the Black Sea and Hydrate Ridge, respectively. Consequently, the rates observed *in vitro* at only 1·1 MPa must still represent substantial underestimations of rates occurring under *in situ* conditions. This inability in reaching environmental methane concentrations might also inhibit attempts to culture these organisms in the laboratory (see below).

The pH of the environment might also influence activities, as well as the composition of microbial communities. For example, Nauhaus *et al.* (2005) showed that ANME-2 from Hydrate Ridge sediments had a distinct maximum of AOM rates at pH 7·4 (pH 7–7·5), whilst the pH optimum was broader for the ANME-1 community from the Black Sea, ranging from pH 6·8 to 8·1. These small differences in environmental preferences might contribute to the development of either ANME-1- or -2-dominated communities.

Besides the pH, temperature is another important environmental factor influencing micro-organisms. Despite a difference of only 4 °C between *in situ* temperatures of habitats investigated in the Black Sea and at Hydrate Ridge, the ANME-2 community of Hydrate Ridge was more active at low temperatures (8–12 °C), whereas the ANME-1 community in the Black Sea was mesophilic, with highest AOM activities between 16 and 24 °C (Nauhaus *et al.*, 2005). These distinct temperature optima for AOM of environmental samples dominated by ANME-1 or ANME-2 may indicate a selective advantage for either population. Further temperature optima for AOM ranged from 4 °C for sediment from the Haakon Mosby Mud Volcano (−1·5 °C *in situ*) to 25 °C in the Baltic Sea (between 4 and 16 °C *in situ*) (Krüger *et al.*, 2005). So far, the different temperature optima of AOM reflected the different *in situ* temperatures of the habitat. Especially at shallow sites, seasonal temperature changes might also cause changes in AOM activity. Indeed, such seasonality of AOM activity has been reported from studies in Eckernförde Bay, Baltic Sea (Treude *et al.*, 2005b) and Cape Lookout Bight, USA (Hoehler *et al.*, 1994).

So far, reports on the discovery of AOM have been restricted to habitats with sufficient concentrations of sulfate available for the microbial partners of the methanotrophic archaea (Hinrichs & Boetius, 2002 and references therein), predominantly in marine habitats. However, the question is still pending whether, under specific environmental conditions, AOM might also proceed with other alternative electron acceptors. In a recent study with samples from Hydrate Ridge and the Black Sea, Nauhaus *et al.* (2005) observed no AOM activity without sulfate. Instead, both ANME-1 and -2 communities oxidized methane with similar rates at sulfate concentrations ranging from 10 to 100 mM. Other electron acceptors for AOM, such as nitrate, sulfur, ferric iron and manganese oxide, were also reduced in Hydrate Ridge sediments by the indigenous microbial population. However, this reduction was not coupled to AOM.

In summary, the ecological niches occupied more frequently by ANME-1 or ANME-2 seem to be defined mainly by temperature and the simultaneous availability of methane and sulfate. Nevertheless, the factor(s) leading to the dominance of either group remain to be identified.

MECHANISM OF AOM

The invesigation of mechanistic details of AOM is still hindered by the lack of pure cultures of anaerobic methanothophs. Only a few studies are available, which have been conducted with environmental samples naturally enriched in methanotrophic biomass (Hoehler *et al.*, 1994; Nauhaus *et al.*, 2002, 2005). An important question regarding the mechanism of AOM is whether the two reactions involved, i.e. methane oxidation and sulfate reduction, are indeed carried out by a consortium of methanotrophic archaea and associated bacteria (Fig. 3), as indicated by the striking structural features revealed by microscopic analysis (Boetius *et al.*, 2000; Orphan *et al.*, 2001b; Michaelis *et al.*,

2002), or whether the entire process is mediated by a single organism. The latter is indicated by repeated findings of ANME (mainly ANME-1) without contact to sulfate-reducing bacteria (Orphan *et al.*, 2001a; Michaelis *et al.*, 2002; Joye *et al.*, 2004; Treude *et al.*, 2005b).

Besides the microscopic evidence for a consortium of methanotrophic archaea and sulfate-reducing bacteria mediating AOM, further evidence was gained by inhibition experiments. In these experiments on environmental regulation of AOM, molybdate and bromoethanesulfonate (BES) as inhibitors for sulfate reduction and methanogenesis, respectively, have been used (Alperin & Reeburgh, 1985; Hoehler *et al.*, 1994; Hansen *et al.*, 1998; Nauhaus *et al.*, 2005). In both ANME-1- and -2-dominated samples, AOM was inhibited completely by BES. This inhibition was reversible and AOM activity was resumed after removal of BES. Molybdate inhibited AOM completely in ANME-2-dominated samples, but only partially in ANME-1-dominated samples (Nauhaus *et al.*, 2005), which was explained by strong adsorption of molybdate to extracellular polysaccharide. These results are in good agreement with previous studies, in which a partial to complete inhibition of AOM by these compounds was observed (Alperin & Reeburgh, 1985; Hoehler *et al.*, 1994). In conclusion, the application of a specific inhibitor, i.e. BES for methanogens and molybdate for sulfate-reducing bacteria, inhibited AOM. However, this inhibition is not absolute proof for syntrophy or a symbiotic association between ANME and sulfate-reducing bacteria. A complete inhibition would also be observed if methane oxidation and sulfate reduction were carried out by a single organism.

Nevertheless, the inhibition of AOM by a methanogen-specific inhibitor provides important evidence for the hypothesis that AOM is taking place via a reversal of methanogenesis using similar enzymes (Zehnder & Brock, 1980; Hoehler *et al.*, 1994). Recently, an abundant protein closely resembling methyl coenzyme M reductase, the terminal enzyme in methanogenesis (Thauer, 1998), was detected in ANME-1 cells from Black Sea samples (Krüger *et al.*, 2003). This protein represents a likely candidate for the initial step in AOM (Fig. 4). Furthermore, Hallam *et al.* (2003) found gene sequences of the same enzyme in sediments with AOM activity, which they assigned to ANME-2. In a subsequent paper, even the almost-complete methanogenic enzymic system was attributed to ANME, based on the analysis of a metagenomic library from an AOM site (Fig. 4) (Hallam *et al.*, 2004).

If, indeed, a syntrophic mechanism is necessary, the question arises as to what the exact mechanism of this process might be and which intermediates are exchanged within the consortium (Fig. 3). The nature of the cooperation between the archaeal and bacterial partners in AOM has not yet been elucidated (Sørensen *et al.*, 2001; Nauhaus *et al.*, 2002).

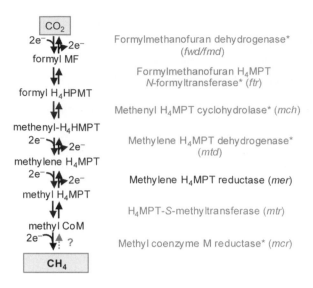

Fig. 4. Activities (marked with an asterisk) and genes for methanogenic enzymes found in methanotrophic archaea (ANME). Activities are after Krüger *et al.* (2003) and genes are after Hallam *et al.* (2004).

The addition of exogenous electron donors in the form of methanogenic substrates or other C_1–C_3 compounds, including acetate, formate, hydrogen and methanol, did not stimulate sulfate reduction in the absence of methane (Nauhaus *et al.*, 2002, 2005). In theory, if the sulfate-reducing bacteria are adapted to the substrate supplied by the methane-oxidizing partner, they should respond immediately to the addition of potential AOM intermediates, with higher activity.

As an alternative to hydrogen or carbon compounds, the possibility of electron transfer between the archaeal and sulfate-reducing partners has been discussed for AOM consortia (Sørensen *et al.*, 2001; Nauhaus *et al.*, 2002). Such an electron transfer to an external acceptor has been reported previously for different anaerobic reactions (Seeliger *et al.*, 1998; Schink & Stams, 2001). However, the addition of several compounds able to capture electrons – phenazines, AQDS (anthroquinone disulfonate) and humic acids – did not replace the function of the sulfate-reducing bacterium. Neither of the added compounds induced AOM (Nauhaus *et al.*, 2005). One explanation for this might be that, besides their important role in membrane-bound electron transport, phenazines might also have toxic effects on micro-organisms (Ingram & Blackwood, 1970; Abken *et al.*, 1998). Even though Straub & Schink (2003) showed that AQDS may serve as an electron shuttle for iron-reducing bacteria, the redox conditions in the incubations might change upon the addition of AQDS or humic acids, perhaps suppressing AOM (Hernandez & Newman, 2001). In summary, these experiments did not provide direct evidence for a methanogenic or sulfidogenic substrate or electrons as an intermediate in AOM. Consequently, despite the conspicuous aggregation of archaea and sulfate-reducing bacteria in the environment, it might be possible that AOM is carried out by one organism alone (see below).

GROWTH OF AOM MICRO-ORGANISMS IN THE LABORATORY

Pure cultures of anaerobic methanotrophic archaea have not yet been isolated. Consequently, physiological studies have only been possible on environmental samples naturally enriched in methanotrophic archaea and their sulfate-reducing partners (Hoehler *et al.*, 1994; Blumenberg *et al.*, 2004, 2005; Nauhaus *et al.*, 2002, 2005). However, sample availability and quality have limited the scope of these experiments. For major questions regarding the mechanism and regulation of AOM, it seems – despite substantial progress – inevitable to work with pure cultures. Nevertheless, recent studies on microbial communities with a limited diversity have shown that there is the possibility to work on these aspects of AOM by using metagenomic (Hallam *et al.*, 2004) or biochemical (Krüger *et al.*, 2003) approaches.

Recently, Girguis *et al.* (2003) developed a novel continuous-flow system to study AOM, which simulates the *in situ* conditions and could support the growth of anaerobic, methanotrophic archaea. The major limitation of this system was that it only operated under 0·1 MPa methane pressure and did not resemble *in situ* conditions on the sea floor. Consequently, thermodynamic calculations showed that the Gibbs free-energy yield for AOM at 1 atm methane pressure was low. In contrast, high *in situ* methane concentrations and a close physical association between AOM partners can produce energy yields sufficient to support biosynthesis (Sørensen *et al.*, 2001; Nauhaus *et al.*, 2002). Therefore, it would be advantageous to use high-pressure systems, as described by Nauhaus *et al.* (2002), for future research on the growth of AOM micro-organisms, thus allowing the application of elevated methane partial pressures.

The means by which methanotrophic archaea are capable of growing at atmospheric methane pressures are difficult to understand. High rates of AOM observed in the deep oceans require high dissolved-methane concentrations (Nauhaus *et al.*, 2002), which can only be sustained at high pressures. Nevertheless, there is evidence for AOM at atmospheric methane pressure in different shallow-water ecosystems (Iversen & Jørgensen, 1985; Martens *et al.*, 1999; Krüger *et al.*, 2005).

CONCLUSIONS

Microbially mediated AOM significantly influences biological and biogeochemical processes on local to global scales. The process reduces methane flux into the water column, stimulates subsurface microbial metabolism and also supports rich deep-sea chemosynthetic communities that derive energy from one of its by-products, hydrogen sulfide. However, advective systems such as methane seeps might still represent a significant source of methane emission from the ocean to the hydrosphere, as recent data have shown that AOM is only able to inhibit methane emission into the water column completely in diffusive systems (Table 1; Treude, 2003).

The range of datasets available today (Orphan *et al*., 2001b; Hinrichs & Boetius, 2002; Teske *et al*., 2002; Knittel *et al*., 2003; Treude *et al*., 2003; Boetius & Suess, 2004; Kallmeyer & Boetius, 2004) suggests that the co-occurrence of methane and sulfate is the major environmental factor defining the ecological niche occupied by AOM communities. So far, no environment is known in which only one of the ANME groups occurs. The significant dominance of either group points to the presence of defined environmental niches within AOM zones, which have not been distinguished so far. Further experimental studies of ANME-enriched samples from different environments with a range of habitat characteristics are needed to answer the question of niche selection.

Finally, it seems that, although a lot is known about AOM and the respective microbial consortia, there are still many questions to be answered. For example, the intermediate that is exchanged between archaea and sulfate-reducing bacteria is still unknown. Thermodynamic considerations show that the archaea alone might not gain free energy by this process, thus depending on cooperation with the sulfate-reducing bacteria. Despite recent progress in the fields of biochemistry and metagenomics, this and other questions will probably not be answered until pure cultures become available.

ACKNOWLEDGEMENTS

We especially thank Dr Gundula Eller for critical reading of the manuscript. This is publication no. GEOTECH-129 of the programme GEOTECHNOLOGIEN of the Bundesministerium für Bildung und Forschung and Deutsche Forschungsgemeinschaft. Within this programme, the study is part of projects MUMM-1, -2 and GHOSTDABS. Further support came from the Max Planck Society.

REFERENCES

Abegg, F. & Anderson, A. L. (1997). The acoustic turbid layer in muddy sediments of Eckernfoerde Bay, Western Baltic: methane concentration, saturation and bubble characteristics. *Mar Geol* **137**, 137–147.

Abken, H.-J., Tietze, M., Brodersen, J., Bäumer, S., Beifuss, U. & Deppenmeier, U. (1998). Isolation and characterization of methanophenazine and function of phenazines in membrane-bound electron transport of *Methanosarcina mazei* Gö1. *J Bacteriol* **180**, 2027–2032.

Alperin, M. J. & Reeburgh, W. S. (1985). Inhibition experiments on anaerobic methane oxidation. *Appl Environ Microbiol* **50**, 940–945.

Blumenberg, M., Seifert, R., Reitner, J., Pape, T. & Michaelis, W. (2004). Membrane lipid patterns typify distinct anaerobic methanotrophic consortia. *Proc Natl Acad Sci U S A* **101**, 11111–11116.

Blumenberg, M., Seifert, R., Nauhaus, K., Widdel, F., Pape, T. & Michaelis, W. (2005). In vitro study of lipid biosynthesis in an anaerobically methane oxidizing microbial mat. *Appl Environ Microbiol* (in press).

Boetius, A. & Suess, E. (2004). Hydrate Ridge: a natural laboratory for the study of microbial life fueled by methane from near-surface gas hydrates. *Chem Geol* **205**, 291–310.

Boetius, A., Ravenschlag, K., Schubert, C. J. & 7 other authors (2000). A marine microbial consortium apparently mediating anaerobic oxidation of methane. *Nature* **407**, 623–626.

Borowski, W. S., Hoehler, T. M., Alperin, M. J., Rodriguez, N. M. & Paull, C. K. (2000). Significance of anaerobic methane oxidation in methane-rich sediments overlying the Blake Ridge gas hydrates. In *Proceedings of the Ocean Drilling Program, Scientific Results*, vol. 164, pp. 87–99. Edited by C. K. Paull, R. Matsumoto, P. J. Wallace & W. P. Dillon. College Station, TX: Texas A&M University.

Devol, A. H. (1983). Methane oxidation rates in the anaerobic sediments of Saanich Inlet. *Limnol Oceanogr* **28**, 738–742.

D'Hondt, S., Rutherford, S. & Spivack, A. J. (2002). Metabolic activity of subsurface life in deep-sea sediments. *Science* **295**, 2067–2070.

Elvert, M., Suess, E. & Whiticar, M. J. (1999). Anaerobic methane oxidation associated with marine gas hydrates: superlight C-isotopes from saturated and unsaturated C_{20} and C_{25} irregular isoprenoids. *Naturwissenschaften* **86**, 295–300.

Elvert, M., Boetius, A., Knittel, K. & Jørgensen, B. B. (2003). Characterization of specific membrane fatty acids as chemotaxonomic markers for sulfate-reducing bacteria involved in anaerobic oxidation of methane. *Geomicrobiol J* **20**, 403–419.

Gage, J. D. & Tyler, P. A. (1996). *Deep-Sea Biology: a Natural History of Organisms at the Deep-Sea Floor*. Cambridge: Cambridge University Press.

Girguis, P. R., Orphan, V. J., Hallam, S. J. & DeLong, E. F. (2003). Growth and methane oxidation rates of anaerobic methanotrophic archaea in a continuous-flow bioreactor. *Appl Environ Microbiol* **69**, 5472–5482.

Hallam, S. J., Girguis, P. R., Preston, C. M., Richardson, P. M. & DeLong, E. F. (2003). Identification of methyl coenzyme M reductase A (*mcrA*) genes associated with methane-oxidizing archaea. *Appl Environ Microbiol* **69**, 5483–5491.

Hallam, S. J., Putnam, N., Preston, C. M., Detter, J. C., Rokhsar, D., Richardson, P. M. & DeLong, E. F. (2004). Reverse methanogenesis: testing the hypothesis with environmental genomics. *Science* **305**, 1457–1462.

Hansen, L. B., Finster, K., Fossing, H. & Iversen, N. (1998). Anaerobic methane oxidation in sulfate depleted sediments: effects of sulfate and molybdate additions. *Aquat Microb Ecol* **14**, 195–204.

Harder, J. (1997). Anaerobic methane oxidation by bacteria employing ^{14}C-methane uncontaminated with ^{14}C-carbon monoxide. *Mar Geol* **137**, 13–23.

Hernandez, M. E. & Newman, D. K. (2001). Extracellular electron transfer. *Cell Mol Life Sci* **58**, 1562–1571.

Hinrichs, K.-U. & Boetius, A. (2002). The anaerobic oxidation of methane: new insights in microbial ecology and biogeochemistry. In *Ocean Margin Systems*, pp. 457–477. Edited by G. Wefer, D. Billett, D. Hebbeln, B.B. Jørgensen, M. Schlüter & T. van Weering. Berlin: Springer.

Hinrichs, K.-U., Hayes, J. M., Sylva, S. P., Brewer, P. G. & DeLong, E. F. (1999). Methane-consuming archaebacteria in marine sediments. *Nature* **398**, 802–805.

Hinrichs, K.-U., Summons, R. E., Orphan, V., Sylva, S. P. & Hayes, J. M. (2000). Molecular and isotopic analysis of anaerobic methane-oxidizing communities in marine sediments. *Organic Geochem* **31**, 1685–1701.

Hoehler, T. M., Alperin, M. J., Albert, D. B. & Martens, C. S. (1994). Field and laboratory studies of methane oxidation in an anoxic marine sediment: evidence for a methanogen-sulfate reducer consortium. *Global Biogeochem Cycles* **8**, 451–463.

Holmes, A. J., Owens, N. J. P. & Murrell, J. C. (1996). Molecular analysis of enrichment cultures of marine methane oxidising bacteria. *J Exp Mar Biol Ecol* **203**, 27–28.

Ingram, J. M. & Blackwood, A. C. (1970). Microbial production of phenazines. *Adv Appl Microbiol* **13**, 267–282.

IPCC (1994). *Climate Change 1994: Radiative Forcing of Climate Change and an Evaluation of the IPCC IS92 Emission Scenarios*. Cambridge: Cambridge University Press.

Ivanov, M. V., Polikarpov, G. G., Lein, A. Y. & 7 other authors (1991). Biogeochemistry of the carbon cycle in the zone of Black Sea methane seeps. *Dokl Akad Nauk SSSR* **320**, 1235–1240.

Iversen, N. & Blackburn, T. H. (1981). Seasonal rates of methane oxidation in anoxic marine sediments. *Appl Environ Microbiol* **41**, 1295–1300.

Iversen, N. & Jørgensen, B. B. (1985). Anaerobic methane oxidation rates at the sulfate-methane transition in marine sediments from Kattegat and Skagerrak (Denmark). *Limnol Oceanogr* **30**, 944–955.

Jørgensen, B. B. (1982). Mineralization of organic matter in the sea bed – the role of sulfate reduction. *Nature* **296**, 643–645.

Jørgensen, B. B. & Fenchel, T. (1974). The sulfur cycle of a marine sediment model system. *Mar Biol* **24**, 189–210.

Joye, S. B., Boetius, A., Orcutt, B. N., Montoya, J. P., Schulz, H. N., Erickson, M. J. & Lugo, S. K. (2004). The anaerobic oxidation of methane and sulfate reduction in sediments from Gulf of Mexico cold seeps. *Chem Geol* **205**, 219–238.

Kallmeyer, J. & Boetius, A. (2004). Effects of temperature and pressure on sulfate reduction and anaerobic oxidation of methane in hydrothermal sediments of Guaymas Basin. *Appl Environ Microbiol* **70**, 1231–1233.

Knittel, K., Boetius, A., Lemke, A., Eilers, H., Lochte, K., Pfannkuche, O., Linke, P. & Amann, R. (2003). Activity, distribution, and diversity of sulfate reducers and other bacteria in sediments above gas hydrates (Cascadia Margin, Oregon). *Geomicrobiol J* **20**, 269–294.

Knittel, K.,, Lösekann, T., Boetius, A., Kort, R. & Amann, R. (2005). Diversity and distribution of methanotrophic archaea at cold seeps. *Appl Environ Microbiol* **71**, 467–479.

Krüger, M., Meyerdierks, A., Glöckner, F. O. & 8 other authors (2003). A conspicuous nickel protein in microbial mats that oxidize methane anaerobically. *Nature* **426**, 878–881.

Krüger, M., Treude, T., Wolters, H., Nauhaus, K. & Boetius, A. (2005). Microbial methane turnover in different marine habitats. *Palaeogeogr Palaeoclimatol Palaeoecol* (in press).

Lein, A. Y., Ivanov, M. V., Pimenov, N. V. & Gulin, M. B. (2002). Geochemical characteristics of the carbonate constructions formed during microbial oxidation of methane under anaerobic conditions. *Mikrobiologiia* **71**, 89–102 (in Russian).

Lidstrom, M. E. (1988). Isolation and characterization of marine methanotrophs. *Antonie van Leeuwenhoek* **54**, 189–199.

Martens, C. S. & Berner, R. A. (1974). Methane production in the interstitial waters of sulfate-depleted marine sediments. *Science* **185**, 1167–1169.

Martens, C. S., Albert, D. B. & Alperin, M. J. (1999). Stable isotope tracing of anaerobic methane oxidation in the gassy sediments of Eckernförde Bay, German Baltic Sea. *Am J Sci* **299**, 589–610.

Michaelis, W., Seifert, R., Nauhaus, K. & 14 other authors (2002). Microbial reefs in the Black Sea fueled by anaerobic oxidation of methane. *Science* **297**, 1013–1015.

Milkov, A. V. (2000). Worldwide distribution of submarine mud volcanoes and associated gas hydrates. *Mar Geol* **167**, 29–42.

Nauhaus, K., Boetius, A., Krüger, M. & Widdel, F. (2002). *In vitro* demonstration of anaerobic oxidation of methane coupled to sulfate reduction in sediment from a marine gas hydrate area. *Environ Microbiol* **4**, 296–305.

Nauhaus, K., Treude, T., Boetius, A. & Krüger, M. (2005). Environmental regulation of the anaerobic oxidation of methane: a comparison of ANME-I and ANME-II communities. *Environ Microbiol* **7**, 98–106.

Niewöhner, C., Hensen, C., Kasten, S., Zabel, M. & Schulz, H. D. (1998). Deep sulfate reduction completely mediated by anaerobic methane oxidation in sediments of the upwelling area off Namibia. *Geochim Cosmochim Acta* **62**, 455–464.

Orphan, V. J., House, C. H., Hinrichs, K.-U., McKeegan, K. D. & DeLong, E. F. (2001a). Methane-consuming archaea revealed by directly coupled isotopic and phylogenetic analysis. *Science* **293**, 484–487.

Orphan, V. J., Hinrichs, K.-U., Ussler, W., III, Paull, C. K., Taylor, L. T., Sylva, S. P., Hayes, J. M. & DeLong, E. F. (2001b). Comparative analysis of methane-oxidizing archaea and sulfate-reducing bacteria in anoxic marine sediments. *Appl Environ Microbiol* **67**, 1922–1934.

Pancost, R. D., Sinninghe Damsté, J. S., de Lint, S., van der Maarel, M. J. E. C. & Gottschal, J. C. (2000). Biomarker evidence for widespread anaerobic methane oxidation in Mediterranean sediments by a consortium of methanogenic archaea and bacteria. *Appl Environ Microbiol* **66**, 1126–1132.

Pimenov, N. V., Rusanov, I. I., Poglazova, M. N., Mityushina, L. L., Sorokin, D. Y., Khmelenina, V. N. & Trotsenko, Yu. A. (1997). Bacterial mats on coral-like structures at methane seeps in the Black Sea. *Mikrobiologiia* **66**, 354–360 (in Russian).

Reeburgh, W. S. (1976). Methane consumption in Cariaco Trench waters and sediments. *Earth Planet Sci Lett* **28**, 337–344.

Reeburgh, W. S. (1980). Anaerobic methane oxidation: rate depth distributions in Skan Bay sediments. *Earth Planet Sci Lett* **47**, 345–352.

Reeburgh, W. S. (1996). "Soft spots" in the global methane budget. In *Eighth International Symposium on Microbial Growth on C_1 Compounds*, pp. 334–342. Edited by M. E. Lidstrom & F. R. Tabita. Dordrecht: Kluwer.

Sahling, H., Rickert, D., Lee, R. W., Linke, P. & Suess, E. (2002). Macrofaunal community structure and sulfide flux at gas hydrate deposits from the Cascadia convergent margin, NE Pacific. *Mar Ecol Prog Ser* **231**, 121–138.

Schink, B. & Stams, A. J. M. (2001). Syntrophism among prokaryotes. In *The Prokaryotes: an Evolving Electronic Resource for the Microbiological Community*. Edited by M. Dworkin, S. Falkow, E. Rosenberg, K.-H. Schleifer & E. Stackebrandt. Heidelberg: Springer (http://www.prokaryotes.com).

Schouten, S., Wakeham, S. G., Hopmans, E. C. & Sinninghe Damsté, J. S. (2003). Biogeochemical evidence that thermophilic archaea mediate the anaerobic oxidation of methane. *Appl Environ Microbiol* **69**, 1680–1686.

Seeliger, S., Cord-Ruwisch, R. & Schink, B. (1998). A periplasmic and extracellular c-type cytochrome of *Geobacter sulfurreducens* acts as a ferric iron reductase and as an electron carrier to other acceptors or to partner bacteria. *J Bacteriol* **180**, 3686–3691.

Sørensen, K. B., Finster, K. & Ramsing, N. B. (2001). Thermodynamic and kinetic requirements in anaerobic methane oxidizing consortia exclude hydrogen, acetate and methanol as possible electron shuttles. *Microb Ecol* **42**, 1–10.

Straub, K. L. & Schink, B. (2003). Evaluation of electron-shuttling compounds in microbial ferric iron reduction. *FEMS Microbiol Lett* **220**, 229–233.

Suess, E., Torres, M. E., Bohrmann, G. & 8 other authors (1999). Gas hydrate destabilization: enhanced dewatering, benthic material turnover and large methane plumes at the Cascadia convergent margin. *Earth Planet Sci Lett* **170**, 1–15.

Teske, A., Hinrichs, K.-U., Edgcomb, V., Gomez, A. de V., Kysela, D., Sylva, S. P., Sogin, M. L. & Jannasch, H. W. (2002). Microbial diversity of hydrothermal sediments in the Guaymas Basin: evidence for anaerobic methanotrophic communities. *Appl Environ Microbiol* **68**, 1994–2007.

Thauer, R. K. (1998). Biochemistry of methanogenesis: a tribute to Marjory Stephenson. *Microbiology* **144**, 2377–2406.

Thiel, V., Peckmann, J., Richnow, H. H., Luth, U., Reitner, J. & Michaelis, W. (2001). Molecular signals for anaerobic methane oxidation in Black Sea seep carbonates and a microbial mat. *Mar Chem* **73**, 97–112.

Torres, M. E., McManus, J., Hammond, D. E., de Angelis, M. A., Heeschen, K. U., Colbert, S. L., Tryon, M. D., Brown, K. M. & Suess, E. (2002). Fluid and chemical fluxes in and out of sediments hosting methane hydrate deposits on Hydrate Ridge, OR, I: Hydrological provinces. *Earth Planet Sci Lett* **201**, 525–540.

Treude, T. (2003). *Anaerobic oxidation of methane in marine sediments*. PhD thesis, Universität Bremen, Bremen, Germany (http://elib.suub.uni-bremen.de/publications/dissertations/E-Diss845_treude.pdf).

Treude, T., Boetius, A., Knittel, K., Wallmann, K. & Jørgensen, B. B. (2003). Anaerobic oxidation of methane above gas hydrates at Hydrate Ridge, NE Pacific Ocean. *Mar Ecol Prog Ser* **264**, 1–14.

Treude, T., Niggemann, J., Kallmeyer, J., Wintersteller, P., Schubert, C. J., Boetius, A. & Jørgensen, B. B. (2005a). Anaerobic oxidation of methane and sulfate reduction along the Chilean continental margin. *Geochim Cosmochim Acta* **69**, 2767–2779.

Treude, T., Krüger, M., Boetius, A. & Jørgensen, B. B. (2005b). Environmental control on anaerobic oxidation of methane in the gassy sediments of Eckernförde Bay (German Baltic). *Limnol Oceanogr* (in press).

Valentine, D. L. & Reeburgh, W. S. (2000). New perspectives on anaerobic methane oxidation. *Environ Microbiol* **2**, 477–484.

Valentine, D. L., Blanton, D. C., Reeburgh, W. S. & Kastner, M. (2001). Water column methane oxidation adjacent to an area of active hydrate dissociation, Eel River Basin. *Geochim Cosmochim Acta* **65**, 2633–2640.

Yamamoto, S., Alcauskas, J. B. & Crozier, T. E. (1976). Solubility of methane in distilled water and seawater. *J Chem Eng Data* **21**, 78–80.

Zehnder, A. J. B. (1988). *Biology of Anaerobic Microorganisms*. New York: Wiley.

Zehnder, A. J. B. & Brock, T. D. (1979). Methane formation and methane oxidation by methanogenic bacteria. *J Bacteriol* **137**, 420–432.

Zehnder, A. J. B. & Brock, T. D. (1980). Anaerobic methane oxidation: occurrence and ecology. *Appl Environ Microbiol* **39**, 194–204.

Biogeochemical roles of fungi in marine and estuarine habitats

Nicholas Clipson,[1] Eleanor Landy[2] and Marinus Otte[3]

[1,3]Department of Industrial Microbiology[1] and Department of Botany[3], University College Dublin, Belfield, Dublin 4, Ireland

[2]School of Biomedical and Molecular Sciences, University of Surrey, Guildford GU2 7XH, UK

INTRODUCTION

A fungal component of the marine biota was only recognized as recently as 1944 (Barghoorn & Linder, 1944), and it was not until the 1960s that studies commenced to assess the extent and diversity of fungi in marine systems. Since this time, considerable effort has been exerted to uncover marine fungal diversity, with high decadal discovery indices in the 1970s and 80s (Hawksworth, 1991), resulting in around 1000 fungal species known today from marine environments. Nevertheless, it is hardly surprising that, with the extent of marine environments globally, we probably have a very incomplete view of fungal diversity, together with their frequency and function in these ecosystems. The objective of this review is to assess the extent of our present knowledge and to highlight future directions to further elucidate their biology and ecology.

THE NATURE OF MARINE ENVIRONMENTS

Marine ecosystems are globally extensive, and account for around 70 % of global surface area. They can be defined generally as aquatic systems influenced by substantial concentrations of salts, particularly sodium chloride, from existing oceanic systems. Seas and oceans divide between regions bordering and influenced by terrestrial regions and the open ocean, which is strongly zoned through the water column. These broad boundaries are illustrated in Fig. 1, which also details linkages between marine compartments. At the sea surface, light is an important factor driving primary productivity in the photic zone, to a depth of around 200 m (Dring, 1992), with carbon and other nutrients moving downwards through the water column to stimulate food webs within benthic zones, finally adding to sediments.

SGM symposium 65: Micro-organisms and Earth systems – advances in geomicrobiology.
Editors G. M. Gadd, K. T. Semple & H. M. Lappin-Scott. Cambridge University Press. ISBN 0 521 86222 1 ©SGM 2005

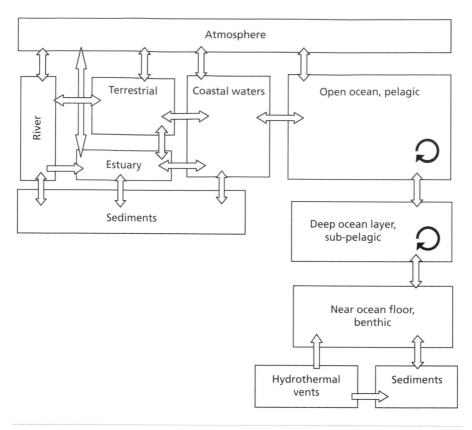

Fig. 1. Interactions between and within marine ecosystems. While most interactions are bi-directional, interactions between rivers and estuaries and between hydrothermal vents and their surroundings are unidirectional. Circular arrows within oceanic compartments indicate cycles that are largely confined within those compartments (after Wangersky, 1980; Salomons & Förstner, 1984).

Salinity concentrations within open ocean systems are characteristically very stable, at around 500 mM NaCl, although these are lower in enclosed seas such as the Baltic and Black Seas, where NaCl concentrations are diluted by the effects of freshwater inputs from riverine discharge and ice melt and may be in the region of 50–200 mM. Typical open ocean sea water ion concentrations are given in Table 1. The open ocean water column changes with depth, with productivity declining dramatically as light and temperature decrease. The ocean floor is an area of low primary productivity and low biomass density (Nybakken, 1997) except for deep-sea hydrothermal vent and cold seep communities, which are characterized by temperatures in the range 8–16 °C (as opposed to the normal 2 °C of most of the deep sea) and high pressures exerted by the overlying water column (Ballard, 1977). These environments are generally found at the edges of tectonic plates and are typically very rich in reduced sulfur compounds. Primary productivity is driven by prokaryal chemolithoautotrophs, which drive

Table 1. Concentrations of major and selected minor constituents of sea water

Data were compiled from Martin (1970) and Dring (1992).

Constituent	Concentration		Constituent	Concentration
	(g kg⁻¹)	**(mM)**		**(µg l⁻¹)**
Chloride	19·35	548	Silicon	0–4900
Sodium	10·76	470	Nitrogen (combined)	0–560
Sulfate	2·71	28	Phosphorus	0–90
Magnesium	1·29	54	Aluminium	0–10
Calcium	0·413	10	Iron	0·1–62
Potassium	0·387	10	Zinc	1–48
Bicarbonate	0·142	2	Iodine	48–80
Bromide	0·067	0·8	Copper	0·5–27
Strontium	0·008	1·5	Manganese	0·2–8·6
Boron	0·0045	0·4	Cobalt	0·005–4·1
Fluoride	0·001	0·07		

relatively rich food webs with high productivity (Nybakken, 1997). Cold seeps are often associated with hypersaline brines or hydrocarbon seeps.

Near-shore marine ecosystems are much more variable and dynamic, particularly being affected by terrestrial and meteorological influences. The most extensive and productive of these environments are saline wetlands, which characteristically form as salt marshes in polar and temperate zones dominated by low-growing herbs and mangrove swamps in tropical regions dominated by salt-tolerant trees (Packham & Willis, 1997). Both systems are affected by diurnal tidal inundation (much less so in enclosed seas, where diurnal variation in tidal height is very small), which makes soil and sediment environmental parameters highly variable, particularly for soil salinity, oxygenation and drainage (Jefferies *et al.*, 1979; Armstrong *et al.*, 1985). Terrestrial influences can also be important, resulting in inputs of nutrients and freshwater as a result of riverine deposition and run-off. Other salt-affected coastal environments might include dune and shingle systems and rocky foreshores, which are influenced as spray communities from sea water (Packham & Willis, 1997). Adjacent to coastal regions, and where continental shelves are shallow, coastal sea communities form, including coral reefs, which are found in both tropical and cold seas. Marine communities can also be associated with former oceanic activity, generally where seas have become separated from present oceanic areas. These can either form highly saline inland seas such as the Dead Sea, where saline concentrations can reach 4–5 M (Buchalo *et al.*, 1998), or evaporate completely to form salt pans and salt deserts. Many of these highly saline environments are dominated by ions other than sodium and chloride, including magnesium, calcium and potassium, with the counter ions carbonate, bicarbonate and sulfate (Flowers *et al.*, 1986). Perhaps 10 % of global land surface area is dominated by these hypersaline environments.

For a detailed view of the elemental contents of the major environments associated with marine systems, see Hutzinger (1980) and Salomons & Förstner (1984). In general terms, open oceans are more dilute than areas above continental shelves, adjacent to land masses or estuarine. Terrestrial and near-shore systems are usually well connected via rivers, leading to rapid elemental transfer from terrestrial to marine ecosystems. Transport to the open ocean and its underlying layers is much slower. In both cases, many elements are immobilized in sediments, which act as elemental sinks after deposition. These can be remobilized back slowly to the water column, particularly for phosphorus and metals, where transfer to the atmosphere is negligible (Elmsley, 1980; Salomons & Förstner, 1984). Some elements have a large atmospheric component, such as carbon, nitrogen and sulfur, which are much more rapidly cycled in these systems. This may also include some metals and metalloids such as mercury, arsenic and selenium, which can be methylated and lost to the atmosphere.

Although open ocean systems are more stable than marine systems at terrestrial boundaries, important dynamics do occur through the water column, largely reflecting oceanic mixing by circulatory systems. Through the oceans there are local and depth-delineated changes in factors such as temperature, salinity, oxygen and nutrient levels. Examples are water-column transect studies across the Atlantic Ocean (Lavin *et al.*, 2003) and near-shore studies across the Strait of Georgia (British Columbia) (Masson, 2002). In the open ocean North Atlantic, Lavin *et al.* (2003) found a mean salinity of around 3·5 %, varying between 3·45 and 3·7 % between 6000 m depth and the surface. Mean surface oxygen was 200 μmol kg^{-1}, declining to around 150 μmol kg^{-1} at 1000 m but increasing to around 250 μmol kg^{-1} at 6000 m. Silicates, nitrates and phosphates were all depleted in the surface oceanic layers, increasing to average levels at around 1000 m and below. Oxygen and nutrient levels are also heavily influenced by the circulation of currents through these systems, leading to localized variations.

MARINE FUNGAL DIVERSITY IN MARINE ECOSYSTEMS

The presence of fungi in marine ecosystems was first recognized by Barghoorn & Linder (1944), who proposed the term 'marine fungi' for fungal species living in these systems. Although there has been considerable effort since the early 1960s to elucidate marine fungal biodiversity, present estimates probably grossly underrepresent its true extent. This is for a number of reasons. Firstly, the oceans are exceptionally extensive, both on the basis of their global surface area and as they are compartmentalized through the water column. Many of these environments are inaccessible and a challenge to meaningful sampling. Secondly, total fungal diversity is itself grossly underestimated, with many geographical areas undersampled, many species lying hidden in interactions with other organisms such as insects or plants and the problems of isolating and culturing most species. Overall, Hawksworth (2001) estimates that around 80 000 species have been identified and classified in a background of possibly in excess of 1·5 million species. It is within this background that marine fungal diversity is probably heavily underestimated. Generally, two approaches are used to assess marine fungal diversity. Firstly, direct observation of fungal structures on natural substrates or baited substrates such as wood blocks or litter bags can be made. Secondly, isolations can be made onto traditional agar-based cultural media. Certainly, in the latter case, culturability of environmental isolates of fungi probably falls into the region of that for bacteria, in the region of 1–5 %. There have been very few studies (e.g. Buchan *et al.*, 2003) to test whether conventionally made diversity estimates in a given location correspond to molecular estimates. In the case of fungal diversity during the degradation of the salt-marsh halophyte *Spartina alterniflora*, Buchan *et al.* (2003) found *Phaeosphaeria spartinicola*, *Phaeosphaeria halima* and a *Mycosphaerella* sp. to be dominant using terminal restriction fragment length polymorphism.

There are also problems in defining what a marine fungus is. Kohlmeyer & Kohlmeyer (1979) advanced three ecological groupings of fungi occurring in marine environments, including obligate marine fungi, facultative marine fungi and terrestrial fungi (see also Kohlmeyer & Volkmann-Kohlmeyer, 2003). Obligate marine fungi were defined by their inability to complete their life cycle outside the marine environment, whereas facultative species could complete life cycles in either marine or terrestrial systems but were of terrestrial origin. Terrestrial fungi do not have the ability to complete life cycles in marine environments and, although commonly isolated, might just exist in a dormant state until more favourable conditions prevail for spore germination (Kohlmeyer & Kohlmeyer, 1979). On a physiological basis, it is unclear what confers the marine lifestyle of fungal organisms. The principal adaptation required by marine fungi is probably salt-tolerance. Many terrestrial species, for example most aspergilli or penicillia (e.g. Beever & Laracy, 1986), are exceptionally salt-tolerant both vegetatively and for sporulation, but are not generally found to be active in marine ecosystems. Some terrestrial *Aspergillus* species have been found in marine ecosystems, such as *Aspergillus sydowii* and *Aspergillus fumigatus*, which cause disease in the Caribbean sea fan *Gorgonia ventalina*. Disease-causing populations are probably maintained by continual influx of vegetative material from neighbouring terrestrial environments (Smith *et al.*, 1996; Geiser *et al.*, 1998). The question concerning the nature of the marine fungal lifestyle remains. It might relate to the substrate conditions existing in marine ecosystems, sensitivity at a certain lifestyle stage (for example sporulation or spore germination), the oligotrophic nature of the marine environment, the selection of appropriate media for physiological studies or the ability to attach to substrate.

Biogeographically, and in terms of marine ecosystem-type, fungi appear to be ubiquitous. Nevertheless, some caution must be expressed with this view, as there are substantial areas of the globe which have not been surveyed for this group. Marine fungi appear to be found worldwide, including polar regions (Pugh & Jones, 1986; Grasso *et al.*, 1997), although probably the highest levels of diversity so far have been found in the tropics associated with coastal ecosystems, especially mangroves. This may reflect that most effort has been exerted within these regions. Additionally, coastal and near-shore habitats are much better studied than oceanic habitats because of the expense of boat time. Benthic and deep-sea fungal communities have only been sampled on a very few occasions (e.g. Alongi, 1987; Nagahama *et al.*, 2001). Another important issue leading to diversity underestimates is the issue of culturability. Traditionally, marine fungal diversity has been assessed almost exclusively by culture-based isolation approaches or direct observation of natural substrates.

The most comprehensive assessment of known marine fungal diversity to date was made by Jones & Mitchell (1996), taking into account those that had been described and estimating species waiting for full description and publication, including thraustochytrids and lower fungi. They advanced a total of around 1400 species. Another approach to assessing marine fungal diversity has been to compile checklists of species occurring either on particular substrates or in known areas. Schmit & Shearer (2003) compiled a checklist from available literature of mangrove-associated fungi, recording a total of 625 species, with both marine and terrestrial species recorded. Advances in producing checklists of fungal species illustrate how knowledge of their diversity has been extended. Kohlmeyer (1969) could list 76 mangrove-fungi species, Hyde & Jones (1988) listed 90 species, with Jones & Alias (1997) extending this to 268 species. These latter two studies speculated that mangrove-fungal diversity is most extensive in South-East Asia, reflecting greater substrate diversity, although this probably also reflects greater numbers of isolation studies.

Two studies have aimed to produce checklists on an area basis. Gonzalez *et al.* (2001) have produced a checklist of higher marine fungi of Mexico, listing 62 species. They believed that this number significantly underestimates the true diversity of Mexican marine fungi. In a literature search of marine fungi identified within the seas and coastal regions of the European Union (EU), Clipson *et al.* (2001) found 318 species described. Interestingly, all three checklists, together with that of Jones & Mitchell (1996), found that marine fungi were predominantly either ascomycetes (particularly the Halosphaeriales) or mitosporic fungi, with details given in Table 2. Many of the

Table 2. Marine fungal diversity, based upon a global estimate and literature searches for the European Union area, Mexico and on mangrove substrates

Data were taken from Jones & Mitchell (1996) (global), Clipson *et al.* (2001) (EU marine), Gonzalez *et al.* (2001) (Mexico marine) and Schmit & Shearer (2003) (mangrove). Species awaiting description and publication are indicated in parentheses.

Fungal group	Global	EU marine	Mexico marine	Mangrove
Lower fungi	100 (+32)	12		14
Thraustochytrids	40			
Ascomycetes		214	47	278
Hemiascomycetes	50			
Euascomycetes	305 (+143)			
Deuteromycotina	79 (+200)	84	14	277
Basidiomycotina	7 (+7)	10	1	30
Trichomycetes	23			
Lichens	18 (+410)			

genera found have common characteristics (reproduction in aquatic habitats; thin-walled, unitunicate, deliquescing asci; central pseudoparenchyma in immature asci; hyaline, bicelled ascospores with polar or equatorial appendages) which Jones (1995) considers as adaptations to aquatic habitats. The large number of mitosporic fungi found is problematic because little is known about their relationship with their sexual stages. Currently, the EU list is being updated, with 338 marine species being recognized by 2005 (E. Landy, unpublished).

Schmit & Shearer (2003) also explored substrate preferences of mangrove fungi, with marine ascomycetes being intertidal or submerged species and mitosporic species being associated with sediments and possibly of terrestrial origin. Most marine ascomycetes possess appendages or gelatinous sheaths to aid attachment, making them particularly adapted to the forces operating in intertidal zones. Also, there is a further problem in that marine fungi are defined only imprecisely, with single mangrove trees possessing populations of both marine and terrestrial mangrove fungi. At what position on a single tree, and exposed to what outside environmental conditions, is a fungal species marine? At best, this must represent a rather arbitrary division. As Schmit & Shearer (2003) point out, we need to know much more about the evolution and ecology (related to their physiology and genetics) to understand the determinants of distribution of mangrove fungi. The same is true for marine fungal distribution in all other marine habitats. To date, only coastal habitats have been explored in any detail, leaving fungal diversity in the vast bulk of the open oceans and their depths largely undetermined. It is known that yeasts are frequent in open seas (e.g. Van Uden & Fell, 1968). In a study based in the Atlantic south of Portugal, Gadanho *et al.* (2003) identified 31 yeast taxa using microsatellite-primed PCR. Yeast cell densities typically decline with increased depth and distance from land (Gadanho *et al.*, 2003), presumably reflecting substrate availability and favourable environmental conditions. Yeasts are also known in the deep sea, with Nagahama *et al.* (2002) reporting the presence of red yeasts from abyssal zones. There are also some reports of fungi from hypersaline habitats. From the Dead Sea, Kritzman (1973) reported an osmophilic yeast, and Buchalo *et al.* (1998) made the first reports of filamentous fungi, including species of *Gymnascella*, *Ulocladium* and *Penicillium*, for this hypersaline habitat. A number of species of melanized yeast-like fungi have been reported from hypersaline salterns (Gunde-Cimermann *et al.*, 2000).

FUNGAL ADAPTATION TO MARINE ENVIRONMENTS

That a specialized mycoflora has become representative of marine ecosystems would suggest that specialized adaptations are necessary for the marine fungal habit. Clearly, adaptation to a combination of factors such as salinity, redox potential, pressure, substrate availability and quality and temperature, throughout the fungal life cycle, determines fungal diversity and activity in individual marine ecosystems.

The best-studied fungal response to such factors is that to salinity, which is ubiquitous in marine ecosystems. Much of our understanding of the mechanisms by which marine fungi adapt to salt comes from the physiological studies on the marine hyphomycete *Dendryphiella salina* by Jennings and co-workers in the 1980s and 1990s and from subsequent genetic studies on this and other non-marine fungi (e.g. Clement *et al.*, 1999; and see Bohnert *et al.*, 2001; Hooley *et al.*, 2003). Integrating information between these marine species and other osmotolerant but non-marine species has led to a model of fungal cellular adaptation to salinity. Central to osmotic adaptation by fungi is maintenance of osmotic gradients across the hyphal membrane to maintain inwardly directed water fluxes. Water (=osmotic) potential of sea water is around -2.4 MPa, meaning that cellular water potentials have to be maintained more negative. Fungal cells have rigid cell walls and possess turgor pressure generated as the difference between cellular water potential and cellular osmotic potential (generated by osmotically active solutes), turgor pressure being an important component of apical growth and expansion processes in fungi (Money, 1997). To generate hyphal water potentials, fungal cells have to maintain substantial concentrations of intracellular solutes. The fungal cell has two potential sources of osmotically active solutes, ions taken up from the external environment and organic solutes either synthesized from metabolism or taken up externally if available. Both these strategies are potentially problematic to fungal growth; many ions, particularly sodium and chloride, are toxic to cellular metabolic processes, and organic solutes for osmotic purposes can divert carbon from growth. *In vitro* studies have shown that many enzymes involved in fungal primary metabolism are inhibited substantially by sodium chloride levels well below (approx.

100 mM) those of sea water (Paton & Jennings, 1988), although this may be altered *in vivo* by either upregulation of protein systems or by effects of the counter-anion environment (Nilsson & Adler, 1990; Hooley *et al.*, 2003).

In *Dendryphiella salina*, a balance of these two approaches has been found to generate hyphal osmotic potential (Clipson & Jennings, 1992). X-ray microanalysis studies of hyphal cells growing at sea-water concentrations (500 mM NaCl) indicated cytoplasmic concentrations of around 170 mM sodium (Clipson *et al.*, 1990). This is probably still compatible with cytosolic enzyme activity, albeit with some inhibition of more sensitive enzymes. There was also no evidence of preferential accumulation of ions in vacuoles, which are much less involved in metabolism than is the cytoplasm (Clipson *et al.*, 1990). Many algal and plant halophiles have large vacuoles which are used as ion stores for osmotic adjustment, in conjunction with low cytoplasmic ion levels (Flowers *et al.*, 1986). Although *Dendryphiella salina* does not have large vacuoles, vacuolar localization of ions may be important in those fungal species that do.

In *Dendryphiella salina*, the balance of osmotic potential is generated through compatible solutes, either accumulated or synthesized. These solutes are 'compatible' in that they do not interfere with cellular metabolism. In fungi, the principal compatible solutes are polyols, including glycerol, mannitol, arabinitol and erythritol (Blomberg & Adler, 1993). In most fungi, it is mannitol which is the primary osmoresponsive polyol, with others generally increasing as cultures age. For species such as *Dendryphiella salina* and the osmophilic yeasts *Debaryomyces hansenii* and *Zygosaccharomyces rouxii*, where polyol concentrations have been measured under saline conditions *in vitro*, polyols generated between 36 and 75 % of cellular osmotic potential (Hooley *et al.*, 2003). Other compatible solutes may also be important in marine fungi, including proline and trehalose (Jennings & Burke, 1990; Davis *et al.*, 2000).

Clearly, these osmotic mechanisms are under genetic control, although precise details are not understood at present due to the difficulty of genetic manipulation in marine fungal species. The genetics of osmotolerance in yeast is quite well understood and has been reviewed recently by Yale & Bohnert (2001) and Hooley *et al.* (2003). *Saccharomyces cerevisiae* is not a good model for halophilic fungi as it is relatively salt-sensitive, with the halophilic yeast *Debaryomyces hansenii* being more representative. A number of genes involved in stress responses by the marine yeast *Debaryomyces hansenii* have been characterized, including the cloning of the genes for superoxide dismutase (Hernandez-Saavedro & Romero-Geraldo, 2001), a homologue of the high-osmolarity glycerol response gene *HOG1* encoding a MAP kinase (Bansal & Mondal, 2000) and the identification of several ion-transport-related genes including *DhENA1* and *DhENA2* for sodium ATPases (Almagro *et al.*, 2001). Nevertheless, there is still a very incomplete view of the genetics of salt tolerance in halophilic fungi in general and filamentous marine fungi in particular.

So far, an overview of fungal salt tolerance has been presented. In most marine environments, particularly the open ocean, salt concentrations do not fluctuate appreciably or rapidly, and adaptation is probably a continuous and gradual process presenting little problem to fungal organisms. In coastal regions, the influence of tidal effects, evapotranspiration and freshwater inputs makes salt concentrations much more unstable (Armstrong *et al.*, 1985). Although the mechanisms are not known, presumably fungal species living in these environments are adapted to adjust osmotically rather quickly in response to such fluctuations. Such an ability might differentiate these species from terrestrial fungi. Another important issue is tolerance to marine environmental factors through the full fungal life cycle, particularly spore germination and early hyphal development, and sporulation processes. These have not been examined to any extent, although it is well known that many non-marine osmophilic fungi, particularly aspergilli and penicillia, are tolerant through these stages and certainly in the laboratory can complete life cycles on highly saline media (Hooley *et al.*, 2003).

Marine fungi, dependent upon the habitat they occupy, may have to be adapted to other adverse environmental factors. On salt marshes, sediments and soils tend to be anoxic, with highly negative redox potentials. Most fungal species are aerobic (Carlile *et al.*, 2001) and salt-marsh soils probably do not form an important habitat for fungi. Although fungal species can be cultured from marine soils, these are probably transient species only present as spores and not vegetationally active.

Open oceans tend to be relatively oligotrophic except where ocean circulatory systems cause upwelling of nutrients, for example the Benguela System off south-west Africa. This may lead to planktonic blooms in the photic zone, giving localized regions with higher potential substrate concentrations for degrading organisms such as fungi and bacteria. Within the water column, it might have been thought that low oxygen contents might be inhibitory to fungal activity, although recent oceanological studies would indicate that deep-sea water oxygen concentrations are probably close to saturated (Lavin *et al.*, 2003). Lorenz & Molitoris (1997) tested the effects of simulated deep-sea conditions on a number of marine yeasts, finding that many grew under pressures equivalent to a depth of 4000 m.

MARINE BIOGEOCHEMICAL PROCESSES

Our view of the role of fungi in biogeochemical processes remains rather limited, perhaps with the exception of some coastal ecosystems. This mirrors diversity estimates, particularly with regard to fungal distribution in the open ocean away from coastal regions. In terms of substrate availability, there are radical differences between the marine–terrestrial interface, perhaps one of the most productive ecosystems on Earth (Newell, 1996), and the open ocean, which is relatively oligotrophic. Fungi are heterotrophs and their abundance is likely to reflect availability of carbon-based substrates. Knowledge concerning the fungal contribution to marine cycling processes also mirrors research effort and ease of study. Coastal environments are generally close to research centres and require little specialist equipment to sample. In contrast, studies performed at sea require expensive boat time, detailed planning and, particularly for studies in the benthic and abyssal zones, sophisticated sampling equipment. Nevertheless, this has been carried out widely and successfully for marine bacteria (e.g. Eilers *et al.*, 2000).

Biogeochemical cycling

Biogeochemical cycling refers to all processes and pathways involved in the cycling and turnover of elements. This includes not only chemical transformations, but also physical processes such as adsorption/desorption and biological transfer within (compartmentation) and between organisms. The best studied and most relevant elements important in marine biocycling are carbon, nitrogen, phosphorus and sulfur, for which overviews are given in Fig. 2. For all four elements, the fluxes between and within the compartments are very small compared with the sizes of the terrestrial and oceanic pools. The fluxes of carbon and nitrogen between the oceans and the atmosphere are bi-directional and of equal size, while the fluxes of phosphorus and sulfur show a net deposition. The oceanic pools of nitrogen and phosphorus are of the same order of magnitude as the terrestrial pools. In contrast, the oceanic pools for carbon and sulfur are several orders of magnitude larger than the terrestrial pools. While data available for carbon indicated that there is a distinct difference in the sizes of the carbon pools between the surface layer and deep ocean, no such data are available for the other elements. From a global perspective, all compartments are connected, but this does not apply at the organismal (fungal) level (Hutzinger, 1980), with little connectedness between open ocean near-surface ecosystems and the ocean floor. Within the marine environment, a number of more or less distinct systems can be identified that each have their own biogeochemical cycles (see Fig. 1). The focus of the ensuing sections will be on biogeochemical cycling of individual elements, particularly carbon, nitrogen, phosphorus and sulfur. Although these will be considered separately for clarity, it must be noted that they are intrinsically linked within biogeochemical cycling processes, all being basic constituents of organic matter.

Carbon, nitrogen and phosphorus. The quality and availability of organic compounds based upon carbon skeletons is central to fungal distribution and function within marine ecosystems. Fungi develop a range of responses to organic matter within

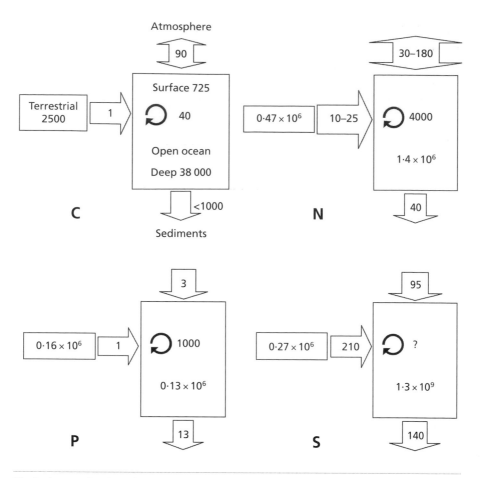

Fig. 2. Fluxes and cycling of carbon, nitrogen, phosphorus and sulfur through the oceans. Pools (boxes) are expressed in 10^{15} g for carbon and 10^{12} g for nitrogen, phosphorus and sulfur; fluxes (arrows) are in 10^{15} g year^{-1} for carbon and 10^{12} g year^{-1} for nitrogen, phosphorus and sulfur. Circular arrows within oceanic compartments indicate cycling through biota, but no such data were available for sulfur (after Pierrou, 1976; Richey, 1983; Rosswall, 1983; Freney *et al.*, 1983).

ecosystems, dependent upon the condition of organic matter, and are capable of deriving nutrition saprotrophically from non-living organic matter or either biotrophically (symbiosis) or pathogenically from other living organisms. Certainly, we know most about how fungi derive nutrition saprotrophically; nevertheless, there are many examples of fish, invertebrate and plant pathogens in both open ocean and coastal systems (see Porter, 1986, for overview) and examples of symbionts, particularly as mycorrhizas and endophytes of marine plants (Cooke *et al.*, 1993; Cornick *et al.*, 2005). Availability of hosts will be important in determining the distribution of such fungal interactions, although this has not been studied to any great extent.

Table 3. Production of exoenzymes for the breakdown of substrate polymers demonstrated in isolates of filamentous marine fungi

Data were taken from Torzilli (1982), Schaumann *et al.* (1986), Rohrmann *et al.* (1992) and Bucher *et al.* (2004).

Marine fungal species	Exoenzyme produced
Lulworthia spp.	Carrageenase
Halocyphina villosa, *Lulworthia* spp.	Laminarinase
Halocyphina villosa	Gelatinase
Halocyphina villosa, *Lulworthia* spp., *Corollospora maritima*, *Digitatispora marina*, *Cirrenalia pygmea*, *Varicosporina ramulosa*	Caseinase
Lulworthia spp.	Alginase
Pleospora pelagica, *Pleospora vagans*, *Phaeosphaeria typharum*	Pectinase
Halocyphina villosa, *Pleospora pelagica*, *Pleospora vagans*, *Phaeosphaeria typharum*	Xylanase
Lulworthia spp.	Agarase
Lulworthia spp.	Lipase
Lulworthia spp.	Amylase
Lulworthia spp.	Chitinase
Corollospora maritima, *Cirrenalia pygmea*, *Varicosporina ramulosa*, *Pleospora pelagica*, *Pleospora vagans*, *Phaeosphaeria typharum*	Cellulase complex
Lulworthia spp.	Laccase
Lulworthia spp.	Tyrosinase
Ascocratera manglicola, *Astrosphaeriella striatispora*, *Cryptovalsa halosarceicola*, *Linocarpon bipolaris*, *Rhizophila marina*	Ligninase

Central to fungal nutrition is the production and extracellular secretion of degradatory enzymes, making fungi particularly effective in the breakdown of complex polymeric compounds. A number of studies has examined the range of exoenzymes produced by marine fungal saprotrophs *in vitro* (Torzilli, 1982; Torzilli & Andrykovitch, 1986; Molitoris & Schaumann, 1986; Schaumann *et al.*, 1986; Lorenz & Molitoris, 1992; Rohrmann *et al.*, 1992), which are summarized in Table 3.

Although there is little quantitative information except for salt-marsh systems, fungal biomass will reflect substrate availability. This differs markedly between substrate-rich coastal ecosystems and the generally oligotrophic open ocean. In the open ocean, fungal communities are largely yeast-dominated, as opposed to comprising filamentous species (Sieburth, 1979). In a study in the Pacific, Van Uden & Fell (1968) found viable yeast cell numbers varying between 13 and 274 cells l^{-1}. In general, yeast cell densities decline with increased depth and distance from land (Gadanho *et al.*, 2003), presumably reflecting substrate availability and favourable environmental conditions.

Substrate differences do occur in the open ocean, particularly where the upwelling of colder waters creates regions of high productivity. Presumably, the activity of recycling organisms such as fungi will be higher in these regions. Nevertheless, we know very little about the role of yeasts in biogeochemical cycling in such systems.

Coastal zones are highly productive and rich in vegetation, which provides complex polymeric material for recycling by fungi and other degraders. Fungi are highly effective degraders of lignocellulosic material and may be central in commencing mixed degradative processes for this material (see Newell, 1996). Many fungal isolates from marine systems have been classified as white rot (lignin-degrading) organisms (Molitoris & Schaumann, 1986; Bucher *et al.*, 2004), particularly where isolates have been made from woody material. Overall, it is clear that fungi of coastal zones are able to produce a very extensive battery of exoenzymes, making them important components of recycling activity.

Nitrogen is essential to fungal growth and its availability is central to fungal activity in all environments, including the marine environment (Sguros & Simms, 1963; Jennings, 1989; Feeney *et al.*, 1992; Edwards *et al.*, 1998; Ahammed & Prema, 2002). In general, fungi can utilize both inorganic and organic forms of nitrogen, depending upon species and environment (Carlile *et al.*, 2001). Little is known about how marine fungi derive nitrogen from their environments or contribute to nitrogen cycling, although both generalist and specialist utilizers will be present. In marine systems, total inorganic nitrogen (as combined nitrogen) ranges between 0 and 560 μg l^{-1} (Dring, 1992). This is likely to limit fungal growth very much in the same way that marine primary productivity is limited by oceanic nitrogen and phosphorus concentrations (Dring, 1992). Fungi probably derive the bulk of their nitrogen requirements from organic substrate resulting from primary production, probably as reduced nitrogen, with close linkage between factors affecting primary productivity and fungal biomass.

There are some examples of fungal species in marine habitats deriving nitrogen from quite specific organic sources. For example, the trichomycete fungus *Entromyces callianassae* occurs exclusively in the foregut lining of the callianassid shrimp *Nihonotrypaea harmandi* and appears to be involved in the hydrolysis of certain nitrogen-containing compounds (Kimura *et al.*, 2002). Like their terrestrial counterparts, some marine fungi have been found to degrade nitrogen-rich materials such as chitin which are not easily utilized by other organisms. Kirchner (1995) found that the yeast-like fungus *Aureobasidium pullulans* was involved in the degradation of chitin in the moults and carcasses of the marine copepod *Tisbe holothuriae*. In general, marine fungi may play a role in the degradation of more recalcitrant nitrogen-containing compounds, which may be only slowly degraded by other recyclers. Although quantifying the contribution of marine fungi to nitrogen cycling is extremely difficult, Newell (1996) considered that, on salt marshes, fungi could account for nearly all the nitrogen present in decaying standing biomass of salt-marsh plants. Certainly, the role of fungi in the marine nitrogen cycle may have been grossly underestimated, particularly at the marine–terrestrial interface; our knowledge of the fungal contribution in the open ocean is almost entirely absent.

Open ocean phosphorus concentrations are very low, ranging from 0 to 90 µg l^{-1} (Dring, 1992). Similarly to nitrogen, these are likely to be too low to support appreciable fungal growth, with most phosphorus derived from organic substrates related to primary productivity. However, very few reports exist on marine fungi in relation to phosphorus. Bongiorni & Dini (2002) described the habitat, seasonality and species distribution of thraustochytrids, marine fungoid protists, and found that, among the nutrients assessed in the water column, only total phosphorus was related to variations in thraustochytrid densities.

On salt marshes, fungi may be involved in phosphorus cycling as mycorrhizas, with such associations reported in mangroves (Sengupta & Chaudhuri, 2002) and some salt-marsh plant species (Carvalho *et al.*, 2001). Mycorrhizal colonization of salt-marsh halophytes tends to be quite poorly distributed, with some of the most widespread salt-marsh plant species such as the Chenopodiaceae and many of the Graminae not being strongly mycorrhizal. In contrast, halophytic members of the Asteraceae do tend to be functionally mycorrhizal (Rozema *et al.*, 1986; Carvalho *et al.*, 2001). In the open ocean, a much more oligotrophic environment than coastal waters, phosphorus availability is known to limit bacterial productivity (Van Wambeke *et al.*, 2002), although effects on yeast populations are not known.

Sulfur. While sulfur plays an important role in global biogeochemical cycling in general, and in marine environments in particular, little is known about fungal involvement in sulfur cycling. Sulfate is a major constituent of sea water, which typically contains around 2·6 g l^{-1} (Dring, 1992). Also, sediments and hydrothermal vent regions tend to be rich in sulfides (Dring, 1992). Fungi are known to be able to oxidize inorganic sulfur (Wainwright, 1989) and to produce compounds such as dimethyl sulfide (Slaughter, 1989), which is an important compound in the cycling of sulfur in marine environments. Fungi are therefore likely to be involved in biogeochemical cycling of sulfur, particularly in marine sediments.

A few studies report on fungal sulfur metabolism of inorganic and organic sources. Phae & Shoda (1991) reported a fungal species able to degrade hydrogen sulfide, methanethiol, dimethyl sulfide and dimethyl disulfide, while Faison *et al.* (1991)

reported on a coal-solubilizing *Paecilomyces* sp. able to degrade organic sulfur compounds such as ethyl phenyl sulfide, diphenyl sulfide and dibenzyl sulfide. Reports on the involvement of fungi associated with marine organisms in relation to sulfur are scant. Zande (1999) described an ascomycete on the gills of the gastropod *Bathynerita naticoidea*, which the author suggested could be involved in detoxification of the abundant sulfide compounds in its habitat. Also associated with sulfide-rich marine environments is the fungus *Fusarium lateritium*, which is able to degrade dimethyl-sulfoniopropionate derived from algae and the salt-marsh grass *Spartina alterniflora* (Bacic & Yoch, 1998).

FUNGAL CONTRIBUTION TO CYCLING IN MARINE ECOSYSTEMS

For most marine ecosystems, the contribution by fungi to biogeochemical cycling can at best only be speculated upon. As considered above in the discussion of the effect of carbon availability on marine fungal processes, fungal biomass appears to be maximal (and presumably most active) where organic substrates are at high levels. This gives a fundamental difference between the generally oligotrophic open ocean systems and the land–ocean interface, where some of the most globally productive ecosystems exist. For the latter, salt-marsh ecosystems are well studied and much is known about the role of fungal activity in these systems (see Newell, 1993a, 1996).

Salt-marsh ecosystems have high rates of primary productivity resulting from nutrient recycling, riverine external loading, tidally mediated net export of organic matter, balancing of nutrient inputs and outputs as well as aerobic and anaerobic microbial activity (Adam, 1990). The most intensively studied of these systems are those of the eastern seaboard of the United States, where a sophisticated overview of the role of fungal activity is now available (Gessner & Kohlmeyer, 1976; Gessner, 1977; Newell *et al.*, 2000), particularly for the salt-marsh cordgrass *Spartina alterniflora* Loisel. *Spartina alterniflora* is an abundant and highly productive plant species in this system and has frequently been used as the model for understanding fungal–plant interactions in this unusual environment (Newell, 1993a, b).

Spartina alterniflora is highly lignocellulosic (>70 %) and is extensively colonized by fungi during decomposition (Newell *et al.*, 1996). Newell (1996) considered that there are four main strategies by which marine microbes capture metabolizable organic carbon: (i) maximized use of surface area coupled with high substrate affinity and suitable enzymes that are capable of diffusing into solid particles; (ii) penetration by tunnelling or surface erosion; (iii) penetration by absorptive ectoplasmic nets or rhizoids; and (iv) pervasion via networks of self-extending tubular flowing cytoplasm within rigid microfibrillar tubes of chitin laminaran or cellulose laminaran. Marine fungi contribute to the degradation of *Spartina alterniflora* through the production of lignocellulosic enzymes, which commence the breakdown of this material as plants senesce leading to a subsequent mixed degradation. The life cycle of *Spartina alterniflora* involves shoot growth from perennial rhizomes from April to July, maturation and seed production from July to October, frost-mediated death in October and an ensuing decomposition period aided by weathering, shearing and tidal action. Fungi appear to be central in commencing the decomposition process (Newell, 1993a), efficiently breaking down lignin and cellulose into smaller, low-molecular-mass fragments available either for conversion to fungal biomass or for further decomposition by bacteria over the ensuing months. A number of distinct laccase sequence types have been identified from a group of fungi isolated from this salt-marsh decay system (Lyons *et al.*, 2003).

Some of the fungi reported to be involved in this process include *Dreschslera halodes*, *Halosphaeria hamata*, *Phaeosphaeria typharum*, *Stagonospora* sp., *Buergenerula spartinae*, *Alternaria alternata*, *Epicoccum nigrum*, *Claviceps purpurea*, *Leptosphaeria obiones*, *Leptosphaeria pelagica*, *Pleospora pelagica*, *Pleospora vegans* and *Lulworthia* sp. (Gessner, 1977). More recently, ascomycetous fungi have been reported as the major secondary producers within standing decaying leaves of *S. alterniflora* L. throughout its geographical range (Newell *et al.*, 2000; Newell, 2001). This has been further supported by work on the fungal internal transcribed spacer sequences of these species by Buchan *et al.* (2003). The four ascomycetes that appear to be most involved in this system are *Phaeosphaeria spartinicola* Leuchtmann, *Mycosphaerella* sp. 2 (of Kohlmeyer & Kohlmeyer, 1979), *Phaeosphaeria halima* Johnson and *Buergenerula spartinae* Kohlm. et Gessner (Newell, 2001).

The fungal signal molecule ergosterol has also been used extensively to detail fungal biomass in *Spartina alterniflora*-dominated salt-marsh systems. Significant differences in ergosterol content of decaying leaves occurred, which were interpreted to indicate differences in nitrogen (but not phosphorus) availability (Newell, 2001). Nitrogen is thought to be a limiting nutrient for both *Spartina alterniflora* (Mendelssohn & Morris, 2000) and its associated fungal populations (Newell *et al.*, 1996). Looking for seasonal differences, Newell's group also found that the ratio between fungal productivity and mean ergosterol contents was significantly higher in winter/spring than in summer/autumn.

The situation in open oceans differs greatly from that in coastal regions. Organic matter is rarely abundant, except where planktonic blooms occur, often resulting from nutrient upwelling. Low substrate availability generally leads to very low fungal biomass, of yeasts rather than the filamentous fungi characteristic of coastal ecosystems.

There are at present no quantitative data linking primary production to fungal numbers in these systems. Beneath the photic zones in the open ocean, degraders presumably become more important as organic material sediments through the water column. It is known that bacterial numbers reduce with depth, with numbers at 5000 m being around 10 % of those at 75 m (Dring, 1992), together with much lower bacterial growth rates. It might be expected that, at some depth, particularly below the photic zone, micro-organisms and fungi are the most important organisms (relative to total bio-mass), growing on the detritus precipitating down from the ocean surface layers. At great depths, a surprising abundance of life can be found near 'smokers' (deep-ocean hydrothermal vents) and it is very likely that fungi form a significant component of these systems (Duhig *et al.*, 1992; Burgath & Von Stackelberg, 1995; Nagahama *et al.*, 2001).

CONCLUSIONS

In comparison with terrestrial environments, we have a rather restricted view of the role of fungi in marine ecosystems. This partially reflects the extent of marine environments on a global basis, together with their remoteness and difficulty for sampling. There is a better understanding of land–sea interfaces, particularly from the studies of Newell and co-workers in the eastern USA, although these studies need to be taken up for other salt marsh/mangrove systems in other parts of the world. Although there has been considerable effort to uncover fungal diversity, marine mycologists have been slow to apply molecular methods to explore diversity. Whereas these would be routine for marine bacteriologists, fungal molecular studies are few (e.g. Buchan *et al.*, 2003). Our knowledge of fungal diversity in open ocean systems is even more restricted, but could more frequently be incorporated into programmes examining marine bacterial populations, particularly as the predominant fungi in these systems are yeasts. We urgently need to know more about fungi existing in benthic and deep-sea (hydro-thermal) zones. Although we know something about marine fungal diversity in general, little is known about their function except in salt marshes. We need to know to what extent fungi contribute to food webs and fluxes of organic matter in marine systems other than salt marshes, in order to give fuller views of these systems in general. Exploration of new marine environments and determining the role of fungi in these environments is the next challenge for marine mycology.

REFERENCES

Adam, P. (1990). *Saltmarsh Ecology*. Cambridge: Cambridge University Press.
Ahammed, S. & Prema, P. (2002). Influence of media nutrients on synthesis of lignin peroxidase from *Aspergillus* sp. *Appl Biochem Biotechnol* **102/103**, 327–336.

Almagro, A., Prista, C., Benito, B., Loureiro-Dias, M. C. & Ramos, J. (2001). Cloning and expression of two genes coding for sodium pumps in the salt-tolerant yeast *Debaryomyces hansenii*. *J Bacteriol* **183**, 3251–3255.

Alongi, D. M. (1987). The distribution and composition of deep-sea microbenthos in a bathyal region of the western Coral Sea. *Deep Sea Res* **34**, 1245–1254.

Armstrong, W., Wright, E. J., Lythe, S. & Gaynard, T. J. (1985). Plant zonation and the effects of the spring-neap tide cycle on soil aeration in a Humber salt marsh. *J Ecol* **73**, 323–339.

Bacic, M. K. & Yoch, D. C. (1998). *In vivo* characterization of dimethylsulfoniopropionate lyase in the fungus *Fusarium lateritium*. *Appl Environ Microbiol* **64**, 106–111.

Ballard, R. D. (1977). Notes on a major oceanographic find (marine animals near hot-water vents at ocean bottom). *Oceanus* **20**, 35–44.

Bansal, P. K. & Mondal, A. K. (2000). Isolation and sequence of the HOG1 homologue from *Debaryomyces hansenii* by complementation of the *hog1Delta* strain of *Saccharomyces cerevisiae*. *Yeast* **16**, 81–88.

Barghoorn, E. S. & Linder, D. H. (1944). Marine fungi: their taxonomy and biology. *Farlowia* I, 395–467.

Beever, R. E. & Laracy, E. P. (1986). Osmotic adjustment in the filamentous fungus *Aspergillus nidulans*. *J Bacteriol* **168**, 1358–1365.

Blomberg, A. & Adler, L. (1993). Tolerance of fungi to NaCl. In *Stress Tolerance of Fungi*, pp. 233–256. Edited by D. H. Jennings. New York: Marcel Dekker.

Bohnert, H. J., Ayoubi, P., Borchert, C. & 23 other authors (2001). A genomics approach towards salt stress tolerance. *Plant Physiol Biochem* **39**, 295–311.

Bongiorni, L. & Dini, F. (2002). Distribution and abundance of thraustochytrids in different Mediterranean coastal habitats. *Aquat Microb Ecol* **30**, 49–56.

Buchalo, A. S., Nevo, E., Wasser, S. P., Oren, A. & Molitoris, H. P. (1998). Fungal life in the extremely hypersaline water of the Dead Sea: first records. *Proc Biol Sci* **265**, 1461–1465.

Buchan, A., Newell, S. Y., Butler, M., Biers, E. J., Hollibaugh, J. T. & Moran, M. A. (2003). Dynamics of bacterial and fungal communities on decaying salt marsh grass. *Appl Environ Microbiol* **69**, 6676–6687.

Bucher, V. V. C., Hyde, K. D., Pointing, S. B. & Reddy, C. A. (2004). Production of wood decay enzymes, loss of mass and lignin solubilization in wood by diverse marine fungi. *Fungal Divers* **15**, 1–14.

Burgath, K. P. & Von Stackelberg, U. (1995). Sulfide-impregnated volcanics and ferro-manganese incrustations from the southern Lau basin (Southwest Pacific). *Mar Georesour Geotechnol* **13**, 263–308.

Carlile, M. J., Watkinson, S. C. & Gooday, G. W. (2001). *The Fungi*, 2nd edn. San Diego: Academic Press.

Carvalho, L. M., Caçador, I. & Martins-Loução, M. (2001). Temporal and spatial variation of arbuscular mycorrhizas in salt marsh plants of the Tagus estuary (Portugal). *Mycorrhiza* **11**, 303–309.

Clement, D. J., Stanley, M. S., O'Neil, J., Woodcock, N. A., Fincham, D. A., Clipson, N. J. W. & Hooley, P. (1999). Complementation cloning of salt tolerance determinants from the marine hyphomycete *Dendryphiella salina* in *Aspergillus nidulans*. *Mycol Res* **103**, 1252–1258.

Clipson, N. J. W. & Jennings, D. H. (1992). *Dendryphiella salina* and *Debaryomyces hansenii*: models for ecophysiological adaptation to salinity by fungi that grow in the sea. *Can J Bot* **70**, 2097–2105.

Clipson, N. J. W., Hajibagheri, H. A. & Jennings, D. H. (1990). X-ray microanalysis of the marine fungus *Dendryphiella salina* at different salinities. *J Exp Bot* **41**, 199–202.

Clipson, N. J. W., Landy, E. T. & Otte, M. L. (2001). Fungi. In *European Register of Marine Species: a Checklist of the Marine Species in Europe and a Bibliography of Identification Guides*, Collection Patrimoines Naturels, vol. 50, pp. 15–19. Edited by M. J. Costello, C. S. Emblow & R. White. Paris: Publications Scientifiques du MNHN.

Cooke, J. C., Butler, R. H. & Madole, G. (1993). Some observations on the vertical distribution of vesicular arbuscular mycorrhizae in roots of salt marsh grasses growing in saturated soils. *Mycologia* **85**, 547–550.

Cornick, J., Standwerth, A. & Fisher, P. J. (2005). A preliminary study of fungal endophyte diversity in a stable and declining bed of *Spartina anglica* Hubbard. *Mycologist* **19**, 24–29.

Davis, D. J., Burlak, C. & Money, N. P. (2000). Osmotic pressure of fungal compatible osmolytes. *Mycol Res* **104**, 800–804.

Dring, M. J. (1992). *The Biology of Marine Plants*, 2nd edn. Cambridge: Cambridge University Press.

Duhig, N. C., Davidson, G. J. & Stolz, J. (1992). Microbial involvement in the formation of Cambrian sea-floor silica-iron oxide deposits, Australia. *Geology* **20**, 511–514.

Edwards, J., Chamberlain, D., Brosnan, G., West, D., Stanley, M. S., Clipson, N. J. W. & Hooley, P. (1998). A comparative physiological and morphological study of *Dendryphiella salina* and *D. arenaria* in relation to adaptation to life in the sea. *Mycol Res* **102**, 1198–1202.

Eilers, H., Pernthaler, J., Glockner, F. O. & Amann, R. (2000). Culturability and *in situ* abundance of pelagic bacteria from the North Sea. *Appl Environ Microbiol* **66**, 3044–3051.

Elmsley, J. (1980). The phosphorus cycle. In *The Handbook of Environmental Chemistry*, vol. 1, part A, *The Natural Environment and the Biogeochemical Cycles*, pp. 147–167. Edited by O. Hutzinger. Berlin: Springer.

Faison, B. D., Clark, T. M., Lewis, S. N., Ma, C. Y., Sharkey, D. M. & Woodward, C. A. (1991). Degradation of organic sulfur compounds by a coal-solubilizing fungus. *Appl Biochem Biotechnol* **28/29**, 237–251.

Feeney, N., Curran, P. M. T. & O'Muircheartaigh, I. G. (1992). Biodeterioration of woods by marine fungi and *Chaetomium globosum* in response to an external nitrogen source. *Int Biodeterior Biodegrad* **29**, 123–133.

Flowers, T. J., Hajibagheri, M. A. & Clipson, N. J. W. (1986). Halophytes. *Q Rev Biol* **6**, 313–337.

Freney, J. R., Ivanov, M. V. & Rodie, H. (1983). The sulphur cycle. In *The Major Biogeochemical Cycles and their Interactions*, pp. 46–50. Edited by B. Bolin & R. B. Cook. Chichester: Wiley.

Gadanho, M., Almeida, J. M. & Sampaio, J. P. (2003). Assessment of yeast diversity in a marine environment in the south of Portugal by microsatellite-primed PCR. *Antonie van Leeuwenhoek* **84**, 217–227.

Geiser, D. M., Taylor, J. W., Ritchie, K. B. & Smith, G. W. (1998). Cause of sea fan death in the West Indies. *Nature* **394**, 137–138.

Gessner, R. V. (1977). Seasonal occurrence and distribution of fungi associated with *Spartina alterniflora* from a Rhode Island estuary. *Mycologia* **69**, 477–491.

Gessner, R. V. & Kohlmeyer, J. (1976). Geographical distribution and taxonomy of fungi from salt marsh *Spartina*. *Can J Bot* **54**, 2023–2037.

Gonzalez, M. C., Hanlin, R. T. & Ulloa, M. (2001). A checklist of higher marine fungi of Mexico. *Mycotaxon* **80**, 241–253.

Grasso, S., Bruni, V. & Maio, G. (1997). Marine fungi in Terra Nova Bay (Ross Sea, Antarctica). *New Microbiol* **20**, 371–376.

Gunde-Cimermann, N., Zalarb, P., de Hoogc, S. & Plemenitasd, A. (2000). Hypersaline waters in salterns – natural ecological niches for halophilic black yeasts. *FEMS Microbiol Ecol* **32**, 235–240.

Hawksworth, D. L. (1991). The fungal dimension of biodiversity: magnitude, significance, and conservation. *Mycol Res* **95**, 641–655.

Hawksworth, D. L. (2001). The magnitude of fungal diversity: the 1·5 million species estimate revisited. *Mycol Res* **105**, 1422–1432.

Hernandez-Saavedra, N. Y. & Romero-Geraldo, R. (2001). Cloning and sequencing the genomic encoding region of copper-zinc superoxide dismutase enzyme from several marine strains of the genus *Debaryomyces* (Lodder & Kreger-van Rij). *Yeast* **18**, 1227–1238.

Hooley, P., Fincham, D. A., Whitehead, M. P. & Clipson, N. J. W. (2003). Fungal osmotolerance. *Adv Appl Microbiol* **53**, 177–211.

Hutzinger, O. (editor) (1980). *The Handbook of Environmental Chemistry*, vol. 1, part A, *The Natural Environment and the Biogeochemical Cycles*. Berlin: Springer.

Hyde, K. D. & Jones, E. B. G. (1988). Marine mangrove fungi. *Mar Ecol* **9**, 15–33.

Jefferies, R. L., Davy, A. J. & Rudmik, T. (1979). The growth strategies of coastal halophytes. In *Ecological Processes in Coastal Environments*, pp. 243–268. Edited by R. L. Jefferies & A. J. Davy. Oxford: Blackwell.

Jennings, D. H. (1989). Some perspectives on nitrogen and phosphorus metabolism in fungi. In *Nitrogen, Phosphorus and Sulphur Utilization by Fungi*, pp. 1–32. Edited by L. Boddy, R. Marchant & D. J. Read. Cambridge: Cambridge University Press.

Jennings, D. H. & Burke, R. M. (1990). Compatible solutes – the mycological dimension and their role as physiological buffering agents. *New Phytol* **116**, 277–283.

Jones, E. B. G. (1995). Ultrastructure and taxonomy of the aquatic ascomycetous order Halosphaeriales. *Can J Bot* **73**, S790–S801.

Jones, E. B. G. & Alias, S. A. (1997). Biodiversity of mangrove fungi. In *Diversity of Tropical Fungi*, pp. 177–186. Edited by K. D. Hyde. Hong Kong: Hong Kong University Press.

Jones, E. B. G. & Mitchell, J. I. (1996). Biodiversity of marine fungi. In *Biodiversity, International Biodiversity Seminar*, pp. 31–42. Edited by A. Cimermann & N. Gunde-Cimermann. Ljubljana: National Institute of Chemistry and Slovenia National Commission for UNESCO.

Kimura, H., Harada, K., Hara, K. & Tamaki, A. (2002). Enzymatic approach to fungal association with arthropod guts: a case study for the crustacean host, *Nihonotrypaea harmandi*, and its foregut fungus, *Enteromyces callianassae*. *Mar Ecol* **23**, 157–183.

Kirchner, M. (1995). Microbial colonization of copepod body surfaces and chitin degradation in the sea. *Helgol Meeresunters* **49**, 2001–2012.

Kohlmeyer, J. (1969). Ecological notes on fungi in mangrove forests. *Trans Br Mycol Soc* **53**, 237–250.

Kohlmeyer, J. & Kohlmeyer, E. (1979). *Marine Mycology: the Higher Fungi*. New York: Academic Press.

Kohlmeyer, J. & Volkmann-Kohlmeyer, B. (2003). Fungi from coral reefs: a commentary. *Mycol Res* **107**, 386–387.

Kritzman, G. (1973). *Observations on the microorganisms in the Dead Sea*. MSc thesis, Hebrew University of Jerusalem (in Hebrew).

Lavin, A. M., Bryden, H. L. & Parilla, G. (2003). Mechanisms of heat, freshwater, oxygen and nutrient transports and budgets at 24·5° N in the subtropical North Atlantic. *Deep Sea Res Part I Oceanogr Res Pap* **50**, 1099–1128.

Lorenz, R. & Molitoris, H.-P. (1992). Combined influence of salinity and temperature (*Phoma* pattern) on growth of marine fungi. *Can J Bot* **70**, 2111–2115.

Lorenz, R. & Molitoris, H.-P. (1997). Cultivation of fungi under simulated deep sea conditions. *Mycol Res* **101**, 1355–1365.

Lyons, J. I., Newell, S. Y., Buchan, A. & Moran, M. A. (2003). Diversity of ascomycete laccase gene sequences in a southeastern US salt marsh. *Microb Ecol* **45**, 270–281.

Martin, D. F. (1970). *Marine Chemistry*, vol. 2. New York: Marcel Dekker.

Masson, D. (2002). Deep water renewal in the Strait of Georgia. *Estuar Coast Shelf Sci* **54**, 115–126.

Mendelssohn, I. A. & Morris, J. T. (2000). Eco-physiological controls on the primary productivity of *Spartina alterniflora*. In *Concepts and Controversies in Tidal Marsh Ecology*, pp. 59–80. Edited by M. P. Weinstein & D. A. Kreeger. Dordrecht: Kluwer.

Molitoris, H.-P. & Schaumann, K. (1986). Physiology of marine fungi: a screening programme for growth and enzyme production. In *The Biology of Marine Fungi*, pp. 35–48. Edited by S. T. Moss. Cambridge: Cambridge University Press.

Money, N. P. (1997). Wishful thinking of turgor revisited: the mechanics of fungal growth. *Fungal Genet Biol* **21**, 173–187.

Nagahama, T., Hamamoto, M., Nakase, T., Takami, H. & Horikoshi, K. (2001). Distribution and identification of red yeasts in deep-sea environments around the northwest Pacific Ocean. *Antonie van Leeuwenhoek* **80**, 101–110.

Newell, S. Y. (1993a). Decomposition of shoots of a saltmarsh grass methods and dynamics of microbial assemblages. *Adv Microb Ecol* **13**, 301–326.

Newell, S. Y. (1993b). Membrane containing fungal mass and fungal specific growth rate in natural samples. In *Handbook of Methods in Aquatic Microbial Ecology*, pp. 579–586. Edited by P. F. Kemp, B. F. Sherr, E. B. Sherr & J. J. Cole. Boca Raton, FL: Lewis Publishers.

Newell, S. Y. (1996). Established and potential impacts of eukaryotic mycelial decomposers in marine/terrestrial ecotones. *J Exp Mar Biol Ecol* **200**, 187–206.

Newell, S. Y. (2001). Multiyear patterns of fungal biomass dynamics and productivity within naturally decaying smooth cordgrass shoots. *Limnol Oceanogr* **46**, 573–583.

Newell, S. Y., Porter, D. & Lingle, W. L. (1996). Lignocellulosis by ascomycetes (fungi) on a saltmarsh grass (smooth cordgrass). *Microsc Res Tech* **33**, 32–46.

Newell, S. Y., Blum, L. K., Crawford, R. E., Dai, T. & Dionne, M. (2000). Autumnal biomass and potential productivity of salt marsh fungi from 29° to 43° north latitude along the United States Atlantic coast. *Appl Environ Microbiol* **66**, 180–185.

Nilsson, A. & Adler, L. (1990). Purification and characterisation of glycerol-3-phosphate dehydrogenase (NAD$^+$) in the salt-tolerant yeast *Debaryomyces hansenii*. *Biochim Biophys Acta* **1034**, 180–185.

Nybakken, J. W. (1997). *Marine Biology – an Ecological Approach*. Boston: Addison–Wesley.

Packham, J. R. & Willis, A. J. (1997). *Ecology of Dunes, Salt Marsh and Shingle*. London: Chapman & Hall.

Paton, F. M. & Jennings, D. H. (1988). Effect of sodium and potassium chloride and polyols on malate and glucose 6-phosphate dehydrogenase from the marine fungus *Dendryphiella salina*. *Trans Br Mycol Soc* **91**, 205–215.

Phae, C. G. & Shoda, M. (1991). A new fungus which degrades hydrogen sulfide, methanethiol, dimethyl sulfide and dimethyl disulfide. *Biotechnol Lett* **13**, 375–380.

Pierrou, U. (1976). The global phosphorus cycle. In *The Major Biogeochemical Cycles and their Interactions*, pp. 46–50. Edited by B. Bolin & R. B. Cook. Chichester: Wiley.

Porter, D. (1986). Mycoses of marine organisms: an overview of pathogenic fungi. In *The Biology of Marine Fungi*, pp. 141–154. Edited by S. T. Moss. Cambridge: Cambridge University Press.

Pugh, G. J. F. & Jones, E. B. G. (1986). Antarctic marine fungi: a preliminary account. In *The Biology of Marine Fungi*, pp. 323–330. Edited by S. T. Moss. Cambridge: Cambridge University Press.

Richey, J. E. (1983). The phosphorus cycle. In *The Major Biogeochemical Cycles and Their Interactions*, pp. 51–56. Edited by B. Bolin & R. B. Cook. Chichester: Wiley.

Rohrmann, S., Lorenz, R. & Molitoris, H. P. (1992). Use of natural and artificial seawater for the investigation of growth, fruit body production, and enzyme activities in marine fungi. *Can J Bot* **70**, 2106–2110.

Rosswall, T. (1983). The nitrogen cycle. In *The Major Biogeochemical Cycles and Their Interactions*, pp. 46–50. Edited by B. Bolin & R. B. Cook. Chichester: Wiley.

Rozema, J., Arp, W., Van Diggelen, J., Van Esbroek, M., Broekman, R. & Punte, H. (1986). Occurrence and ecological significance of vesicular-arbuscular mycorrhiza in the salt marsh environment. *Acta Bot Neerl* **35**, 457–467.

Salomons, W. & Förstner, U. (1984). *Metals in the Hydrocycle*. Berlin: Springer.

Schaumann, K., Mulach, W. & Molitoris, H.-P. (1986). Comparative studies on growth and exoenzyme production of different *Lulworthia* isolates. In *The Biology of Marine Fungi*, pp. 49–60. Edited by S. T. Moss. Cambridge: Cambridge University Press.

Schmit, J. P. & Shearer, C. A. (2003). A checklist of mangrove-associated fungi, their geographical distribution and known host plants. *Mycotaxon* **85**, 423–477.

Sengupta, A. & Chaudhuri, S. (2002). Arbuscular mycorrhizal relations of mangrove plant community at the Ganges river estuary in India. *Mycorrhiza* **12**, 169–174.

Sguros, P. L. & Simms, J. (1963). Role of marine fungi in biochemistry of oceans. 2. Effect of glucose, inorganic nitrogen, and tris (hydroxymethyl) amino methane on growth and pH changes in synthetic media. *Mycologia* **55**, 728–741.

Sieburth, J. M. (1979). *Sea Microbes*. Oxford: Oxford University Press.

Slaughter, J. C. (1989). Sulphur compounds in fungi. In *Nitrogen, Phosphorus and Sulphur Utilization by Fungi*, pp. 91–105. Edited by L. Boddy, R. Marchant & D. J. Read. Cambridge: Cambridge University Press.

Smith, G. W., Ives, L. D., Nagelkerken, I. A. & Ritchie, K. B. (1996). Caribbean sea-fan mortalities. *Nature* **383**, 487.

Torzilli, A. P. (1982). Polysaccharidase production and cell wall degradation by several salt marsh fungi. *Mycologia* **74**, 297–302.

Torzilli, A. P. & Andrykovitch, G. (1986). Degradation of *Spartina* lignocellulose by individual and mixed cultures of salt-marsh fungi. *Can J Bot* **64**, 2211–2215.

Van Uden, N. & Fell, J. W. (1968). Marine yeasts. In *Advances in the Microbiology of the Sea*, vol. 1, pp. 167–202. Edited by M. R. Droop & E. J. F. Wood. New York: Academic Press.

Van Wambeke, F., Christaki, U., Giannakourou, A., Moutin, T. & Souvemerzoglou, K. (2002). Longitudinal and vertical trends of bacterial limitation by phosphorus and carbon in the Mediterranean Sea. *Microb Ecol* **43**, 119–133.

Wainwright, M. (1989). Inorganic sulphur oxidation by fungi. In *Nitrogen, Phosphorus and Sulphur Utilization by Fungi*, pp. 73–90. Edited by L. Boddy, R. Marchant & D. J. Read. Cambridge: Cambridge University Press.

Wangersky, P. J. (1980). Chemical oceanography. In *The Handbook of Environment Chemistry*, vol. 1, part A, *The Natural Environment and the Biogeochemical Cycles*, pp. 51–68. Edited by O. Hutzinger. Berlin: Springer.

Yale, J. & Bohnert, H. J. (2001). Transcript expression in *Saccharomyces cerevisiae* at high salinity. *J Biol Chem* **276**, 15996–16007.

Zande, J. M. (1999). An ascomycete commensal on the gills of *Bathynerita naticoidea*, the dominant gastropod at Gulf of Mexico hydrocarbon seeps. *Invertebr Biol* **118**, 57–62.

Role of micro-organisms in karstification

Philip C. Bennett[1] and Annette Summers Engel[2]

[1]Department of Geological Sciences, The University of Texas at Austin, Austin, TX 78712, USA

[2]Department of Geology and Geophysics, Louisiana State University, Baton Rouge, LA 70803, USA

INTRODUCTION

Whilst chemolithoautotrophic micro-organisms are found in nearly every environment on Earth, they are more abundant in dark habitats where competition by photosynthetic organisms is eliminated. Caves, particularly, represent dark but accessible subsurface habitats where the importance of microbial chemolithoautotrophy to biogeochemical and geological processes can be examined directly. At Lower Kane Cave, WY, USA, hydrogen sulfide-rich springs provide a rich energy source for chemolithoautotrophic micro-organisms, supporting a surprisingly complex consortium of micro-organisms, dominated by sulfur-oxidizing bacteria. Several evolutionary lineages within the class '*Epsilonproteobacteria*' dominate the biovolume of subaqueous microbial mats, and these microbes support the cave ecosystem through chemolithoautotrophic carbon fixation. The anaerobic interior of the cave microbial mats is a habitat for anaerobic metabolic guilds, dominated by sulfate-reducing and -fermenting bacteria. Biological controls of speleogenesis had not been considered previously and it was found that cycling of carbon and sulfur through the different microbial groups directly affects sulfuric acid speleogenesis and accelerates limestone dissolution. This new recognition of the contribution of microbial processes to geological processes provides a better understanding of the causal factors for porosity development in sulfidic groundwater systems.

Karst landscapes form where soluble carbonate rocks dissolve by chemical solution (karstification), resulting in numerous geomorphic features, including caves and subterranean-conduit drainage systems (e.g. White, 1988; Ford & Williams, 1989). This

SGM symposium 65: Micro-organisms and Earth systems – advances in geomicrobiology.
Editors G. M. Gadd, K. T. Semple & H. M. Lappin-Scott. Cambridge University Press. ISBN 0 521 86222 1 ©SGM 2005

has traditionally been viewed as an abiotic, chemical process that occurs near the water table, with biologically produced CO_2 as the principal reactive component.

This model fits well with the classic view of terrestrial subsurface microbial environments as a 'top-downward' model, where photosynthetically fixed, reduced carbon (C) supplies the subsurface community with substrates for cell mass and energy (Kinkle & Kane, 2000), because microbial processes occurring in the absence of light energy have generally been considered insufficient to support ecosystem-level processes. In the past, the subsurface community has been portrayed as a mixed community of heterotrophs, with only hydrogen-oxidizing methanogens typically recognized as a significant autotrophic population (Stevens & McKinely, 1995; Chapelle *et al.*, 2002). However, the absence of light energy does not preclude life, as subsurface environments couple relatively constant temperatures and protection from surface conditions with an abundance of dissolved inorganic solutes and exposed mineral surfaces that can serve as sources of energy and nutrients (Maher & Stevenson, 1988; Stevens & McKinley, 1995; Ghiorse, 1997; Bachofen *et al.*, 1998; Rogers *et al.*, 1999; Bennett *et al.*, 2000; Ben-Ari, 2002). Consequently, in subsurface environments, chemosynthetic microorganisms are vitally important to global chemical and ecosystem processes because they gain cellular energy from the chemical oxidation of inorganic compounds and fixing inorganic C, and serve as catalysts for reactions that would otherwise not occur or would proceed slowly over geological time.

Sulfur-based microbial ecosystems include communities that oxidize energy-rich, non-surface-derived, reduced S substrates [S(−II)], coupled to molecular-oxygen reduction, to fix CO_2. In near-surface environments, such as caves and mines in shallow bedrock (< 200 m depth), these communities can be examined directly and chemolithoautotrophic S oxidizers form isolated but diverse and highly productive microbial-mat communities that support higher trophic levels within the subsurface ecosystems (e.g. Sarbu *et al.*, 1996; Angert *et al.*, 1998; Hose *et al.*, 2000; Sarbu *et al.*, 2000; Engel *et al.*, 2001). The ubiquity of reduced S compounds in anaerobic groundwater suggests that microbial communities that utilize reduced S are dispersed more widely in the terrestrial subsurface than is currently recognized. Equally important, S(−II) oxidation to sulfate generates excess protons that accelerate rock weathering, porosity development and mobilization of toxic metals, demonstrating a geological significance of terrestrial S-based microbial ecosystems beyond the descriptive characterization of novel populations. Understanding the dynamics of S-based subsurface communities and the coupling of the S and C cycles is a step toward linking microbial ecology to geological phenomena.

Karst aquifers

Caves act as portals to subsurface karst environments, with the typical habitat characteristics being complete darkness, nearly constant air and water temperatures and usually oligotrophic nutrient conditions, due to a limited supply of allochthonous organic material (Kinkle & Kane, 2000; Poulson & Lavoie, 2000). The classic model for karst development (speleogenesis) involves carbonic acid dissolution, usually at and rarely below the water table. More recently, sulfuric acid speleogenesis was proposed from work in Lower Kane Cave, WY, USA (Egemeier, 1981). Based on observations of H_2S-bearing thermal springs, extensive gypsum mineral deposits and gypsum-replaced carbonate rock walls, Egemeier (1981) put forth the sulfuric acid speleogenesis model, which included volatilization of H_2S from the sulfidic groundwater to the cave atmosphere and H_2S autoxidation to sulfuric acid on the moist cave walls:

$$H_2S + 2O_2 \rightarrow H_2SO_4 \qquad \text{(equation 1)}$$

where the acid then reacted with and replaced the carbonate with gypsum:

$$H_2SO_4 + CaCO_3 + H_2O \rightarrow CaSO_4 . 2H_2O + CO_2 \qquad \text{(equation 2)}$$

Gypsum is dissolved easily into groundwater and the net result is mass removal and an increase in void volume. Cave formation due to sulfuric acid has now been recognized in several active sulfidic cave systems (e.g. Hose *et al.*, 2000; Sarbu *et al.*, 2000) and has been a process linked to the development of at least 10 % of carbonate caves worldwide (Palmer, 1991).

Sulfur-based cave ecosystems

Microbial communities colonizing sulfidic cave habitats have received recent attention due to their chemolithoautotrophic metabolism (e.g. Sarbu *et al.*, 1996) and their geological impact due to acid production (Vlasceanu *et al.*, 2000; Engel *et al.*, 2001; Northup & Lavoie, 2001). Early studies of sulfuric acid speleogenesis (Egemeier, 1981) assumed that abiotic H_2S autoxidation was the important process for cave formation, and did not consider microbial S(–II) oxidation. In some of the active systems, e.g. in the Movile Cave (Romania), the Frasassi Caves (Italy) and Parker Cave (USA), filamentous aerobic to microaerophilic S-oxidizing bacteria (SOB) dominate subaqueous microbial mats (Angert *et al.*, 1998; Hose *et al.*, 2000) and were found to fix inorganic C, providing energy to sustain complex cave ecosystems (Sarbu *et al.*, 1996, 2000). Movile Cave provides us with an extreme example of a highly evolved, terrestrial, chemolithoautotrophically based ecosystem (Sarbu *et al.*, 1996), where 33 endemic, cave-adapted, invertebrate taxa have been identified (from 48 total taxa), including 24 terrestrial and nine aquatic animal species.

Culture-independent and -dependent approaches have been used to characterize the microbial communities, and 16S rRNA gene-based phylogenetic analyses of the microbial mats from Movile Cave, Parker Cave, Cueva de Villa Luz (Mexico), the Frasassi Caves and Cesspool Cave (USA) show diverse SOB, belonging to distinct lineages within different classes of the proteobacteria and including the genera *Thiothrix, Thiobacillus, Thiomonas, Thiomicrospira, Thiovulum* and *Achromatium* (Sarbu *et al.*, 1996; Angert *et al.*, 1998; Hose *et al.*, 2000; Vlasceanu *et al.*, 2000; Engel *et al.*, 2001). The predominant microbial groups in many of the investigated sulfidic caves were found to belong to novel lineages within the class 'Epsilonproteobacteria' (Angert *et al.*, 1998; Engel *et al.*, 2001, 2003) and, in addition to caves, closely related organisms have been described from many S-rich, oligotrophic natural habitats, including sulfidic springs, groundwater associated with oilfields, marine waters and sediments, deep-sea hydrothermal-vent sites and vent-associated metazoans (Engel *et al.*, 2004b). Despite the unique S-based microbial diversity of these habitats, there are few detailed studies describing the occurrence of epsilonproteobacteria from terrestrial environments. Moreover, little is known about either the ecophysiology of most epsilonproteobacteria, as many assemblages have not been cultured; from the work that has been done, most of these bacteria cycle S as SOB (e.g. Takai *et al.*, 2003). The SOB communities, including epsilonproteobacteria, occupy the redox boundary where reduced S mixes with aerobic water, and these microbes can take advantage of a potential energy gradient to produce sulfate and protons as the ultimate end products of their metabolism.

Colonization of the sulfidic caves by non-SOB microbial groups has rarely been addressed. Heterotrophic and anaerobic micro-organisms related to the phylum 'Bacteroidetes' have been reported from some of the caves from 16S rRNA gene-clone libraries (Angert *et al.*, 1998; Engel *et al.*, 2001), and dissimilatory sulfate-reducing bacteria (SRB) and fermenting bacteria have been characterized from molecular investigations of few sulfidic aquifers, including those associated with oilfields (Voor-douw *et al.*, 1996; Ulrich *et al.*, 1998). Based on geochemical data, SRB and SOB have been characterized from Yucatan cenotes open to phototrophic activity (Stoessell *et al.*, 1993). However, the importance of SRB for recycling S compounds by generating supplemental H_2S that SOB can use (e.g. Widdel & Bak, 1992) has only recently been addressed (Engel *et al.*, 2004b).

Coupled C and S metabolism

Whilst the microbiology and biochemistry of an S-based system are complex, only a few aspects relevant to $S(-II)$ oxidation in a non-photosynthesizing environment are reviewed here. SOB use reduced S compounds as electron donors (Madigan *et al.*, 1997):

$$H_2S + 2O_2 \rightarrow SO_4^{2-} + 2H^+ \; (-798 \, \text{kJ mol}^{-1}) \qquad \qquad \text{(equation 3)}$$

$$HS^- + {}^1\!/_2O_2 + H^+ \rightarrow S^0 + H_2O \; (-209 \, \text{kJ mol}^{-1}) \qquad \qquad \text{(equation 4)}$$

$$S^0 + H_2O + 1{}^1\!/_2O_2 \rightarrow SO_4^{2-} + 2H^+ \; (-587 \, \text{kJ mol}^{-1}) \qquad \qquad \text{(equation 5)}$$

$$S_2O_3^{2-} + H_2O + 2O_2 \rightarrow 2SO_4^{2-} + 2H^+ \; (-823 \, \text{kJ mol}^{-1}) \qquad \qquad \text{(equation 6)}$$

These general reactions proceed both biotically and abiotically, with abiotic autoxidation of H_2S proceeding spontaneously in aerobic aqueous systems (Millero *et al.*, 1987). Almost all non-photosynthesizing, chemolithoautotrophic SOB are obligate aerobes, requiring O_2 as the electron acceptor for S(–II) oxidation (Ehrlich, 1996). The mechanisms of biological S oxidation are still being elucidated. For example, whilst thiobacilli oxidize sulfide to sulfate directly (i.e. via equation 3), many SOB initially form S^0 as an intermediate species (equation 4) that is then stored intracellularly and further oxidized (equation 5) during periods of limiting sulfide (Ehrlich, 1996). Others oxidize reduced-sulfur intermediates, such as thiosulfate (equation 6). *Thiobacillus denitrificans* (Justin & Kelly, 1978), *Thermothrix thiopara* (Caldwell *et al.*, 1976) and some epsilonproteobacteria (Moyer *et al.*, 1995) couple S oxidation to nitrate reduction in dysoxic environments, whereas *Thiobacillus ferrooxidans* couples the anaerobic oxidation of S^0 to iron reduction (Corbett & Ingledew, 1987). Chemo-organotrophs in dysoxic marine environments conserve energy from S^0 conversion to sulfuric acid and H_2S by electron transfer in the presence of a sulfide scavenger (e.g. Fe^{2+}) without O_2 (sulfur disproportionation) (Thamdrup *et al.*, 1993):

$$4S^0 + 4H_2O \rightarrow SO_4^{2-} + 3H_2S + 2H^+ \qquad \qquad \text{(equation 7)}$$

Anaerobic microbes are common in marine and freshwater habitats, from anoxic sediments and bottom waters. Of the anaerobic microbial guilds, specifically SRB, many have not been studied in detail from groundwater and karst springs. SRB, obligate anaerobes that use molecular hydrogen to reduce sulfate to H_2S, are divided into two broad physiological subgroups: those that oxidize acetate and those that do not. Certain species of each subgroup are capable of chemolithoautotrophic growth with CO_2 as the sole C source. If sulfate concentrations are high, SRB oxidize fermentation by-products completely to CO_2. In low-sulfate environments, however, SRB compete with methanogens for hydrogen and organic compounds (Widdel & Bak, 1992; Scholten *et al.*, 2002).

MICROBIAL GEOCHEMISTRY METHODS

An interdisciplinary approach was used to examine the diversity of cave microbial communities and their roles in rock modification and karstification. By combining detailed characterization of both the geochemical characteristics of the habitat (Engel *et al.*, 2004a) and the composition and structure of the microbial community (Engel *et al.*, 2003, 2004b), the link between microbial metabolism and geochemical interactions can be examined. Recent research on the S-based ecosystem in Lower Kane Cave reveals complex microbial communities that tightly cycle S and C compounds across redox boundaries, whilst generating acidity and accelerating rock weathering (Engel *et al.*, 2004a).

Geochemical characterization

The objective of the geochemical characterization was to define the microbial habitat and to quantify the changes in solute concentration or speciation that are indicators of biogeochemical interactions. Water samples were collected for complete analyses of dissolved constituents, isotopic ratios and geochemical parameters by using standard methods and methods developed specifically for sulfidic cave systems (Engel *et al.*, 2004a). Field parameters included pH, temperature, specific conductance and field alkalinity on a filtered sample. Dissolved oxygen (DO) is a critical habitat constraint and was surveyed by using multiple methods, including electrochemical, microelectrode fluorescence quenching, colorimetric analyses and gas chromatography (GC). Dissolved Fe^{2+} is a sensitive indicator of the presence of dissimilatory iron-reducing bacteria (Lovley & Phillips, 1988) and this was measured immediately in the field by the ferrozine method to prevent sample degradation. Total and dissolved metal concentrations are used to characterize the geochemical consequences of the microbial activity and were determined from a filtered acid-preserved sample by inductively coupled plasma-mass spectrometry (ICP-MS). Dissolved anions, nutrients and organic acids were measured by ion chromatography (IC) and total inorganic and dissolved organic C (DOC) were measured by a carbon analyser. Stream water-dissolved solute speciation and activity, equilibrium gas partial pressure and saturation state with respect to mineral phases were calculated by using the geochemical speciation model PHREEQC (Parkhurst & Appelo, 1999). Sediments and core were characterized by X-ray diffraction (XRD) analysis of dried samples and total elemental composition by dissolution and ICP-MS analysis. Sediment and core samples were also examined by conventional and environmental scanning electron microscopy (CSEM and ESEM, respectively) for the presence of micro-organisms on mineral surfaces, mineral composition by energy-dispersive X-ray spectrometry (EDX) analysis and for evidence of mineral weathering.

Microbial-community analysis

Both culture-dependent and -independent methods were used to characterize the metabolically active microbes in Lower Kane Cave. Typically, only 1 % of environmental bacteria can be cultured (Leff *et al.*, 1995) and standard culturing methods often introduce a selective bias toward micro-organisms that are able to grow quickly and to utilize the substrates provided in the medium (McDougald *et al.*, 1998). However, culturing allows for quantification of metabolically active organisms in the samples and can lead to species identification through phylogenetic analyses (Palleroni, 1997). Laboratory strain isolation is essential to quantify the potential geological significance of microbes from the different habitats and can be used to help target unknown environmental samples. In contrast, molecular methods allow for characterization of a microbial community that may have populations that are difficult, if not impossible, to cultivate (Head *et al.*, 1998).

Biomass determination. Biomass, both viable and non-viable, is a basic attribute for characterizing mineral–microbe interactions and for understanding the implications of the more detailed metabolic and phylogenetic analyses. Biomass of filtered samples was determined by 4,6-diamidino-2-phenylindole (DAPI) cell counts (Gough & Stahl, 2003). This method is excellent for individual cells, but it is less reliable for filamentous organisms where the definition of a single cell is problematic. For filamentous mat samples, total organic C content of an acidified sample (to remove carbonate-mineral material) was analysed by using a carbon analyser and the total cell-count estimate was done by using the conversion factor of 350 fg C per cell (Bratbak & Dundas, 1984). This method offers a maximum biomass, as it assumes that all of the organic C is associated with biomass. Parallel determination of bacterial biomass can also be done by using phosphatidyl ester-linked fatty acid (PLFA) analysis (Vestal & White, 1989). Phospho-lipids are membrane lipids that are turned over rapidly during metabolism (White *et al.*, 1979). Consequently, phospholipids indicate viable microbial biomass at the moment that the microbial mats were extracted and are not subject to the positive artefact of the DAPI cell-count or C biomass-estimate methods.

Molecular methods. Microbial-community composition and structure were assessed by using established methods and a full-cycle rRNA approach (Engel *et al.*, 2003, 2004b). Total microbial-mat environmental DNA was extracted and near-full-length 16S rRNA gene sequences (rDNA) were obtained by PCR amplification using either archaeal and bacterial universal primers or lineage-specific primers (e.g. Ausubel *et al.*, 1990; Lane, 1991; Engel *et al.*, 2003). A TA cloning kit (Invitrogen) was used to facilitate PCR-product transformation and cloning procedures. Sequence inserts from clone plasmids were PCR-amplified with plasmid-specific primers, purified by using Sepha-dex columns and sequenced by using an automated ABI sequencer (e.g. ABI Prism 377X Perkin Elmer sequencer at Brigham Young University, Provo, UT, USA). 16S rDNA clone libraries were constructed from samples throughout Lower Kane Cave and, to facilitate community-composition analyses, clone sequence inserts were screened by using restriction-fragment length-polymorphism (RFLP) analysis, where selected clones from each representative RFLP pattern were sequenced (Engel *et al.*, 2004b). Results were analysed by using several phylogenetic-analysis methods after closely related rRNA gene sequences obtained from the Ribosomal Database Project were aligned by using CLUSTAL_X (Thompson *et al.*, 1997) and then adjusted manually based on conserved primary and secondary gene structure. Phylogenetic methods have included neighbour joining by PHYLIP (Felsenstein, 1993), maximum likelihood, minimum evolution and maximum parsimony by PAUP* (Swofford, 2002) and Bayesian inference in MrBayes (Ronquist & Huelsenbeck, 2003).

To identify and reliably quantify micro-organisms directly from the microbial mats, 16S rRNA-targeted oligonucleotide probes were used for fluorescence *in situ* hybridization (FISH) (Alm *et al.*, 1996). In addition to general archaeal and bacterial probes, gene probes were designed to target two novel groups of epsilonproteobacteria found in Lower Kane Cave (Engel *et al.*, 2003).

Metabolic survey. Microbial groups were enriched in pre-reduced anaerobic media specific for different metabolic groups, including chemolithoautotrophic, lactate/formate- or acetate-utilizing SRB and fermenting bacteria. Enumeration of each metabolic group was done by using the most probable number (MPN) estimates of tenfold serial dilutions of cultivable micro-organisms from each enrichment medium. All enrichments were incubated in the dark at room temperature in a Coy anaerobic chamber with $N_2 : H_2$ mixed gases to maintain anaerobic conditions. Following enrichment, growth was monitored by measuring OD_{590}, evolved gases produced by the mixed cultures (e.g. CH_4 for methanogens) were measured by GC and SRB were screened for by measuring H_2S production by GC.

In situ microcosms

The microbial influence on speleogenesis was evaluated by using the *in situ* microcosm approach that has been used extensively to examine silicate colonization and weathering (Hiebert & Bennett, 1992; Roberts *et al.*, 2004). Sterile and non-sterile field chambers (*in situ* microcosms) containing $0.5–1.0$ cm^3 chips of Iceland Spar calcite (Wards Scientific) and native limestone were deployed in the cave to test whether micro-organisms or the bulk cave stream-water chemistry controlled carbonate dissolution. Microcosms were constructed from 2.5×5 cm PVC pipes with screw caps on both ends. Sterile microcosms had 0.1 μm PVDF hydrophilic filters on the end, whilst non-sterile microcosms had 0.5 mm polyethylene mesh on either end to allow for fluid flow and also for microbial colonization of the chips. Paired sterile and non-sterile microcosms were placed throughout the cave stream and within the microbial mats, and remained in the cave for 2 weeks to 9 months. A technique similar to the buried-slide technique was also used, such that thin, polished wafers (surface areas of $1–2$ cm^2) of limestone and pieces of Iceland Spar were attached to glass slides and then surrounded by a 0.5 mm mesh sack. At each microcosm site, a mesh sack was also deployed. Chips and wafers were examined by ESEM, confocal laser-scanning microscopy (CLSM) for FISH and laser ablation (LA)-ICP-MS. A parallel interaction between the habitat and the microbial communities is the control exerted by the geological matrix, and colonization and weathering patterns from the carbonate surfaces were compared to other mineral substrates, such as chert or insoluble residues specific to the habitat being studied. Results from other environments (Hiebert & Bennett, 1992; Rogers *et al.*, 1998; Bennett *et al.*, 2000; Roberts *et al.*, 2004) suggest that colonization of a particular mineral substratum is influenced directly by the chemical composition of that mineral.

Unpreserved chips were examined by using a Philips XL30 ESEM; chamber conditions varied from 10 to 95 % relative humidity, using a Peltier cooling stage, and variable water pressure [0·9–6·4 torr (120–853 Pa), with corresponding accelerating voltages from 4 to 20 kV]. For FISH, chips and wafers were fixed in two ways within 12 h of collection: (i) with 4 % paraformaldehyde for 3 h; or (ii) with 50 % ice-cold ethanol (Engel *et al.*, 2003). Fixed samples were air-dried and dehydrated by sequential washes with 50, 80 and 100 % ethanol for 3 min prior to hybridization. Gene probes specific for Lower Kane Cave epsilonproteobacterial lineages, as well as probes targeting other microbial groups, were applied to the rock surfaces and surfaces were examined by using CLSM.

The example of Lower Kane Cave

S-oxidizing bacteria are an underappreciated geological weathering force, particularly in subsurface karst systems. In simple microbial systems, such as planktonic organisms in groundwater, the production of acidity can accelerate rock weathering, but in the complex, stratified microbial-mat communities that can attach to rock surfaces, the microbial influences on weathering can be more difficult to elucidate. Influences may involve local production of H_2S or other S gases and transfer to the vapour phase, with the net effect of transferring protons (or dissolution capacity) from aqueous to subaerial environments.

Lower Kane Cave is located in the Bighorn Basin near Lovell, WY, USA, adjacent to oilfields and localized sulfidic thermal and non-thermal springs (Egemeier, 1981). The cave is formed in the Little Sheep Mountain Anticline along the Bighorn River in the Madison Limestone and is ~350 m long. Thick, white, filamentous microbial mats, 3–10 cm thick and interconnected with white, web-like films (Fig. 1), are associated with four sulfidic springs that discharge into the cave, and mats stretch for up to 20 m in the cave stream below the springs (Engel *et al.*, 2003, 2004a, b). The biomass of the filamentous mats was 10^{12} cells ml^{-1}, based on the total organic C-estimate method.

The cave's springs are all of the calcium/bicarbonate/sulfate water type. Although the cave is forming from sulfuric acid speleogenesis, the spring and stream pH is buffered to circumneutral by ongoing carbonate dissolution. The dissolved sulfide concentration of

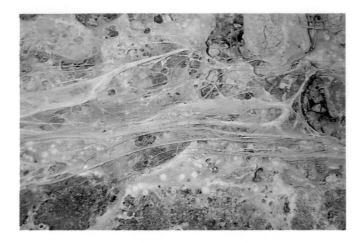

Fig. 1. Photograph of subaqueous microbial mats in Lower Kane Cave. Image width is 50 cm, water depth is 10 cm.

incoming spring water was ~38 µmol l^{-1} (speciated as ~60 % HS$^-$: 40 % H$_2$S, based on the pH of the springs, pK 7·04), with non-detectable DO. The concentration of DO and sulfide changed downstream, such that, at the end of the microbial mats, DO exceeded 40 µmol l^{-1} and sulfide was non-detectable. DOC (including methane) in all of the waters was extremely low, at < 80 µmol l^{-1}.

Microbial-mat diversity, characterized by 16S rDNA clone library construction and sequence analyses, revealed the presence of several different bacterial phyla. The surface of the mats, predominantly of white filament bundles, was oxygenated based on microelectrode DO profiles, whereas the grey mat interior was anoxic ~3 mm beneath the surface of the mat–water interface. The majority of the sequences retrieved from the white filaments belonged to the phylum *Proteobacteria*, specifically the classes 'Epsilonproteobacteria' (68 %), 'Gammaproteobacteria' (12·2 %), 'Betaproteobacteria' (11·7 %) and 'Deltaproteobacteria' (0·8 %), as well as other bacterial phyla, including the class *Acidobacteria* (5·6 %) and the *Bacteroidetes/Chlorobi* groups (1·7 %). In comparison, ~50 % of the retrieved 16S rDNA clone sequences from the interior mats were related to groups of SRB affiliated to the class 'Deltaproteobacteria' and to unculturable members of the phylum *Chloroflexi*. Rarer clones belonged to the phyla 'Gammaproteobacteria', *Planctomycetes*, *Bacteroidetes/Chlorobi*, 'Betaproteobacteria', 'Epsilonproteobacteria', *Verrucomicrobia*, 'Alphaproteobacteria', *Spirochaetes*, *Actinobacteria*, *Acidobacteria* and different OP candidate groups. FISH of the microbial mats revealed that ~70 % of the total bacterial biovolume was dominated by filamentous epsilonproteobacteria and specifically dominated by one novel lineage, referred to as LKC group II (Engel *et al.*, 2003) (Fig. 2). The microbial

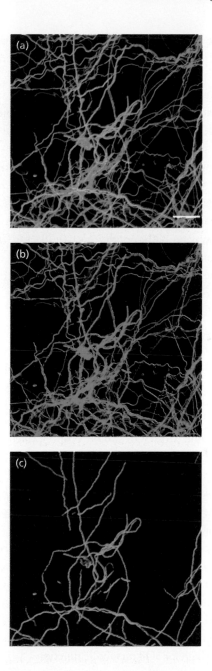

Fig. 2. FISH image of microbial-mat sample. (a) Overlapping probes; green, EUB338I-IIImix probes for all eubacteria; light blue, GAM42a for gammaproteobacteria (e.g. *Thiothrix* species). Bar, 20 µm. (b) EUB338I-IIImix-targeted filaments, being mostly epsilonproteobacteria belonging to the Lower Kane Cave group II evolutionary lineage (Engel *et al.*, 2003). (c) GAM42a probe only. All images were taken by using CLSM. Photograph scale is the same in all images.

mats from Lower Kane Cave represent the first non-marine natural system demonstrably driven by the activity of filamentous epsilonproteobacteria.

The high concentrations of S(–II) in the cave springs provide a rich energy source for SOB, and the concentration of dissolved sulfide decreased rapidly through the microbial mats. Abiotic sulfide autoxidation is extremely slow in disaerobic water at pH ~7·4 [autoxidation half-life was calculated to be > 800 h (Engel *et al.*, 2004a)] and sulfide volatilization from the water to the cave atmosphere accounts for < 8 % of the sulfide loss in the stream, based on gas-flux experiments (Engel *et al.*, 2004a). With no other mechanism for S(–II) loss, there would be, for example, much higher sulfide concentrations at the end of the microbial mats. The only other loss mechanism, microbial consumption under microaerophilic conditions, could cause the observed rapid loss in S(–II). Because epsilonproteobacterial filaments were found to dominate the microbial mats, these organisms are presumed to be SOB. Future culturing of these microbial groups will verify these observations.

The epsilonproteobacteria influence mat redox chemistry and ecosystem function directly by colonizing the initially nutrient-poor habitat, providing an energy source through chemolithoautotrophic C fixation, forming a dense mat and consuming O_2. Chemolithoautotrophy in the cave serves as the base for the cave food web and bulk white mats had a mean $\delta^{13}C$ value of –36‰, demonstrating chemolithoautotrophic fractionation against ^{13}C from an inorganic C source in the cave-stream water of –8·9‰ (Engel *et al.*, 2004b). Most SOB and the epsilonproteobacterial filaments form complex microbial mats and consumption of DO by the SOB creates an anaerobic habitat within the mat interior (Engel *et al.*, 2004b). The MPN method was used to estimate the biomass of anaerobic metabolic guilds; up to 10^6 cells ml^{-1} were found, with SRB and fermenters being the dominant culturable groups; iron reducers, S^0 reducers and methanogens were in low abundance overall (< 10^3 cells ml^{-1}). Redox stratification of the mats spatially separates metabolic guilds, such as SOB from SRB, and nutrients are consequently cycled between the multiple ecosystem components, whilst also being advected downstream. Based on the full-cycle rRNA approach, microbial diversity is greater within the mat interior and downstream portions of the mat, as mat density and organic C availability increase due to chemolithoautotrophic input and heterotrophic C degradation. The mat possibly terminates when H_2S is consumed and O_2 is too high for SOB metabolic function.

The dominant mechanism for S(–II) loss is subaqueous microbial oxidation and most SOB oxidize reduced S compounds completely to sulfate, with a substantial energy yield (e.g. equation 3). As a result of the energetic oxidation reactions, acidity is generated in the form of sulfuric acid, which can attack the geological matrix supporting the

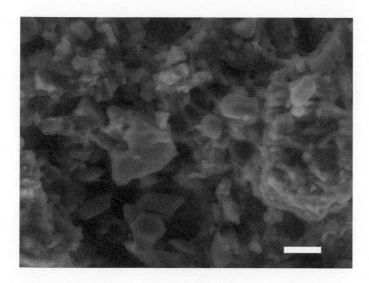

Fig. 3. ESEM image of a limestone fragment collected from the thick microbial-mat region. The surface is deeply corroded, consistent with chemical weathering. Smooth crystals are authigenic gypsum crystals. Filamentous microbial biofilm is barely visible covering the specimen under the imaging conditions used. Bar, 5 μm.

microbial community. Although some SOB can be acidophiles, most are neutrophilic and colonization of carbonate surfaces or habitats may buffer excess acidity and maintain pH homeostasis. Consequently, observed deeply corroded native carbonate rocks in the cave stream (Fig. 3), with dissolution effects only on surfaces exposed to the stream water and filamentous microbial mats, can be attributed to the activity of SOB. Examination of the surfaces by CSEM and ESEM revealed a complex reacting environment, with dissolving carbonate, secondary gypsum mineral deposits and a thick cover of predominantly filamentous microbes in an exopolysaccharide matrix.

The results from the experimental *in situ* microcosm chips show that the filamentous sulfide-oxidizing bacteria colonize the calcite surface (Fig. 4a). Over time, the sulfide oxidation results in chemical corrosion of the surface, first producing shallow, etched trenches along the filament (Fig. 4b) and culminating in a broadly etched surface resembling the native limestone fragments (Fig. 5).

FISH probes were applied to the experimental microcosm carbonate surfaces to determine the identities of the filaments colonizing the surfaces (Engel *et al.*, 2004a). Gene probes targeting the LKC group II ('*Epsilonproteobacteria*'), '*Gammaproteobacteria*' (e.g. *Thiothrix* species) and all eubacteria were used. Positive hybridization signals were observed for filamentous organisms and most of the filaments on the carbonate surfaces hybridized simultaneously with the general eubacterial probe and

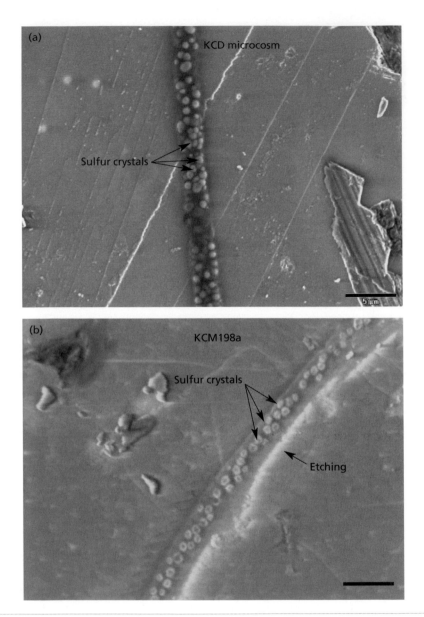

Fig. 4. ESEM image of chips of calcite collected from *in situ* microcosms exposed to microbial colonization for 3 months. (a) Image of S-oxidizing *Thiothrix* sp. filament with abundant stored elemental sulfur. (b) *Thiothrix* sp. filament on calcite surface, with visible etching of the mineral along a filament trench forming in the mineral. Bars, 5 μm.

the LKC group II probe. Exceptionally bright hybridization signals for each of the probes indicated high rRNA content and suggested that the microbes were active when the microcosms were retrieved.

Fig. 5. CSEM photomicrograph of calcite chip from *in situ* microcosm, showing dissolution of the mineral surface (lower left) in association with filamentous micro-organisms that formed a thick biofilm on the mineral surface. Bar, 20 μm.

The rapid loss of sulfide from the cave stream, carbonate dissolution associated with microbial filaments and the dominance of epsilonproteobacteria on the experimental carbonate surfaces support the hypothesis that these organisms, as SOB, have a direct role in sulfuric acid speleogenesis in Lower Kane Cave. The epsilonproteobacteria generate sulfuric acid as a by-product of their metabolism and locally depress the pH at limestone surfaces, which subsequently focuses carbonate dissolution. The previous cave-formation model was based on H_2S volatilization from the cave stream to the atmosphere, but negligible volatilization and abiotic autoxidation of S(–II) were found in the cave. Instead, S(–II) was consumed by subaqueous SOB and cave enlargement occurs via microbially enhanced dissolution of the cave floor.

CONCLUSIONS

The subsurface is a varied habitat for diverse microbial communities, as reactive rock surfaces and mineral-rich groundwater provide a variety of energy sources for micro-organisms, especially chemolithoautotrophs. Stream geochemistry and the spatial relationships of aerobic and anaerobic guilds allow for tight cycling of C and S, constrained by redox boundaries within the mat environment. In Lower Kane Cave, the bulk of the subaqueous microbial mats were dominated by novel evolutionary lineages of the class '*Epsilonproteobacteria*', representing the first non-marine system driven by the activity of this group. Ecologically, the epsilonproteobacteria are chemolitho-autotrophic SOB and their genetic and metabolic diversity creates habitats within the microbial-mat interior for other microbial groups, such as SRB and other anaerobic metabolic guilds.

Nearly all of the S(–II) coming into the Lower Kane Cave is consumed by SOB within the subaqueous microbial mats, and these bacteria drive subaqueous sulfuric acid speleogenesis by attachment to carbonate surfaces and generation of sulfuric acid, which focuses local carbonate undersaturation and dissolution. Prior to this work, sulfuric acid speleogenesis was considered to be an abiotic, subaerial process or to be limited to shallow groundwater depths because of oxygen requirements for abiotic autoxidation of S(–II). However, chemolithoautotrophy linked to S(–II) oxidation under microaerophilic conditions extends the phreatic depths to which porosity and conduit enlargement can occur. The recognition of the geomicrobiological contributions to subaqueous carbonate dissolution fundamentally changes the model for sulfuric acid speleogenesis and supplements how subsurface porosity may develop in karst aquifers.

ACKNOWLEDGEMENTS

We thank the US Bureau of Land Management for permission to access this field site. We are grateful to M. L. Porter, S. A. Engel, T. J. Dogwiler, K. Mabin, M. Edwards, R. Payn, J. Deans and H. H. Hobbs, III, for field assistance. Funding for this work was provided by the Life in Extreme Environments (LExEn) program of the US National Science Foundation (EAR-0085576), the National Speleological Society, the Geological Society of America and the Geology Foundation of the University of Texas.

REFERENCES

Alm, E. W., Oerther, D. B., Larsen, N., Stahl, D. A. & Raskin, L. (1996). The oligonucleotide probe database. *Appl Environ Microbiol* **62**, 3557–3559.

Angert, E. R., Northup, D. E., Reysenbach, A.-L., Peek, A. S., Goebel, B. M. & Pace, N. R. (1998). Molecular phylogenetic analysis of a bacterial community in Sulphur River, Parker Cave, Kentucky. *Am Miner* **83**, 1583–1592.

Ausubel, F., Brent, R., Kingston, R., Moore, D., Seidman, J., Smith, H. & Strujil, K. (1990). *Current Protocols in Molecular Biology*, pp. 1.6.1–1.6.2. New York: Greene Publishing Associates and Wiley-Interscience.

Bachofen, R., Ferloni, P. & Flynn, I. (1998). Microorganisms in the subsurface. *Microbiol Res* **153**, 1–22.

Ben-Ari, E. T. (2002). Microbiology and geology: solid marriage made on Earth. *ASM News* **68**, 13–18.

Bennett, P. C., Hiebert, F. K. & Rogers, J. R. (2000). Microbial control of mineral-groundwater equilibria: macroscale to microscale. *Hydrogeol J* **8**, 47–62.

Bratbak, G. & Dundas, I. (1984). Bacterial dry matter content and biomass estimations. *Appl Environ Microbiol* **48**, 755–757.

Caldwell, D. E., Caldwell, S. J. & Laycock, J. P. (1976). *Thermothrix thiopara* gen. et sp. nov., a facultatively anaerobic facultative chemolithotroph living at neutral pH and high temperature. *Can J Microbiol* **22**, 1509–1517.

Chapelle, F. H., O'Neil, K., Bradley, P. M., Methé, B. A., Ciufo, S. A., Knobel, L. L. &

Lovley, D. R. (2002). A hydrogen-based subsurface microbial community dominated by methanogens. *Nature* **415**, 312–315.

Corbett, C. M. & Ingledew, W. J. (1987). Is $Fe^{3+/2+}$ cycling an intermediate in sulphur oxidation by Fe^{2+}-grown *Thiobacillus ferrooxidans*? *FEMS Microbiol Lett* **41**, 1–6.

Egemeier, S. J. (1981). Cavern development by thermal waters. *Nat Speleol Soc Bull* **43**, 31–51.

Ehrlich, H. L. (1996). *Geomicrobiology*, 3rd edn. New York: Marcel Dekker.

Engel, A. S., Porter, M. L., Kinkle, B. K. & Kane, T. C. (2001). Ecological assessment and geological significance of microbial communities from Cesspool Cave, Virginia. *Geomicrobiol J* **18**, 259–274.

Engel, A. S., Lee, N., Porter, M. L., Stern, L. A., Bennett, P. C. & Wagner, M. (2003). Filamentous "*Epsilonproteobacteria*" dominate microbial mats from sulfidic cave springs. *Appl Environ Microbiol* **69**, 5503–5511.

Engel, A. S., Stern, L. A. & Bennett, P. C. (2004a). Microbial contributions to cave formation: new insights into sulfuric acid speleogenesis. *Geology* **32**, 369–372.

Engel, A. S., Porter, M. L., Stern, L. A., Quinlan, S. & Bennett, P. C. (2004b). Bacterial diversity and ecosystem function of filamentous microbial mats from aphotic (cave) sulfidic springs dominated by chemolithoautotrophic "*Epsilonproteobacteria*". *FEMS Microbiol Ecol* **51**, 31–53.

Felsenstein, J. (1993). PHYLIP (phylogeny inference package), version 3.5c. Department of Genome Sciences, University of Washington, Seattle, USA.

Ford, D. C. & Williams, P. W. (1989). *Karst Geomorphology and Hydrology*. London: Unwin Hyman.

Ghiorse, W. C. (1997). Subterranean life. *Science* **275**, 789–790.

Gough, H. L. & Stahl, D. A. (2003). Optimization of direct cell counting in sediments. *J Microbiol Methods* **52**, 39–56.

Head, I. M., Saunders, J. R. & Pickup, R. W. (1998). Microbial evolution, diversity, and ecology: a decade of ribosomal RNA analysis of uncultivated microorganisms. *Microb Ecol* **35**, 1–21.

Hiebert, F. K. & Bennett, P. C. (1992). Microbial control of silicate weathering in organic-rich ground water. *Science* **258**, 278–281.

Hose, L. D., Palmer, A. N., Palmer, M. V., Northup, D. E., Boston, P. J. & DuChene, H. R. (2000). Microbiology and geochemistry in a hydrogen-sulphide-rich karst environment. *Chem Geol* **169**, 399–423.

Justin, P. & Kelly, D. P. (1978). Growth kinetics of *Thiobacillus denitrificans* in anaerobic and aerobic chemostat culture. *J Gen Microbiol* **107**, 123–130.

Kinkle, B. & Kane, T. C. (2000). Chemolithoautotrophic microorganisms and their potential role in subsurface environments. In *Ecosystems of the World*, vol. 30, pp. 309–318. Edited by H. Wilkens, D. C. Culver & W. F. Humphreys. Amsterdam: Elsevier.

Lane, D. J. (1991). 16S/23S rRNA sequencing. In *Nucleic Acid Techniques in Bacterial Systematics*, pp. 115–175. Edited by E. Stackebrandt & M. Goodfellow. Chichester: Wiley.

Leff, L. G., Dana, J. R., McArthur, J. V. & Shimkets, L. J. (1995). Comparison of methods of DNA extraction from stream sediments. *Appl Environ Microbiol* **61**, 1141–1143.

Lovley, D. R. & Phillips, E. J. P. (1988). Novel mode of microbial energy metabolism: organic carbon oxidation coupled to dissimilatory reduction of iron or manganese. *Appl Environ Microbiol* **54**, 1472–1480.

Madigan, M. T., Martinko, J. M. & Parker, J. (1997). *Brock's Biology of Microorganisms*, 8th edn. Upper Saddle River, NJ: Prentice Hall.

Maher, K. A. & Stevenson, D. J. (1988). Impact frustration of the origin of life. *Nature* **331**, 612–614.

McDougald, D., Rice, S. A., Weichart, D. & Kjelleberg, S. (1998). Nonculturability: adaptation or debilitation? *FEMS Microbiol Ecol* **25**, 1–9.

Millero, F. J., Hubinger, S., Fernandez, M. & Garnett, S. (1987). Oxidation of H_2S in seawater as a function of temperature, pH, and ionic strength. *Environ Sci Technol* **21**, 439–443.

Moyer, C. L., Dobbs, F. C. & Karl, D. M. (1995). Phylogenetic diversity of the bacterial community from a microbial mat at an active, hydrothermal vent system, Loihi Seamount, Hawaii. *Appl Environ Microbiol* **61**, 1555–1562.

Northup, D. E. & Lavoie, K. H. (2001). Geomicrobiology of caves: a review. *Geomicrobiol J* **18**, 199–222.

Palleroni, N. J. (1997). Prokaryotic diversity and the importance of culturing. *Antonie van Leeuwenhoek* **72**, 3–19.

Palmer, A. N. (1991). Origin and morphology of limestone caves. *Geol Soc Am Bull* **103**, 1–21.

Parkhurst, D. L. & Appelo, C. A. J. (1999). Users guide to PHREEQC (version 2) – a computer program for speciation, batch-reaction, one-dimensional transport, and inverse geochemical calculations. In *Water-Resources Investigations Report 99–4259*. Washington, DC: US Geological Survey.

Poulson, T. L. & Lavoie, K. H. (2000). The trophic basis of subsurface ecosystems. In *Ecosystems of the World*, vol. 30, pp. 231–249. Edited by H. Wilkens, D. C. Culver & W. F. Humphreys. Amsterdam: Elsevier.

Roberts, J. A., Bennett, P. C., González, L. A., Macpherson, G. L. & Milliken, K. L. (2004). Microbial precipitation of dolomite in methanogenic groundwater. *Geology* **32**, 277–280.

Rogers, J. R., Bennett, P. C. & Choi, W. J. (1998). Feldspars as a source of nutrients for microorganisms. *Am Miner* **83**, 1532–1540.

Rogers, J. R., Bennett, P. C. & Hiebert, F. K. (1999). Patterns of microbial colonization on silicates. In *US Geological Survey Toxic Substances Hydrology Program: Proceedings of the Technical Meeting*, pp. 237–242. Edited by D. W. Morganwalp & H. T. Buxton. Charleston, SC, USA, 8–12 March 1999.

Ronquist, F. & Huelsenbeck, J. P. (2003). MrBayes 3: Bayesian phylogenetic inference under mixed models. *Bioinformatics* **19**, 1572–1574.

Sarbu, S. M., Kane, T. C. & Kinkle, B. K. (1996). A chemoautotrophically based cave ecosystem. *Science* **272**, 1953–1955.

Sarbu, S. M., Galdenzi, S., Menichetti, M. & Gentile, G. (2000). Geology and biology of Grotte di Frasassi (Frasassi Caves) in central Italy, an ecological multi-disciplinary study of a hypogenic underground karst system. In *Ecosystems of the World*, vol. 30, pp. 361– 381. Edited by H. Wilkens, D. C. Culver & W. S. Humphreys. Amsterdam: Elsevier.

Scholten, J. C. M., van Bodegom, P. M., Vogelaar, J., van Ittersum, A., Hordijk, K., Roelofsen, W. & Stams, A. J. M. (2002). Effect of sulfate and nitrate on acetate conversion by anaerobic microorganisms in a freshwater sediment. *FEMS Microbiol Ecol* **42**, 375–385.

Stevens, T. O. & McKinley, J. P. (1995). Lithoautotrophic microbial ecosystems in deep basalt aquifers. *Science* **270**, 450–455.

Stoessell, R. K., Moore, Y. H. & Coke, J. G. (1993). The occurrence and effect of sulfate reduction and sulfide oxidation on coastal limestone dissolution in Yucatan cenotes. *Ground Water* **31**, 566–575.

Swofford, D. L. (2002). PAUP*: Phylogenetic analysis using parsimony (*and other methods) (version 4). Sunderland, MA: Sinauer Associates.

Takai, K., Inagaki, F., Nakagawa, S., Hirayama, H., Nunoura, T., Sako, Y., Nealson, K. H. & Horikoshi, K. (2003). Isolation and phylogenetic diversity of members of previously uncultivated ε-Proteobacteria in deep-sea hydrothermal fields. *FEMS Microbiol Lett* **218**, 167–174.

Thamdrup, B., Finster, K., Hansen, J. W. & Bak, F. (1993). Bacterial disproportionation of elemental sulfur coupled to chemical reduction of iron or manganese. *Appl Environ Microbiol* **59**, 101–108.

Thompson, J. D., Gibson, T. J., Plewniak, F., Jeanmougin, F. & Higgins, D. G. (1997). The CLUSTAL_X windows interface: flexible strategies for multiple sequence alignment aided by quality analysis tools. *Nucleic Acids Res* **25**, 4876–4882.

Ulrich, G., Martino, D., Burger, K., Routh, J., Grossman, E. L., Ammerman, J. W. & Suflita, J. M. (1998). Sulfur cycling in the terrestrial subsurface: commensal interactions, spatial scales, and microbial heterogeneity. *Microb Ecol* **36**, 141–151.

Vestal, J. R. & White, D. C. (1989). Lipid analysis in microbial ecology: quantitative approaches to the study of microbial communities. *Bioscience* **39**, 535–541.

Vlasceanu, L., Sarbu, S. M., Engel, A. S. & Kinkle, B. K. (2000). Acidic cave-wall biofilms located in the Frasassi Gorge, Italy. *Geomicrobiol J* **17**, 125–139.

Voordouw, G., Armstrong, S. M., Reimer, M. F., Fouts, B., Telang, A. J., Shen, Y. & Gevertz, D. (1996). Characterization of 16S rRNA genes from oil field microbial communities indicates the presence of a variety of sulfate-reducing, fermentative, and sulfide-oxidizing bacteria. *Appl Environ Microbiol* **62**, 1623–1629.

White, W. B. (1988). *Geomorphology and Hydrology of Karst Terrains*. New York: Oxford University Press.

White, D. C., Davis, W. M., Nickels, J. S., King, J. D. & Bobbie, R. J. (1979). Determination of the sedimentary microbial biomass by extractable lipid phosphate. *Oecologia* **40**, 51–62.

Widdel, F. & Bak, F. (1992). Gram-negative mesophilic sulfate-reducing bacteria. In *The Prokaryotes*, pp. 3352–3378. Edited by A. Balows, H. Trüper, M. Dworkin, W. Harder & K.-H. Schleifer. New York: Springer.

INDEX

References to tables/figures are shown in italics

Acetate oxidation 191
acetogenesis 111, 189, 241
Acharax 309
Achromatium 35–37
    A. oxaliferum 36
    A. volutans 53
    calcite deposition 51, *52, 53*
    genetic and ecological diversity 60–61, *62, 63*
    RY5, RYKS, RY8 spp. 61
    sulfur cycle 46, 50, 51, 348
acid mine drainage (AMD) 12, 14, 23, 25, 26
Acidobacteria 354
acidolysis, by fungi 207, 208–211, 214, 215, 218
Acinetobacter sp. B0064 *117*, 118
Acremomium-like hyphomycete 215–216
actinides, microbial reduction 286–291
Actinobacteria 354
Aeromonas 277, 279
Ag
    fungal interactions 211, 215
    microbial reduction 282
    microbial resistance 115, 273
aggregation
    diffusion-limited (DLA) 142–143
    reaction-limited (RLA) 142–143
Al, fungal interactions 212, 216
'*Alphaproteobacteria*' 354
Alternaria alternata 337
Alteromonas putrefaciens see Shewanella oneidensis
americium (Am), microbial reduction 289
ammonia
    anaerobic oxidation 152, *164*
    marine cycling 250–251
ammonium, in primary production 181
anaerobic methanotrophs (ANMEs) 308, 310, 311, 312, 313, *314*, 316
Antarctic dry valleys 71–73
    location *72*
    microbial activity in 73, 75–76
    organisms in 73, *74*
    sources of resources 76–81
AOM *see* methane (anaerobic oxidation)
apatite 211
archaeol 307
archaeal 'strain 121' 277
Arthrobacter 113, 115, *117*, 119
As
    fungal interactions 215

groundwater contamination 283–284
    microbial influences 22
    microbial reduction *164*, 283–286
    microbial resistance 273, 284
Ascocratera manglicola 333
ascomycetes 206, 326, *327*, 337
Aspergillus 218, 220, 330
    A. fumigatus 325
    A. nidulans 215
    A. niger 208, 218
    A. sydowii 325
    P37 sp. 215
Astrosphaeriella striatispora 333
ATPases, P_{IB}-type 115–122
Aureobasidium pullulans 334
autunite 212
azo dyes 293–294
Azotobacter vinelandii 120

Bacillus
    B. arseniciselenatis 285
    B. selenitireducens 285
    B. sp. U26 123
    B. sp. V6 123–124
    B. subtilis 87–88, *89–90*
bacterial cell wall 85–104
    cell surface complexity 94–99
    cryo-transmission electron microscopy 86–93
    hydrophobicity studies 99–104
    metal–ion interaction and mineralization 91–93
    potentiometric properties 93–99
'*Bacteroidetes*' 348, 354
basalt 205, 208, 234
Basidiomycotina 327
Beauveria caledonica 209, *210*, *212–213*
Beggiatoa 37–38
    B. alba 45
    carbon cycle 55, 56
    nitrogen cycle 53, *54*
    sulfur cycle 44, 45, 46, 47, 309
    within-genus diversity 57–59
benthic communities
    organic matter content *178*
    organic matter degradation *178, 179, 180, 182–183*, 188–191
    oxygen uptake *182*
    respiration 186
bentonite 141, 217, *218*, 219

beta-imaging 5
'*Betaproteobacteria*' 354
biocatalysts 273, 281, 283–294
biofilms
    existence, structure and function 22–23, *24*
    formation *24*
    in mycotransformation of minerals 205, 206
    intraterrestrial environment 239
    metabolic links to metals 23
    metal geochemistry interactions 11–28
        aqueous, fundamentals 12–18
        emerging technologies 19, 27–28
        environmental aspects 23, *24*, *25*, *26*
        important solid phases 13–15
        microbial functional metabolism 19–22
        role of redox status 18
        sorption reactions 15, *16*, 17–18
biogeochemical cycling, in coastal zones 173–193
        functional groups 188–191
        organic matter degradation and recycling
            182–183
        physico-chemical differences in latitudinal
            regions 173–175
        primary production *174*, 175–182, 183
        role of coastal wetlands 191–193
        temperature–substrate concentration
            interaction 183–188
biogeochemical cycling, marine ecosystems
        fungal contribution 336–338
        role of fungi 330–336
            C, N and P 331–335
            sulfur 331, *332*, 335–336
biomagnetite 160–161
biomarkers, lipid 133, 306–308
biomass, microbial
        in intraterrestrial environment 239
        marine, resource availability 261–265
        subsurface microbial mats 110, 351, 353, 356
biomineralization
        accidental biominerals *156*, 157–158
        bacterial cell wall interactions *see* bacterial
            cell wall
        biofilms 23–26
        biomagnetite 160–161
        eukaryotic 159–160
        prokaryotic 157–159, 160
        purposeful biominerals *158*, 159–160
        see also biosilification, mycotransformation
bioremediation
        fungal role 220
        metal reduction 273, 278, 279, 282, 287, 289,
            291
        metal resistance 112–114
        radioactive waste 280, 286–291

biosensors, Hg 282
biosignatures 165–166, 216
biosilification 131–145
        chemistry of silica 133–138
            opal-A 137, 138
            precipitation rate, calculation 136
            saturation state 135–136, 137
        cyanobacterial biomineralization pathways
            and colloid aggregation 141–144
        cyanobacterial surface properties and
            function 138–141
biotite 208–209, 216
bioweathering, defined 203
        see also weathering
Black Sea
        AOM studies *307*, 309–310, 311, 312
        redox zones 165, *166*
borehole techniques 236–238
Botrytis 218
bromoethanesulfonate (BES) 313
Buergenerula spartinae 337
building stone, deterioration 208–211
butyrate-oxidizing bacteria, identification *4*

C cycle
        biomineralization and 159
        fungal interactions 210
        giant sulfur bacteria 55–56
        in Antarctic dry valleys 76–81
        in coastal zones 187, 189, 192–193
        intraterrestrial environment 241–243
        life detection and definition 162
        marine, fungal role 331–335
        subsurface ecosystems 345, 346, 353, 356
            coupled C and S metabolism 348–349
        see also primary production
Ca
        fungal interactions 209, 210–211, *212*,
            213–214, 214–215, 216, 218, 219
caesium trifluoroacetate density-gradient
        centrifugation 4
calcite
        coccolithophores 158, 262–265
        fungi 214–215
        sulfur-oxidizing bacteria 35, 51, 52, 53,
            214–215, 357–358, *359*
calcium carbonate 210–211
calcium oxalate 209, *212*, 213–214, 215
calcretes 210, 214
Calothrix 95–96
Calyptogena 309
carbonates
        metal geochemistry 13, 14
        precipitation 212, 214–215

carbon-concentrating mechanisms (CCMs) 257–258, 259–261
carbonic acid/carbonate dissolution 347, 352, 353, 357, 359, 360
carbonic anhydrases 257–258, 259, 261
cave ecosytem *see* subsurface, cave ecosystem
Cd
    fungal interactions 211, 214, 220
    in marine primary production 257, 258
    resistance of subsurface bacteria 115
cell wall *see* bacterial cell wall
chasmolithic organisms 203, *204*, 210
chemical gradients 165–166
chemo-autotrophy, hydrogen-driven 111, 112
chemolithoautotrophic metabolism 111–112, 345, 346, 347, 349, 352, 356
chemotropism 207
Chilean coastal sediments 50, 54–55, 58
Chilean continental margin 305, 307
chlorite 208, 216
Chlorobi 354
Chloroflexi 354
Chrysiogenes arsenatis 284
Cirrenalia pygme 333
Citrobacter 290
Cladosporium cladosporioides 218
Claviceps purpurea 337
clay–fungal interactions 216–218, 220
Clostridium 287
    C. pasteurianum 278
Co
    fungal interactions 214, 220
    in marine primary production 257, 259
    microbial reduction 280
CO_2
    biogeochemical cycling in coastal zones 173, 189, 192–193
    in marine primary production 247, 259, 263
    intraterrestrial environment 241
coal-solubilizing fungus 336
coastal wetlands, biogeochemical cycling 191–193
coastal zones, defined 173
    see also biogeochemical cycling in coastal zones
coccolithophores 52, *158*, 159, 262–265
cold-seep environments 55–56
cold water paradigm 183
Comamonas 116, *117*, *118*
    sp. B0173 123
community gene arrays (CGA) 122
community structure, isotope labelling 1–8
complexolysis, by fungi 207, 209, 219
concrete biodeterioration 219
contact angle measurement (CAM) 101–102, 103

contact guidance 207
contamination *see* bioremediation, horizontal gene transfer, radioactive waste
core-drilling 236–238
Corollospora maritima 333
Cortinarius glaucopus 208
Cr
    fungal interactions 220
    metal geochemistry 18, 22
    microbial reduction 279–280
    resistance of subsurface bacteria 113, 114
crassulacean acid metabolism (CAM) 257
crocetane 307
cryo-transmission electron microscopy (cryoTEM)
    bacterial cell wall 86–93
        freeze-substitution 86–88, 90–92
        frozen hydrated thin sections 88–91
        Gram-positive cell walls 87–88, 89–90
        metal–ion interaction and mineralization 91–93
cryptoendolithic organisms 203, *204*, 206, 210
Cryptovalsa halosarceicola 333
Cu
    fungal interactions 211–212, 214, 216, 220
    resistance of subsurface bacteria 114, 115
cyanobacteria
    gene transfer 121
    in biosilification 133
        biomineralization pathways and colloid aggregation 141–144
        sheath 139–140, 141–143
        surface properties and function 138–141
    marine primary productivity 249, 251, 252–253
cytochromes 55–56
    b 279, 286
    c 279, 282–283, 286, 287–288
    c-type 277, 278, 282–283
    d 279

Dating methods
    groundwater 235
    metabolic abilities 161–166
day length, coastal zones *174*, 175, 176
Dead Sea, marine fungi 328
Debaryomyces hansenii 329
defence compounds 262
Deinococcus radiodurans 119, *120*, 121, 287
'*Deltaproteobacteria*' 354
Dendryphiella salina 328, 329
denitrification
    energy yield 47–49
    giant sulfur bacteria 47–49, 53–55

in coastal zones 189–190
desert varnish 215–216
Desulfobacter 191
Desulfobulbus 308
Desulfocapsa 308
Desulfococcus 308, 309, 310
Desulforhopalus 308
Desulfuromonas 191
Desulfosarcina 308, 309, 310
Desulfotalea/Desulforhopalus group 190
Desulfotomaculum 191
    D. auripigmentum 284
Desulfovibrio 190
    D. desulfuricans 114, 278, 280, 285, 287, 288
    D. vulgaris 287
deuteromycetes (Fungi imperfecti) 206
Deuteromycotina 327
diatoms, silicification by 262–265
diazotrophy 250, 253, 255
Digitatispora marina 333
DNA-SIP 3
DNRA, giant sulfur bacteria 47–49, 54–55
dolocretes 210
dolomite 210
Drechslera halodes 337
dry valleys *see* Antarctic dry valleys
dunite 208

Eckernförde Bay *305*, 312
edge effects *16*, 17
electricity generation 273
electrobioreactor 281
electron shuttles 277–278, 292, 314
electrophoresis, denaturing-gradient gel (DGGE)
    4
electrostatic interaction chromatography (ESIC)
    102
endolithic organisms 203, *204*, 206, 209, 210, 219
energy sources
    intraterrestrial environment 241–242
    microbial 153–154, 155, *156*, 157
    solar, in coastal zones 173–175
energy yield
    dentrification 47–49
    sulfur oxidation 47–49
Enterobacter 280
    E. cloacae 279, 285
Entromyces callianassae 334
environmental aspects, biofilm metal geochemistry
    23, *24*, 25, 26
environmental genomics 8
environmental proteomics 8
Epicoccum nigrum 337
epilithic organisms 203, *204*, 206, 209

'*Epsilonproteobacteria*' 37, 43, 56, 345, 348, 349,
    352, 353, 354, *355*, 356, 357–358, 359
ergosterol 337
Escherichia coli 215, 279, 285, 286, 290, 291, 292
estuarine flushing 174–175
Ethmodiscus 264
eukaryotes
    anaerobic 154–155
    extremophily 166–167
    genetic diversity 152, *153*
    metabolic diversity 152, 154–155, 159–160
    mineral formation 159–160
euoendolithic organisms 203, *204*
exchangeable fraction 17–18
exobiology 168
exoenzymes, fungal 333, 334
extremophiles 166–167, 206

Fe
    anaerobic oxidation *164*
    biofilms 20, 21, 22, *24*, 25, 26
    fungal interactions 212, 213, 215, 216, 218
    in marine primary productivity 252–255
        assimilation of inorganic C 259–261
        fate of biomass 261–265
        interactions with N, P, Zn and other resources
        258–265
    limitation, biogeochemical cycling 180–181,
        182
    microbial reduction 152, *164*, 274–278
        degradation of xenobiotics 274
        diversity of organisms 275–278
        in coastal zones 189, 191
        mechanisms 277–278
        subsurface bacteria 111
    redox status 18
    solid phase reactions 13, 14
    sorption reactions 15, *16*, 17
feedback, geochemical 21, 22, 27
feldspar 208, 209, 218
Ferrimonas 277
FISH (fluorescence *in situ* hybridization) 308,
    354–356
    FISH-MAR 5, 6, 7
fission products, microbial reduction 286–291
flavodoxin 253
food reserve concept 187
formate hydrogenlyase 291, *292*
forsterite 216
freeze-substitution 86–88, 90–93
freshwater sediments, giant sulfur bacteria 50, 59
frozen hydrated thin sections 88–91
functional gene array (FGA) 122

fungi, in marine ecosystems
contribution to cycling 336–338
exoenzymes 333–334
fungal adaptation 328–330
fungal diversity 324–325
in biogeochemical processes 330–336
marine fungus, definition 325
osmotic adaptation 328–330
salt tolerance 328–330
fungi in rock, mineral and soil transformations 201–221
as biosorbents 211–212, 220
environmental biotechnology 219–220
bioremediation 220
fungal–clay interactions 216–218, 220
in terrestrial environment 205–206
metal binding and accumulation 211–212
mycogenic mineral formation 212–216
carbonate precipitation 214–215
oxalate precipitation *212*, 213–214
reduction or oxidation of metals/metalloids 215–216
processes influenced by minerals 205
rock and mineral habitats 203–205
tropic responses 207
weathering processes 202–203, 206–208
biochemical 207–208
biomechanical 206–207
clay and silicate 217–218
concrete biodeterioration 219
rock and building stone 208–211
Fungi imperfecti 206
Fusarium lateritium 336

'**G**ammaproteobacteria' 6, 37, 38, 42, 43, 56, 60, 354, *355*, 357
genetic diversity 152, *153*, 240
see also horizontal gene transfer
genome, P content 256
Geobacter 191, *276*, 277–278, 284, 288, 290, 291
G. metallireducens 275, 276, 278, 282–283, 287, 290
G. sulfurreducens 291
strain GS-15 *see G. metallireducens*
Geobacteraceae 111
Geospirillum barnesii strain SES-3 *see Sulfurospirillum barnesii*
Geovibrio ferrireducens 282
giant sulfur bacteria 35–64
descriptions of genera 35–43
energetic considerations of sulfur oxidation 44–49
energy and geochemical significance 43–56
evolutionary and ecological diversity 56–61

in *Achromatium* 60–61, 62, *63*
metagenomics 61–64
within-genus diversity 57–60
geochemical significance 49–56
carbon cycle 55–56
nitrogen cycle 53–55
sulfur cycle 49–53
phylogenetic tree *37*
glass, deterioration 208
gold (Au) reduction 281–282
Gram-negative bacteria
cell surface 101, *102*
cell wall 94, *95*, 99
metal homeostasis genes 115, *118*, 124
see also cyanobacteria
Gram-positive bacteria
cell wall 87–88, *90*, 91–93, 99
metal homeostasis genes 115, 119, *120*, 123, 124
granite 209, 218
grazers, marine 261–262
groundwater 235–236
contamination 112
As 283–284
radioactive 286–287
dating methods 235
mixing processes 235–236
organisms 239–241
sampling techniques 236–238
Gymnascella 328
gypsum 216, 347, *357*

Halocyphina villosa 333
Halosphaeria hamata 337
Halosphaeriales 326
heavy metals
interactions of subsurface bacteria 112–114
subsurface contamination 109, 112–114
bioremediation 112–114
see also specific metals
Heleboma 217
HetCO$_2$-MAR 7
heterotrophic leaching 207
heterotrophic metabolism, subsurface 110–112
Hg
biosensors 282
fungal interactions 215
microbial reduction 282–283
microbial resistance 113, 273, 282–283
homeoviscous model 184
horizontal gene transfer (HGT)
in life detection and definition 163–164
metal homeostasis genes 114–124
evolution 114–115
mechanisms 114–115

microarray 122–124
  P$_{IB}$-type ATPases 115–122
  subsurface environment 114–124
hornblendes 209
hot-spring environment
  biosilification 131–133, 144–145
  substrate sequestration 184
HRABT (high-resolution acid–base titrations)
  93–94, 95–96, 97–98
Hyalodendron 217
Hydrate Ridge *305, 307,* 309, 310, 311, 312
hydrogen
  intraterrestrial environment 241
hydrogen-driven biosphere hypothesis *241*
hydrogenase 3 291, *292*
hydrophilicity, microbial surface 100, 103
hydrophobic interaction chromatography (HIC)
  101, 102–103
hydrophobicity, microbial surface 99–104
hydroxy-archaeol 307
Hymenoscyphus ericae 210, 217
hyperthermophiles 238, 290
hypolithic organisms 203, *204*
Hysterangium crassum 218

Insolation, coastal zones 173–174, 176, 179–180
interfacial reactions *see* sorption reactions
intraterrestrial environment 233–243
  definition 234
  organisms 238–239
    activity 242–243
    energy sources 241–242
    species diversity 240–241
  range of biomass 239
  strategies for exploration 236–238
  variability 234–236
invertebrates, in Antarctic dry valleys 73, *74*
ion concentrations, marine *323,* 324
ionic strength 15, *16,* 17–18
Iron Mountain ecosystem 62–63
isotope-array approach 5
isotopic labelling methods 1–8, 27–28
  DNA-SIP 3
  FISH-MAR 5, 6, 7
  HetCO$_2$-MAR 7
  intraterrestrial environment 242
  life detection and definition 162
  RNA-SIP 3–5
isotopic signals for soils 77, 78

K
  fungal interactions 209
kaolin 208
kaolinite 217, *218*

karst landscape formation 345–346
kinetic biosignatures 165–166

Laccaria laccata 210
laccase-like multicopper oxidase 216
Lactococcus lactis 115
layered communities 165–166
legacy model 76, 79
Legionella pneumophila 119, *120*
Leptosphaeria
  L. obiones 337
  L. pelagica 337
Leucothrix 41–42
Lichenothelia 215
lichens 73, 201, 207, 208, 214, 216
  marine 327
life detection and definition 161–166
lignin-degrading organisms 334
lignocellulosic enzymes 337
limestone 209, 210, 211, 214–215, 352–353,
  357–358
Linocarpon bipolaris 333
Lower Kane Cave 350, 352, 353–359
Lulworthia 333, 337

Magnetite 160–161, 274, *275, 276*
magnetosomes *160,* 161
Magnetospirillum magnetotacticum 160
magnets, 'designer' 275
mangroves 191–192, 323, 326, 327, 335
marble 209
marine ecosystems
  characteristics 321–324
    ion concentrations *323,* 324
  fungal adaptation 328–330
  fungal contribution to cycling 336–338
  fungal diversity 324–328
  fungi in biogeochemical processes 330–336
  interactions 321, *322*
  primary productivity 322–323, 336–338
marine snow 248, 264
mass spectrometry, secondary-ion 308
mats, microbial
  methanotrophic 309–310
  subsurface ecosystems 345, 346, 347–348, 352,
    353–359
    biomass determination 351, 353
    see also Beggiatoa, Thioploca
melanin 207, 211, 212
membrane fluidity 184
mercury-resistance operon 282
mesophiles 184, 186
metabolic diversity 151–168
  extremophily 166–167

kinetic biosignatures and layered communities 165–166
microbial diversity 152–157
mineral formation 157–161
relevance to exobiology 152, 168
timing emergence of metabolic abilities 161–166
metabolism, microbial
functional 19–22
subsurface 110–112
metagenomics, giant sulfur bacteria 61–64
metal accumulation, by fungi 207, 209, 211–212, 217, 220
metal geochemistry
aqueous, fundamentals 12–18
important solid phases 13–15
role of redox status 18
sorption reactions 15, 16, 17–18
bacterial cell wall interactions see bacterial cell wall
biofilms see biofilms – metal geochemistry interactions
overlap with microbiology 19–27
biofilm metabolic links 23
environmental aspects 23, 24, 25, 26
microbial functional metabolism 19–22
metal homeostasis genes, in subsurface microbial communities 114–124
metalloids, reduction 283–286
metal reduction see specific metals
metal resistance, subsurface bacteria 112–114
metal sequestration, fundamentals 12, 13–18, 25
metatorbenite 211–212
metazeunerite 211
methane, aerobic oxidation 304
methane, anaerobic oxidation (AOM) 152, 159, 164, 303–316
consortium 306, 307, 308, 312, 313
environmental regulation 310–312
history 305–306
laboratory growth 315
mechanism 312–315
micro-organisms 306–308
ANME groups 308, 309, 310, 311, 312, 313, 314, 316
study environments 308–310
Black Sea 307, 309–310, 311, 312
Hydrate Ridge 305, 307, 309, 310, 311, 312
methane metabolism, giant sulfur bacteria 56
methane production 303–304
Methanococcoides 190
methanogenesis
AOM and 306
in Antarctic dry valley soils 80–81

in coastal zones 189, 191
intraterrestrial 241
subsurface bacteria 111–112
methanogenic enzymic system 313, 314
Methanomicrobiales 190, 308
Methanosarcinales 190, 308
Mg
fungal interactions 208–209, 216, 218
mica 216, 218
microarray
community gene arrays (CGA) 122
functional gene array (FGA) 122
horizontal gene transfer (HGT) 122–124
phylogenetic oligonucleotide array (POA) 122
microautoradiography see FISH-MAR, HeCO₂-MAR
microbial communities, isotopic labelling 1–8
'*Microbulbifer flagellatus*' 120
microcline 208–209
Micrococcus 280
'*M. lactilyticus*' 278
microcolonial fungi 206, 209
microelectrophoresis see zeta-potential analysis
microfossils 85–86
formation 145
in life detection and definition 162–163
mineralized component 14
minerals
influence on microbial processes 205
mycotransformation 201–221
biochemical deterioration 202, 207–211
environmental biotechnology 219–220
fungal–clay interactions 216–218
mycogenic formation 212–216
secondary 202–203
see also biomineralization
mitosporic fungi 326, 327
Mn
biofilms 24
fungal interactions 214, 215–216
in marine primary production 258
microbial reduction 152, 164, 189, 191, 274–275, 277–278
redox status 18
solid phase reactions 13, 14
sorption reactions 16, 17
Mo, microbial reduction 280
molybdate 313
montmorillonite 217
Mucor 218
muscovite 208
Mycena galopus 208
Mycosphaerella 325
sp. 2 337

N cycle
    giant sulfur bacteria 47–49, 53–55
    in Antarctic dry valleys 76, 77, 78–81
    in coastal zones 187, 188, 189–190, 191–192
    in marine primary productivity 249–252
        assimilation of inorganic C 259–261
        fate of biomass 261–265
        fungal role 331–335
        interactions with Fe, P, Zn and other
            resources 258–265
        recycled production 250–251
Namibian coastal sediments 49–50
natural organic matter (NOM)
    biogeochemical cycling in coastal zones 176,
        178, 179, 182–183, 188
    in Antarctic dry valleys 73, 74, 75–76, 78–81
    in metal geochemistry 13, 15, 18, 25
nematodes, in Antarctic dry valleys 73, 74
nepheline 208
neptunium (Np) 287, 289–290
Ni, fungal interactions 211, 220
nitrates
    ammonification *see* DNRA
    biogeochemical cycling in coastal zones
        181–182, 189–190, 192
    reduction *see* denitrification
    storage by giant sulfur bacteria 38, 39, 53–55,
        57–58
nitrification, Antarctic dry valleys 78, 79
nuclear magnetic resonance (NMR) 96, 97, 99
nutrient fluxes, coastal zones 174–175, 176
nutrient sources, coastal wetlands 191–192

Oidiodendron maius *210*
olivine 208, 216
organic acids, fungal production 207–211, 214,
    215, 218
organic matter *see* natural organic matter
osmotic adaptation by fungi 328–330
outwelling, coastal wetlands 191
oxalate precipitation *212*, 213–214
oxalic acid 209, *210*, 214, 215, 218
oxidants, use by life forms *154*
oxygen concentration/flux, biofilms 20–21, *24*
oxyhydroxide minerals 13, 14, *16*, 17, 18, *20*, 22

P cycle
    in coastal zones 187, 188, 192
    in marine primary productivity 255–257
        assimilation of inorganic C 259–261
        fate of biomass 261–265
        interactions with Fe, N, Zn and other
            resources 258–265
    marine, fungal role 331–335

Pacific coastal sediments 49
Paecilomyces 218, 336
palladium (Pd) 281
palygorskite 217, *218*
Paracoccus denitrificans 290
parasitism, marine 248, 262
PAR (photosynthetically active radiation) 247, 260
Paxillus involutus 210, 211
Pb
    fungal interactions 209, *210*, 211, 214, 220
    resistance of subsurface bacteria 113, 114,
        115, 116
pelagic primary production 176–179, 179–180, *181*
Pelobacter 191, 276
Penicillium 218, 220, 325, 328, 330
    P. expansum 208
    P. frequentans 218
    P. simplicissimum 208
perchlorate reduction *164*
periplasmic space 90, 91, 93
pH
    effects on AOM 311
    in trace-metal sorption 17
    metal binding by fungi 211
Phaeosphaeria
    P. halima 325, 337
    P. spartinicola 325, 337
    P. typharum 333, 337
phosphite oxidation *164*
phosphorimaging technique 290
photosynthesis 155, *156*, 157, 176, 243
    marine primary production 247, 248, 251,
        259, 260
phototrophic component *24*, *25*
phylogenetic analysis
    microbial mats 348, 352
phylogenetic oligonucleotide array (POA) 122
phylogenetic tree
    life detection and definition 164–165
    microbial diversity 152, *153*, 164–165
phytoextraction 220
phytoremediation 220
pili, in electron transfer 278
Piloderma 208, 209
Planctomycetes 354
platinum-group metals (PDMs) 281
Pleospora
    P. pelagica 333, 337
    P. vegans 333, 337
plutonium (Pu) 287, 289–290
poikilophilic organisms 216
poikilotrophic organisms 205
polar coastal zones
    benthic oxygen uptake *182*

biogeochemical cycling 173–193
  primary production rates *174*, 175–182
  seasonal variation in characteristics 173–175,
    176–180
polyols 329
potentiometric properties of cell surfaces 93–99,
  99
primary production
  in coastal zones 173, *174*, 175–182, 183
  marine
    assimilation of inorganic C 259–261
    constraints 249–258
    fate of biomass 261–265
    Fe, N, P and Zn cycling 247–265
    interactions among resources 258–265
    marine ecosystems 322–323
      fungal contribution 336–338
    seasonal variation 176–180
Prochlorococcus 251, 252, 253, 256
prokaryotes
  extremophily 166–167
  genetic diversity 152, *153*
  metabolic abilities *164*
  metabolic diversity 152–154, 157–159, 160
  mineral formation 157–159, 160
Proteobacteria 354
  metal homeostasis genes 113, 116, *117*, *118*,
    119, *120*
protonation *see* acidolysis
Pseudomonas *118*, 279, 290
  P. aeruginosa 94, 95, 96–97, *120*
  P. ambigua 279
  P. guillermondii 280
  P. isachenkovii 278
  P. putida 279, 287
  P. stutzeri 285
  P. vanadiumreductans 278
psychrophiles 184–185, 186
psychrotolerance 184
Pyrobaculum
  P. aerophilum *120*, 121, 282
  P. islandicum 282, 290
Pyrococcus
  '*P. abyssi*' *120*, 121
  P. furiosus 282
pyromorphite 209, *210*

Quartz 208

Radioactive waste 219
  microbial degradation 280, 286–291
radiotropism 219
Ralstonia
  R. B0665 116

  R. metallidurans 115
redox chemistry 157
  metabolic diversity 151, 153–154
redox ladder *242*, 243
redoxolysis, by fungi 207, 211
redox status 18
  Achromatium communities 61
redox zones, Black Sea 165, *166*
refractory component 14
remazol black B *293*
Rhizoctonia solani 217
rhizoferrin 208
Rhizophila marina 333
Rhizosolenia 264
Rhodobacter sphaeroides 286, 290
rhodochrosite 215
Rhodoferax ferrireducens 111
rhodopsin 155
RNA, as biomarker 3–5
RNA-SIP 3–5
rock
  hard 234, 239, 242
  intraterrestrial environment 234–243
  microbial habitats 203–206
  mycotransformation of minerals 201–221
    weathering processes 202–203, 206–211
  sedimentary 234, 239, 242
rosette formation 41–42, 57
Rubisco 258, 261

Saccharomyces cerevisiae 215, 329
salinity 322, 323–324
  fungal adaptation 328–330
Salmonella typhimurium *120*
salt marshes 191–192
  marine fungi 323, 326, 335, 336, 337
salt-tolerant fungi 325, 328–330
sandstone 209, 211, 218
Sargasso Sea, metagenomic analysis 62, 63
Scopulariopsis 218
  S. brevicaulis 208
Scottnema lindsayae 73, 74
scytonemin 139
Se
  fungal interactions 215
  microbial reduction 285, 286
sedimentary pools 12, 13–15
seep sites, giant sulfur bacteria 56
selenate reduction *164*
serpentine 208
Shewanella
  cell surface properties 96, 97, 98–99, 101–102,
    103
  metal homeostasis genes 114

metal reduction 277
S. alga BrY 97, 98–99, 102
S. algae 282
S. oneidensis 98–99, 101, 102, 275, 277, 278, 279, 287
S. putrefaciens 99, 103, 157–158, 287, 288, 290
    see also S. oneidensis
strain J18143 293
Si cycle
    biomineralization 159
    fungal interactions 213
siderite 215, 274, 275
siderophores 207, 208, 252–253
silica
    amorphous (defined) 133
    chemistry 133–138
silicates
    clay–fungal interactions 216–218
    fungal deterioration 208
silicification by diatoms 262–265
silification *see* biosilification
sinter formation *see* biosilification
soil
    amelioration, fungi 220
    Antarctic dry valleys 73, 75–81
        microbial activity 73, 75–76
        organisms in 73, 74, 206
        properties 73, 74, 75–76
        resources 76–81
        respiration 75–76
    clay–fungal interactions 216–217
    radioactive contamination 286–287
sorption reactions in biofilms 12–13, 15, *16*, 17–18
Spartina alterniflora 325, 336–337
spatial-subsidy model 79, *80*, 81
speleogenesis *see* subsurface, cave ecosystems
Sphingomonas 118
    S. sp. F199 *118*
Spirochaetes 354
spodumene 208
Sr, fungal interactions 214
Stagonospora 37
Staphylococcus aureus 115
Stenotrophomonas maltophilia 115, 119, *120*
stress avoidance, fungal 207
string-of-pearls morphology 42
substrate utilization rate 186–187
subsurface, cave ecosystem 345–360
    coupled C and S metabolism 348–349
    karst aquifers 347
    microbial-community analysis 350–352
        microbial mat diversity 354–356
    microbial geochemistry methods 349–359
        geochemical characterization 350

in situ microcosms 352–353, 357–358, 359
        Lower Kane Cave 353–359
    sulfur-based 347–348
subsurface, deep terrestrial 109–110
    interactions of bacteria with heavy metals 112–114
    metal homeostasis genes in microbial communities 114–124
    microbial biomass and diversity 110
    microbial metabolism in 110–112
Suillus
    S. bovinus 210
    S. granulatus 211
    S. luteus 210
sulfate
    in AOM 304–306, 307, 309, 310, 312–315, 316
    reduction
        coastal zones 178
        in coastal zones 189, 190
        subsurface bacteria 111
sulfides, crystalline metal 161
sulfidic minerals
    fundamental reactions 13, 14
    in biofilm formation *24*
    weathering and oxidation 25, 26
sulfites
    reduction, in coastal zones 191
sulfur cycle
    in life detection and definition 162, 163
    marine, fungal role 331, *332*, 335–336
    subsurface ecosystems 345–360
        coupled C and S metabolism 348–349
    see also giant sulfur bacteria
Sulfurihydrogenibium subterraneum 112
Sulfurospirillum
    S. arsenophilum 284
    S. barnesii 284, 285
    strain MIT-13 *see S. arsenophilum*
sulfur-oxidizing bacteria
    energy considerations 44–49
    giant sulfur bacteria 44–49
    subsurface ecosystems 347–349, 356–357, 358, 359
sulfur-reducing bacteria
    subsurface ecosystems 348, 349, 352, 354, 359
sunlight, as energy source 155, 156, 157
superoxide dismutases 257, 329
Synechococcus 251, 252, 253
syntrophism 153

Tailings 25, 26
Tc, microbial reduction 290–291
Te, microbial reduction 215, 285–286

Telephora terrestris 210
temperate coastal zones
    benthic oxygen uptake *182*
    biogeochemical cycling 173–193
    primary production rates *174*, 175–182
    seasonal variation in characteristics 173–175,
        176–180
temperature
    effects on AOM 311–312
    intraterrestrial environment 238
temperature–substrate concentration interaction
    183–188
TEM (transmission electron microscopy) *see*
    cryo-transmission electron microscopy
tepius 216
textile dyes 293–294
Thalassiosira
    T. pseudonana 253
    T. weissflogii 257–258
Thauera selenatis 285
Thermobacillus ferrooxidans 349
thermophiles 184, 186, 238
Thermotoga maritima 282
Thermothrix thiopara 349
thigmotropism 207
Thiobacillus 348, 349
    T. denitrificans 45, 349
    T. ferrooxidans 26, 282, 290, 349
    T. thiooxidans 290
Thiomargarita 42–43, 45, 50, 57, 58, 59–60
    T. namibiensis 57, 59
Thiomicrospora 348
Thiomonas 348
Thioploca 38–39
    carbon cycle 55, 56
    nitrogen cycle 54
    short-cell morphotype (SCM) 58
    sulfur cycle 45, 47, 49, 51
    T. araucae 58
    T. chilieae 58
    within-genus diversity 57, 58
Thiothrix 40–42, 47, 60, 348, 357–358
    T. defluvii 41
    T. eikelboomii 41
    T. unzii 60
Thiovulum 43, 56, 348
    T. majus 43
    T. minus 43
thorium (Th), microbial reduction 289
thraustochytrids 326, *327*, 335
Tokyo Bay sediments 55, 58
toxic metals *see* specific metals
Trapelia involuta 211–212
Trichoderma 218

Trichodesmium 249, 255
Trichomycetes *327*
tropical coastal zones
    benthic oxygen uptake *182*
    biogeochemical cycling 173–193
    primary production rates *174*, 175–182
    seasonal variation in characteristics 173–175,
        176–180
tropic responses, of fungi 219
tunnelling, feldspar 209

U
    binding by fungi 212
    fungal interactions 211–212
    microbial influences 22
    microbial reduction 287–289
    resistance of subsurface bacteria 114
Ulocladium 328
uncultured micro-organisms, isotopic labelling
    1–8
underground ecosystems *see* intraterrestrial
    environment, subsurface
underground laboratory 236, *238*
Ureaplasma parvum 119, *120*

Vanadium (V), reduction 278
Varicosporina ramulosa 333
veil formation 43
vermiculite 216
Verrucomicrobia 354
vivianite 274, *275*

Waste rock material, weathering 25, 26
water temperature
    coastal zones *174*, 175, *176*, 179–180, 183–188
    temperature–substrate concentration
        interaction 183–188
weathering processes
    carbonates 210–211
    role of fungi
        building stone 208–211
        clay and silicate 217–218
        concrete biodeterioration 219
        rock 202–203, 206–211
        waste rock material 25, 26
weddelite *212*, 213–214
whewelite *212*, 213–214, 215
Wolinella succinogenes 285
wood ash 211

Xenobiotics, microbial degradation 273, 274,
    291–294

Yeasts, marine 327–328, 329, 330, 333–334

Zeta-potential analysis 97, 102
Zn
    fungal interactions 211, *212–213*, 214, 218, 220
    in marine primary productivity 257–258
        assimilation of inorganic C 259–261
        fate of biomass 261–265
        interactions with N, P, Fe and other
          resources 258–265
    resistance of subsurface bacteria 114, 115
Zygosaccharomyces rouxii 329